Energy for Rural and Island Communities II

Energy for Rural and Island Communities II

Proceedings of the Second International Conference,
held at Inverness, Scotland
1-4 September 1981

Edited by

JOHN TWIDELL
Energy Studies Unit and
Department of Applied Physics,
University of Strathclyde,
Glasgow, Scotland

Sponsored and assisted by

U.N.E.S.C.O.

U.K. Department of Energy
British Council
Scottish Solar Energy Group
Occidental Oil Company (Kirkwall)

PERGAMON PRESS

OXFORD · NEW YORK · TORONTO · SYDNEY · PARIS · FRANKFURT

U.K.	Pergamon Press Ltd., Headington Hill Hall, Oxford OX3 0BW, England
U.S.A.	Pergamon Press Inc., Maxwell House, Fairview Park, Elmsford, New York 10523, U.S.A.
CANADA	Pergamon Press Canada Ltd., Suite 104, 150 Consumers Rd., Willowdale, Ontario M2J 1P9, Canada
AUSTRALIA	Pergamon Press (Aust.) Pty. Ltd., P.O. Box 544, Potts Point, N.S.W. 2011, Australia
FRANCE	Pergamon Press SARL, 24 rue des Ecoles, 75240 Paris, Cedex 05, France
FEDERAL REPUBLIC OF GERMANY	Pergamon Press GmbH, 6242 Kronberg-Taunus, Hammerweg 6, Federal Republic of Germany

First edition 1982

Library of Congress Cataloging in Publication Data
Main entry under title:
Energy for rural and island communities, II.
Includes indexes.
1. Power resources—Congresses. I. Twidell, John.
TJ163.15. E533 1982 621.042 82-416
AACR2

British Library Cataloguing in Publication Data
Energy for rural and island communities II
1. Power resources—Congresses
I. Twidell, John
333.79'11'091734 HD9502.A2

ISBN 0-08-027606-7

In order to make this volume available as economically and as rapidly as possible the authors' typescripts have been reproduced in their original forms. This method unfortunately has its typographical limitations but it is hoped that they in no way distract the reader.

Printed in Great Britain by A. Wheaton & Co. Ltd., Exeter

Preface

Renewable energy supplies and increased efficiency of use of energy are two rapidly expanding subjects. The rate of progress between the first conference in this series in 1980 and the second in 1981 is impressive. Some of these developments are reflected in these Proceedings, and some specific examples should be noted:

(a) Generation of power from the wind is now an established industry in Denmark.

(b) Wind generation of electricity into the first U.K. nationalised supply grid has started in Orkney. On these island sites the wind regimes are strong and supplies substitute for expensive oil.

(c) Wave energy devices are now being actively considered for relatively small, near to shore sites. The power would be generated for local consumption on island networks.

(d) Small scale hydro power systems are recognised to be of economic potential. Manufacture of small turbines with electronic controllers has started on a commercial basis.

(e) Water pumping for irrigation is possibly now economic in the tropics, powered by solar engines or photovoltaic arrays.

(f) Solar crop driers, passive solar heating of buildings, and photoelectric generation at the 10 kW level are all present developments with important economic consequences.

(g) Energy surveys and planning for rural and island communities point to the need for careful integration of a range of energy supplies.

(h) As usual with renewable energy supplies, a multidisciplinary approach from biology to social science is obviously necessary to optimise systems and gain greatest benefit. Thus the conference whose participants came from a wide range of backgrounds was considered both beneficial and stimulating.

The 150 participants from 20 countries who attended this year's conference found encouragement that the subject is becoming real. Last year the opportunity and need for special consideration of rural and island communities was recognised. This year the engineering and technology for practical installations was apparent. Of particular importance was the active cooperation between participants from all countries as common problems were realised. This has been the year of the Brandt report for North- South economic and development cooperation and we were pleased to play our part in supporting this aspiration.

My own personal thanks must be expressed to our Sponsors, the Orkney office of the Occidental Oil Company for sponsoring our dinner speaker, U.N.E.S.O. for sponsoring some overseas participants, the Inverness Council for hosting a most successful reception in the Town House, the staff of Eden Court Theatre Conference Centre, and the secretarial staff of the Energy Studies Unit and Department of Applied Physics, University of Strathclyde. Without their help and the added blessings of excellent weather the conference would not have been so successful.

The Third Conference is to be held in September 1983. Emphasis will be on practical case studies and demonstrations of successful programmes that clearly benefit rural and island communities throughout the World.

John Twidell,
Energy Studies Unit,
University of Strathclyde,
Glasgow.

Contents

TOPIC A

Future Strategy and Planning

Review of Renewable Energy Development and Use in Scotland

H. Wyper, F. Riddoch* and J. Twidell****

**Department of Applied Physics and Energy Studies Unit,
University of Strathclyde, Glasgow, UK
*Department of Physics, Heriot-Watt University, Edinburgh, UK

KEYWORDS

Renewable ; energy; review; Scotland.

INTRODUCTION

Our review has three objectives:

(i) To list the names, addresses and brief details of organisations and individuals engaged in renewable energy supply, development and use in Scotland. To list examples of renewable energy installations.

(ii) To discuss, by comparison with the available potential, the relative stage of development for each type of energy source.

(iii) To evaluate the positive and negative factors which influence this development.

Environmental conditions and energy use requirements in Scotland are frequently different from the rest of the U.K. and Europe. Within Scotland, Fig. 1, the Highlands and Islands area is noteworthy with a dispersed population in regions of great renewable energy potential and having to meet high conventional energy costs. The central belt of Scotland is an area of high population density where traditional energy supplies would not be easy to change. In general, business and industry are experienced in engineering, including electronics, and looking for new outlets.

These are some of the reasons why renewable energy options in Scotland must be considered distinctively. Indeed it should be taken as a matter of scientific principle that since renewable supplies draw from the dispersed environment, each locality or region has its own optimum blend of renewable energy supplies and end uses. The planning, development and functioning of such systems must therefore be done with each locality individually considered. It follows moreover that a policy of encouraging renewable energy use will be a policy to encourage rural development with dispersed manufacturing and business activity. Thus even within Scotland very diverse renewable energy policies can be expected.

Until recently the case for emphasizing Scotland's renewable energy resources, especially for the Islands, has made little impact on U.K. central government. However there are encouraging signs that this negative response is changing, (Pooley 1981, D.O.E. 1981b).

We have classified renewable energy sources according to Table 1. Each type and subtype is coded by a number, e.g. wind 7, biogas 1.1. The term 'solar' is not used in the general sense of all energy ultimately arising from the Sun, but as only that energy initially used as solar radiation. Peat, geothermal energy, and certain secondary uses of energy are mentioned, although they are not strictly renewable. However their use is often considered in association with renewable supplies.

TABLE 1

Classification of Renewable Supplies by Source and a Listing of the Appendices.

1. BIOMASS	APPENDICES
1.1 Biogas	
1.2 Seaweed	A. Public bodies and Institutions
1.3 Fermentation	
1.4 Peat	B. Consultancy
1.5 Wood	
1.6 Waste	C. Manufacturers and Agents
2. HYDRO	D. Research and Development
3. GEOTHERMAL	E. Education
4. SOLAR	F. Existing installations
4.1 Passive solar heating	
4.2 Crop Drying	G. Individuals [1].
4.3 Solar water heating	
4.4 Focused solar collectors	
4.5 Photovoltaics	
5. TIDAL	
6. WAVE	
7. WIND	

THE POTENTIAL FOR RENEWABLE ENERGY IN SCOTLAND: INTRODUCTORY SURVEY

Various aspects can be considered:

(i) The Potential of Supply

This must be judged from the properties of the environment. Situated at the North East edge of the Atlantic Ocean, Scotland, has wind and wave power potential as great as any populated part of the Earth. Off the West coast and the major island groups, the power in the wind averages over 300 W per m^2

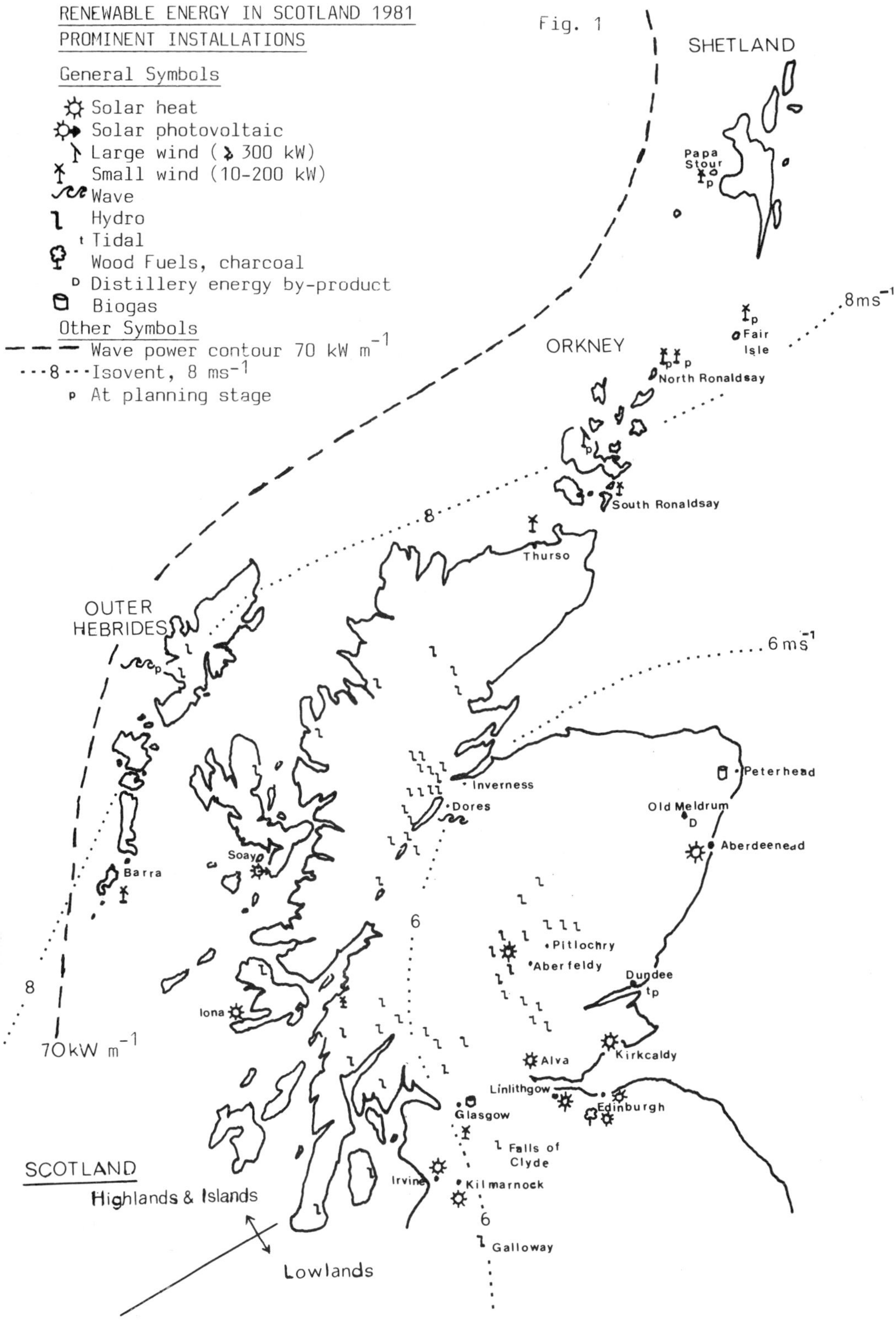
RENEWABLE ENERGY IN SCOTLAND 1981
PROMINENT INSTALLATIONS
General Symbols
Solar heat
Solar photovoltaic
Large wind (≥ 300 kW)
Small wind (10-200 kW)
Wave
Hydro
Tidal
Wood Fuels, charcoal
Distillery energy by-product
Biogas
Other Symbols
Wave power contour 70 kW m-1
Isovent, 8 ms-1
At planning stage
Fig. 1
SHETLAND
Papa Stour
Fair Isle
8ms-1
ORKNEY
North Ronaldsay
South Ronaldsay
Thurso
OUTER HEBRIDES
6 ms-1
Peterhead
Inverness
Dores
Old Meldrum
Aberdeen
Soay
Barra
Pitlochry
Aberfeldy
Dundee
Iona
Kirkcaldy
70kW m-1
Alva
Linlithgow
Edinburgh
Glasgow
Falls of Clyde
SCOTLAND
Highlands & Islands
Irvine
Kilmarnock
Galloway
Lowlands

of wind front (average wind velocity ~7 ms^{-1}), and in the deep water waves off-shore the potential power averages over 70 kW per m of wave front. This may present an over simplistic picture of the theory and technology of wind and wave power extraction, but the importance of these resources is now well documented and accepted (E.T.S.U.).

Hydro power potential is moderate compared with many regions of the World, but nevertheless is the greatest in the U.K. The present installed hydro-electric plant capacity is 1,052 MW (3 x 10^9 kWh $year^{-1}$) by the North of Scotland Hydro Electric Board, 123 MW (0.3 x 10^9 kWH $year^{-1}$) by the South of Scotland Electricity Board and perhaps 1 MW in a range of private schemes. The potential for further development has not been fully assessed. In 1965 the North of Scotland Hydro Electric Board indicated that a further 483 MW of primary hydro supply was being considered for their network, but this did not include the potential for microhydro schemes down to ~5kW capacity (N.S.H.B. 1965).

Unfortunately the full impact of hydro electricity is lessened by treating it as a primary energy resource alongside fossil and nuclear fuels. The latter have very large thermal losses, ~70%, at electrical generation. Whereas the hydro power resource has little loss. Thus Scottish hydro production is often presented as only 4% of the initial primary sources. However the fraction of electricity sent out by hydro is 10% of the total after generation. With full development of Scottish hydropower this could rise to 20%, and with more efficient end use of electricity this fraction could be considerably higher. Even now the 10% contribution of hydroelectricity to the Scottish network with its low generation and works cost is a substantial reason for the lower electricity costs in Scotland compared with England

Solar energy from sunshine has great importance in Scotland despite the small proportion of direct radiation and relatively low yearly energy flux. The major use apart from plant growth is to provide heat for buildings, hot water supply and industrial process heat. Obviously it is sensible to always consider solar input as a fuel saver, for instance to increase background space heat in buildings, to preheat water for conventional heating systems and to partially dry agricultural products. The long space heating season in Scotland (~10 months in the North) allows maximum use of expenditure on solar design. In particular, optimising passive solar input to buildings becomes of equal or greater importance in Scotland than in Southern England, (MacGregor 1981). Space cooling is not needed in Scotland.

Fuels from biological sources must be seriously considered. Far greater use could be made of the traditional supplies of wood and peat, and other developments such as biogas, fuel alcohol, and pyrolytic gases may be important. Since Scotland is a major region for forestry and fermentation industries, it is of the greatest importance to consider the options being developed worldwide in biofuel technology.

The full potential of all renewable sources in Scotland must include these and many other options. Each option is considered in greater detail within this review, and the conclusions summarised.

(ii) The Potential of Use

Scotland is rich in conventional energy resources and present use is dependent on coal, oil, gas and nuclear electricity, Fig. 2. Details of these resources can be found in U.K. government and other publications. The overall picture is that coal is relatively abundant, but the geological

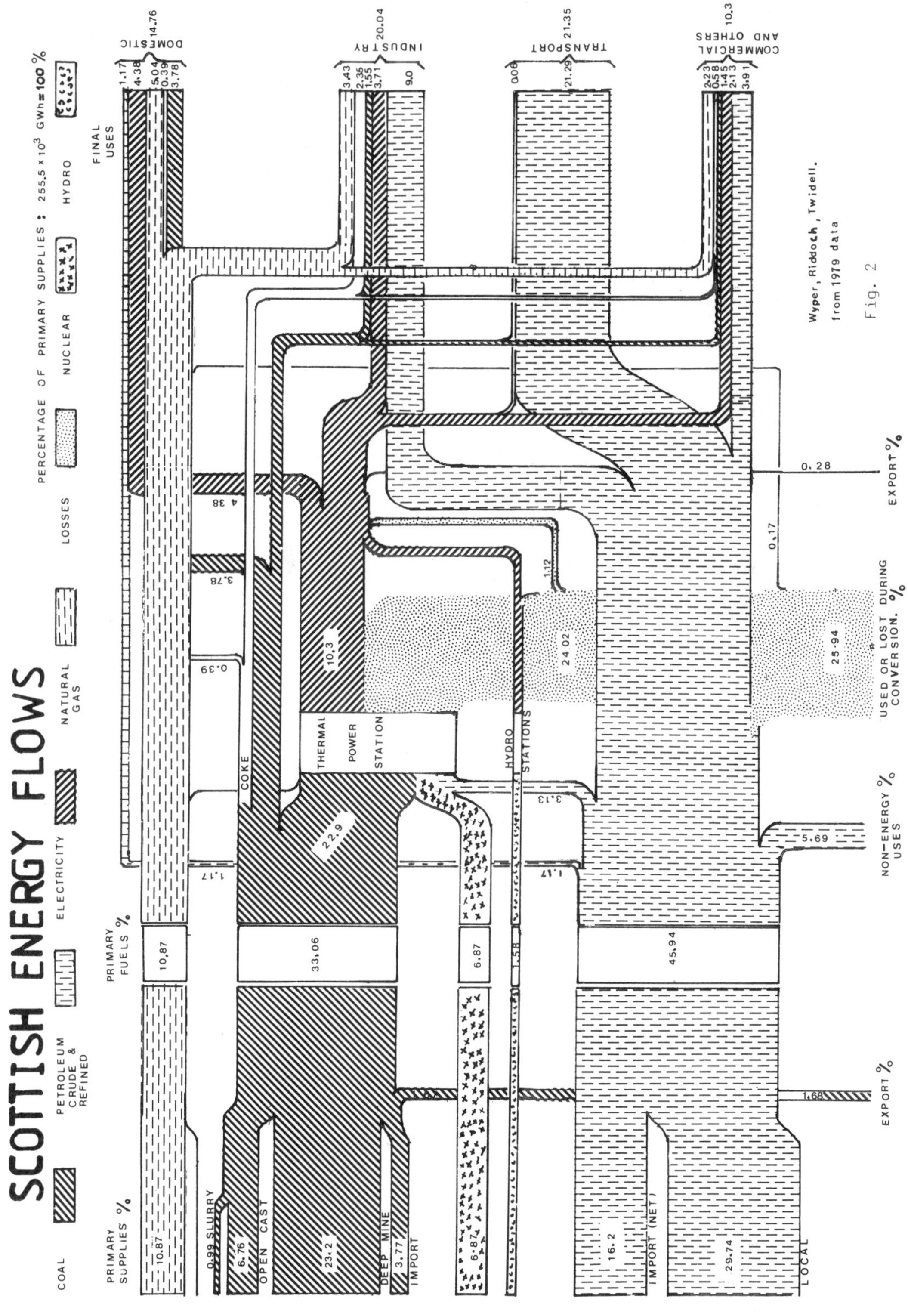
SCOTTISH ENERGY FLOWS
COAL
PETROLEUM CRUDE & REFINED
ELECTRICITY
NATURAL GAS
LOSSES
PERCENTAGE OF PRIMARY SUPPLIES : 255,5 × 10³ GWh = 100 %
NUCLEAR
HYDRO
PRIMARY SUPPLIES %
PRIMARY FUELS %
FINAL USES
DOMESTIC 14.76
INDUSTRY 20.04
TRANSPORT 21.35
COMMERCIAL AND OTHERS 10.3
0.99 SLURRY
6.76 OPEN CAST
23.2
DEEP MINE
3.77 IMPORT
10.87
6.87
16.2 IMPORT (NET)
29.74 LOCAL
33.06
45.94
1.58
THERMAL POWER STATION
HYDRO STATIONS
COKE
22.9
10.3
24.02
25.94
EXPORT %
1.68
0.28
0.17
5.69
NON-ENERGY USES %
USED OR LOST DURING CONVERSION %
Wyper, Riddoch, Twidell.
from 1979 data

Fig. 2

conditions make extraction difficult so the lifetime of the resource will be several hundred years. Oil and gas reserves in the North Sea off Scotland now meet the entire U.K. requirements, but will do so for only the next 20 to 30 years. Electricity production is predominantly from coal with an increasing proportion (now 20% of supply) from nuclear stations.

The present sources of conventional fuels are used predominantly in the U.K. for low temperature heat, <80°C (39%); for high temperature heat, > 80°C, (28%); essential electricity for lighting, machines, etc., but not heat, 9%; for transport (24%), (Leach 1979). It is important to appreciate the relative proportions of these uses. Great importance is attached to the supply of controlled voltage and phase electricity, (~10% of energy use) but frequently the more dominant requirements for heat and transport receive less attention.

The abundant sources of energy supply in Scotland have not led to cheaper prices for the local consumer, since there is a general pressure to equalise energy prices throughout the U.K.. Electricity is available at about 3.6p kWh^{-1}, coal at about 1p kWh^{-1} (£80 per tonne), petroleum transport fuels at about 4p kWh^{-1} (£1.70 per gallon), and heating oil 3p kWh^{-1}. In the Highlands and Islands oil and coal prices tend to be 10% higher than central Scotland (H.I.D.B. 1981).

Renewable energy will only be used in Scotland if it is economically competitive with conventional supplies. The rate of price increase for fossil fuels is above the rate of general inflation, and it is possible that depletion of North Sea and world oil and gas supplies will accelerate this effect. Thus renewable supplies must become increasingly attractive as substitutes for fossil fuels. To summarise present trends for renewable energy use within Scotland we list the following aspects within a framework of economic and technical factors that will be amplified in later sections.

(a) The long term sustainable nature of supply from renewable sources is of significant economic and social importance, leading to greater local self-sufficiency and economic stability.

(b) The outer regions, particularly the Islands, are areas where conventional supply is difficult and expensive, and where the potential of wind and wave power is the greatest.

(c) The demand for building space heat can be helped by passive solar gain and local generation from wind (and occasionally hydro) machines. Such methods are now economically competitive with mains electricity and oil.

(d) Transport presents extreme difficulty. Renewable electricity generation can be channelled to battery and hydrogen vehicles that may be suitable for local transport, especially on the islands. The electrical load that this represents may have benefit for optimising efficiency, e.g. for local control of generating devices by load switching and for night time use.

(e) Electrical grids can be used in various modes. Power can be purchased for an electrical network from dispersed generators, and not all power lines need be maintained at constant voltage and phase, e.g. the proposed wind/diesel system on Fair Isle. One important concept is a grid for lower priced electricity at variable voltage for heat supply. These concepts are most easily applied at the local level.

(f) A number of renewable options can be linked to manufacturing and business developments in Scotland. In general small scale devices and systems encourage economic development in the outer areas, where local manufacture and maintenance services can be based. Larger schemes can be linked with manufacture from central industrial production that is now entirely available within Scotland. The development of hydroelectric power between 1945 and 1965 is an excellent example of such benefit.

1. BIOFUELS

There is a complex and as yet unassessed set of options for energy supply from biological sources. Scottish options are being assessed by individual groups and within a general U.K. assessment sponsored by the Energy Technology Support Unit at Harwell, (e.g. Larkin 1981). Unfortunately the Forestry Commission's terms of reference and activities at a national level do not relate to wood as an energy source. It is reasonable to relate Scottish options for biofuels with those contemplated for countries with similar environments, such as Scandanavia and Canada. The EEC estimates are that within the Community's total energy requirement, 5% could be provided from biological residues and 5% from specific energy crops grown so as not to disrupt present agricultural practice (Palz 1980). Within Scotland we can expect marked variations in supply and use options at the local level.

1.1 BIOGAS

The generation of methane from decomposing biological material and sewage is technically possible and financially attractive in certain circumstances in Scotland, despite the low environmental temperature. Research and development work is being followed in University Departments of Microbiology (e.g. Strathclyde University), Public Corporations (e.g. Glasgow City Council), Research Organisation (e.g. Rowett Research Institute), and commercial organisations (e.g. Biomechanics Ltd. for Scottish distilleries). Further details are given in the appendices. As expected, developments seem most favourable when waste disposal is necessary for reasons of public health. There should be more consideration of biogas production from wet cattle sludge on Scottish farms. Since cattle have to be enclosed and under cover for many months in winter, the sludge forms a regular and viable source for biogas production.

No overall assessment of biogas potential and use has been made for Scotland generally or for specific local regions.

1.2 SEAWEED

Scottish waters contain an estimated stock of 10 million tonnes (wet mass) of marine macrophytic algae. Harvest rates may reach 5 tonnes per year per km of coastline. Traditionally the harvesting of seaweed has been important in island and coastal economies, and it is not unreasonable to consider fuel value for the resource alongside the value for chemical and agricultural use. For instance the digestion of one tonne of wet seaweed (80% moisture content) could produce 4000 kWh^{-1}(14 GJ) of heat from the methane ,(Palz 1980), but in practice yields are likely to be very much less (~3GJ) than this theoretical value (Van Brakel 1980).

1.3 FERMENTATION

The fermentation of grain and distillation of ethyl alcohol for whisky production is of prime economic importance in Scotland. There is a long manufacturing experience and related on-going research by some wealthy companies. Clearly this expertise could be used for the benefit of fuel alcohol production, but there is little indication of this happening. The world potential market for fuel alcohol from biological crops, especially in the tropics, is of great significance and it would be encouraging to see more involvement by Scottish industry. Distilleries in Scotland are however taking increased interest in energy conservation and bye-products (Cockburn 1981).

1.4 PEAT

Peat is a traditional fuel in Scotland still important for domestic heating in some areas of the Highlands and on most Islands. In general it may be considered a non-commercial fuel since most users cut their own supplies, but local marketing firms do exist. An attempt by the N.S.H.E.B. to generate electricity from peat fuel failed some years ago but new developments are being contemplated, for example peat combustion in fluidised beds or part combustion for pyrolytic production of gas. There are indications that the use of peat is now increasing in Scotland after many years of decline. Peat should however not be classed as a renewable resource, especially if harvested intensively.

1.5 WOOD AND FORESTRY

Forestry is extensive in Scotland, with the U.K. headquarters of the Forestry Commission based in Edinburgh. Wood is primarily considered a resource for paper making and building material. There is no official policy or evaluation of wood as an energy source. This is both surprising and unfortunate since in most developing and many developed countries such policies and evaluations exist. However Scottish University Departments and private organisations have considerable experience in the area of wood fuel sources and use (Mitchell 1981, Fraser 1981).

The sales of wood burning stoves have increased rapidly in Scotland over recent years, and there can be little doubt of the desire to use wood fuel when supplies are available.

1.6 BIOLOGICAL WASTE

It is of the greatest importance with all biofuels, including those considered previously, to consider energy supply options from otherwise waste materials (Porteus 1981). Municipal waste in Scotland probably totals 2×10^6 t y^{-1} having a heat of combustion equivalent to 0.4×10^6 t y^{-1} of coal. Energy use options include combustion from thermal and electrical power, and anaerobic digestion in open dumps for methane generation.

1.7 BRACKEN

Serious proposals are being made for the potential of bracken harvesting for biofuels and land improvement (Callaghan 1981).

2. HYDRO

Within the U.K., the only nationalised energy service named with the responsibility for a renewable supply is the North of Scotland Hydro-Electric Board. It was never intended that hydro power should be the sole supply of power by the Board, but the uniqueness of local supply options and use in the Highlands and Islands led to a special Act of Parliament to establish the Board in 1943 (HMSO 1943). Hydro power was then seen to be a major development opportunity for Scotland, and from the onset these developments were linked with ideals for social improvement in the Board's areas. This 1943 Act should be compulsory reading for all those seeking to relate energy supply to development, since it foreshadows many of the development ideals promoted by the United Nations, World Bank, International Labour Office and other world development agencies.

Between 1943 and 1970 the Hydro Board established 54 hydro generating stations each with a capacity to supply from 130 MW to 0.5 MW. The total capacity established was 1052 MW producing a yearly average of 3×10^6 MWh of power to the Scottish grid. The Board's generating cost of hydro electricity is now 0.7 p kWh^{-1}, which is 3 times less than its thermal generation and about 5 times less than its diesel generation. A full account of these activities is given in the Annual Reports of the N.S.H.B. It is unfortunate that a historical account of the Board's work has not been written. With an awakening interest in renewable energy supplies in the 1980s such an account would be of great practical value as well as interest.

The 1965 Annual Report of the Hydro Board listed a further 483 MW of hydro potential which to date has never been developed. The cessation of the hydro programme was due to a combination of causes, including the potential of cheaper oil, aspiration for nuclear power and opposition because of environmental disturbance. The 483 MW potential included sites of medium scale, but did not extend to a comprehensive survey of much smaller sites. It is possible however that further hydro development will occur in the 1980s (N.S.H.E.B. 1981, Francis 1981).

The South of Scotland Electricity Board generates 0.3×10^6 MWh y^{-1} of hydro power from an installed capacity of 123 MW, to produce its cheapest electricity at 0.26p kWh^{-1} (S.S.E.B. 1980) which is 7 times less than from nuclear generation.

The construction of the Scottish hydro system gave valuable experience to a number of engineering firms that later utilised this expertise to gain overseas orders. Sir Alexander Gibb and Partners was one such firm of consultant engineers who continued with major hydro projects in Africa and elsewhere. The role of Scotland as a technical proving ground for renewable supplies that later extended on a World scale is thus recorded in history, and sets a precedent that other options might follow.

There are a number (probably 100) of examples of small scale (5 kW - 500 kW) hydro electric generating plants in Scotland. There is no comprehensive study of these, but the potential for small hydro power in helping those not connected to the grid has been assessed (SRT 1978). Some small private systems sell excess power to the grid, so providing control for the equipment and income for the owner. There has been no comprehensive study of the full potential for small hydro in Scotland (unlike England and Wales, Francis 1981), although discussions are in hand by a University group with the Department of Energy. A recent encouragement has been the manufacture by McKellar Engineering of Granton-on-Spey of a new low head 5 to 15 kW system with electronic load control. This system was devised by Mr. P. Agnew of Glasgow University, and the business enterprise supported by the Highland and Islands Development Board.

3. GEOTHERMAL

Possible geothermal sources of low grade heat have been suggested within the central lowland regions of Scotland and the North Sea (ETSU 1976). However this resource is generally regarded to be of far less potential than other alternatives, despite the great need for low temperature space and process heat. There has been no full investigation of geothermal potential in Scotland.

4. SOLAR

The variation of yearly insolation on a horizontal surface from the North to the South of the U.K. is indicated by Fig. 3. The long daylight hours in Scotland produce a relative increase in summer insolation, but there is a dramatic decrease in winter. This ostensibly depressing factor is somewhat relieved by considering insolation on a vertical surface (of more importance for building space heat) and the long space heating season in Scotland, as discussed below.

The Scottish Solar Energy Group (appendix A) is the first renewable energy association to be formed in Scotland.

4.1 PASSIVE SOLAR ENERGY

Domestic energy requires 28% of primary energy in the U.K. (D.O.E. 1981b). The proportion generally in Scotland may be higher, and in rural and island localities can be 40% (Barbour 1981). MacGregor (1981) has shown that the proportion of the yearly heating load that can be met by passive solar input into buildings increases with increase in latitude across Europe, Fig. 3. It is therefore encouraging that some planning authorities, building firms and individuals are taking an interest in 'solar design'. Even so there are only about 5 building projects designed to maximize solar gain (see appendix for details), and only one of these is in the Highland and Island areas where special conditions including high wind related energy losses can occur.

The importance of developing energy efficient buildings have been analysed in a recent E.T.S.U. sponsored report (Energy Conscious Design, 1981). With minimal design changes 30% energy reductions can be obtained, and this can be increased to 70% with additional expenditure. No final analysis is available to judge the operation of any of the energy and building projects in Scotland.

4.2 CROP DRYING

The drying of grain and hay is an increasing financial and energy burden on farmers. It is therefore encouraging that one national laboratory, the Scottish Institute of Agricultural Engineering at Penicuik, near Edinburgh, has constructed and is testing a solar drying and storage

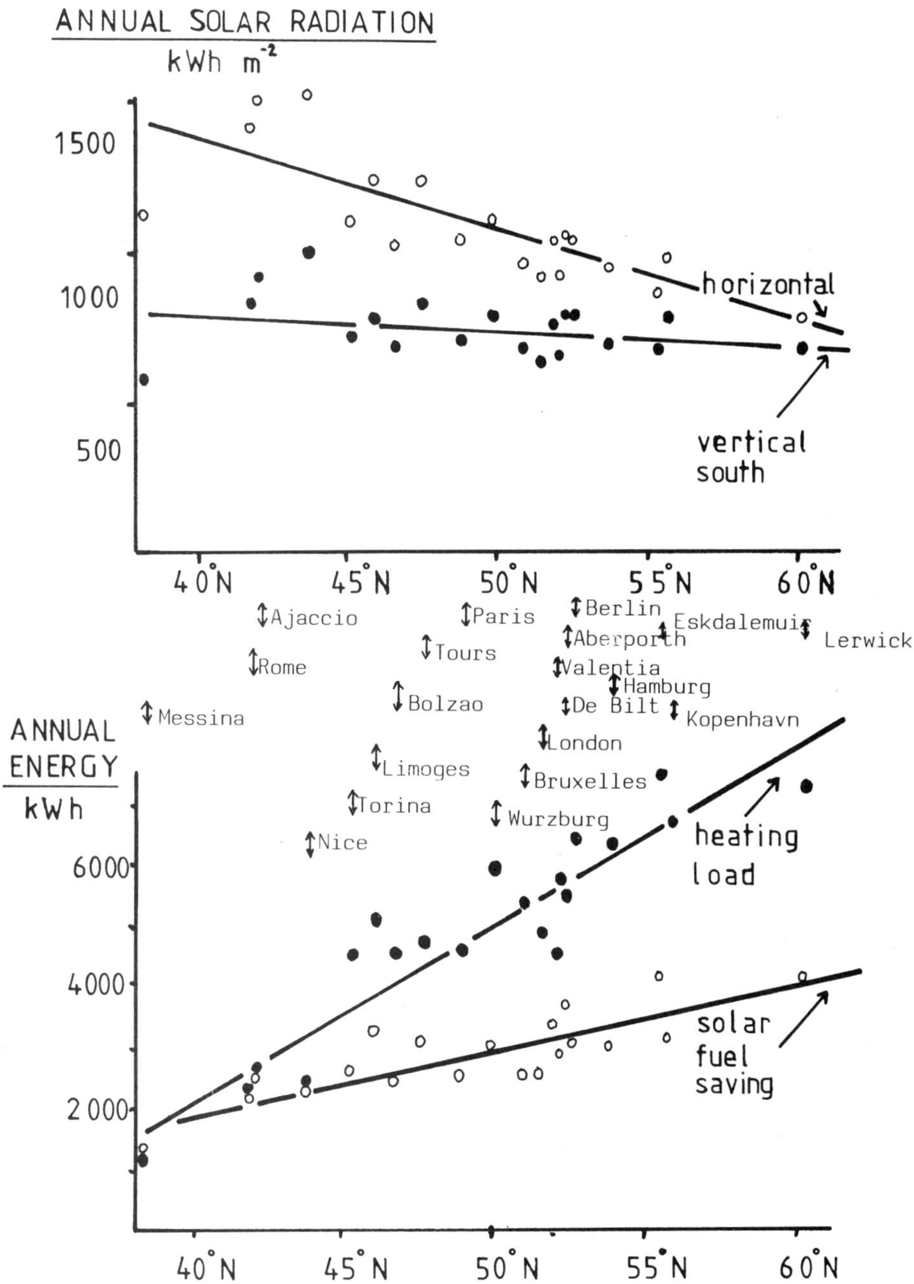

FIG 3 (a) Variation of solar radiation at European sites from Southern to Northern latitude.

(b) Proportion of the annual heating load that could be supplied by solar passive and active energy for standard housing.

From MacGregor (1981b) with permission.

barn (Bailey 1981). It is likely that straw-burning driers - an indirect use of solar energy - will be important for drying at farm level.

4.3 SOLAR WATER HEATING

No independent assessment has been made of the number of solar water heating installations in Scotland. From manufacturer's claims and personal experience, we estimate there are between 500 and 800 installations predominantly purchased during the last 3 years. The U.K. Solar Trade Association has two member organisations based in Scotland, one of whom reports very substantial business. Other firms are engaged in installation. For the U.K. generally, the solar water heating industry is regulated and cooperates with E.T.S.U., the Solar Energy office at Cardiff University and the U.K. Solar Trade Association which has 70 member firms in the U.K.

The economic value of solar water heaters is generally considered to be borderline. Much depends on the operating temperatures, the cost of the displaced heating fuel, and the capital cost of the installation. The other obvious factors are insolation and system efficiency. Most of these variables will be different for Scotland than England (or elsewhere) and so it is most unfortunate that no technical and economic assessment has been undertaken and completed by independent agencies in Scotland.

4.4 FOCUSED SOLAR COLLECTORS

The relatively small proportions of direct solar radiation negates the uses of focused collectors in Scotland.

4.5 PHOTOVOLTAICS

Scottish Physics Departments at Dundee and Heriot-Watt Universities have pioneered research on amorphous silicon devices. A few working photovoltaic silicon crystal systems are used for remote telecommunication stations (Riddoch 1980). A 30kW generating system is planned to be installed on a Scottish island after development at Marchwood Power Station (TREBLE 1980).

5a. TIDAL RANGE (the rise and fall of tides)

U.K. interest in tidal power almost entirely centres on the Severn Estuary, attractive because of the large (~10,000 MW) generating and storage capacity associated with the 9 m average tidal range. The next major U.K. potential is the Solway Firth at the boundary between Scotland and England where the average tidal range is 5 m. In general however the tidal range around the British coast is about 3 m, and estuaries are of considerably less area than those mentioned. As a result generating potential is very considerably reduced.

The main assessment of Scottish tidal potential was made by Davey in 1923, as part of a general U.K. survey. Recently there are indications of revised interest in tidal power generally, including Scotland (B.H.R.A. 1981). If the only criteria for development is large power generation, then clearly no schemes other than the Solway Firth are worth considering. However other factors should be considered

(a) A particular local coastline of successive peninsulars might allow a one or two basin system of up to 10 MW capacity to be constructed with a minimum size of barriers. Such a use of the local coastline might allow generators to be placed on land below water level, Fig. 4.

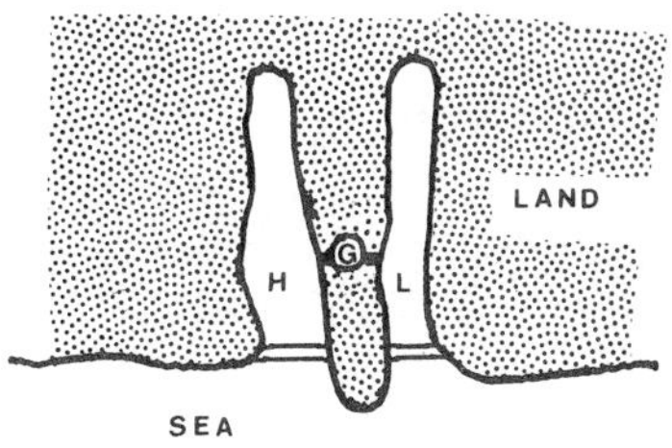

H sea loch at high water level

L sea loch at low water level

G generating plant on land between the two lochs.

Fig.4 Sketch of low cost tidal power system using two neighbouring sea lochs and two short barriers.

(b) The use of the power might be particularly advantageous in the locality (e.g. to substitute for diesel generation).

(c) The barriers might have other valuable use, e.g. for a road or to impound water for fish farming.

(d) The generated power could be kept in phase with the tidal rhythm and still be useful (e.g. absorbed into a larger grid, or used to provide hydro or hydrogen energy storage).

(e) The small scale system could provide test and development facilities for larger schemes, especially for low head turbines. (We note that such a facility will be needed for the development of turbines for the Severn estuary scheme).

Taking all these factors into consideration, we believe it is important that a careful study of the Scottish coastlines should be made to assess the tidal range power potential.

5b. TIDAL FLOW (from tidal currents)

Intense tidal flows occur across the North of Scotland in the Pentland Firth at and around Orkney at the joining of the Atlantic Ocean with the North Sea. This has unassessed generating potential.

Tidal flow power devices are suitable for small scale development at favourable local sites. Development of such systems is being planned in Scotland (Cave 1981, Grant 1981).

6. WAVE

Within its programme for renewable energy, the U.K. Department of Energy has funded wave power research to a greater extent than other systems. This policy and the subsequent developments have been extensively published within an atmosphere of vigorous debate and quiet rivalry. The present state of wave power is very uncertain and may well lead to intense acrimony as future funding is decided.

Interest has focused on Scotland for several reasons:

(a) Wave power potential averages between 50 and 70 kW per meter of wave front in deep water off the Hebrides, Fig. 5. Few locations in the World have greater potential for wave power generation.

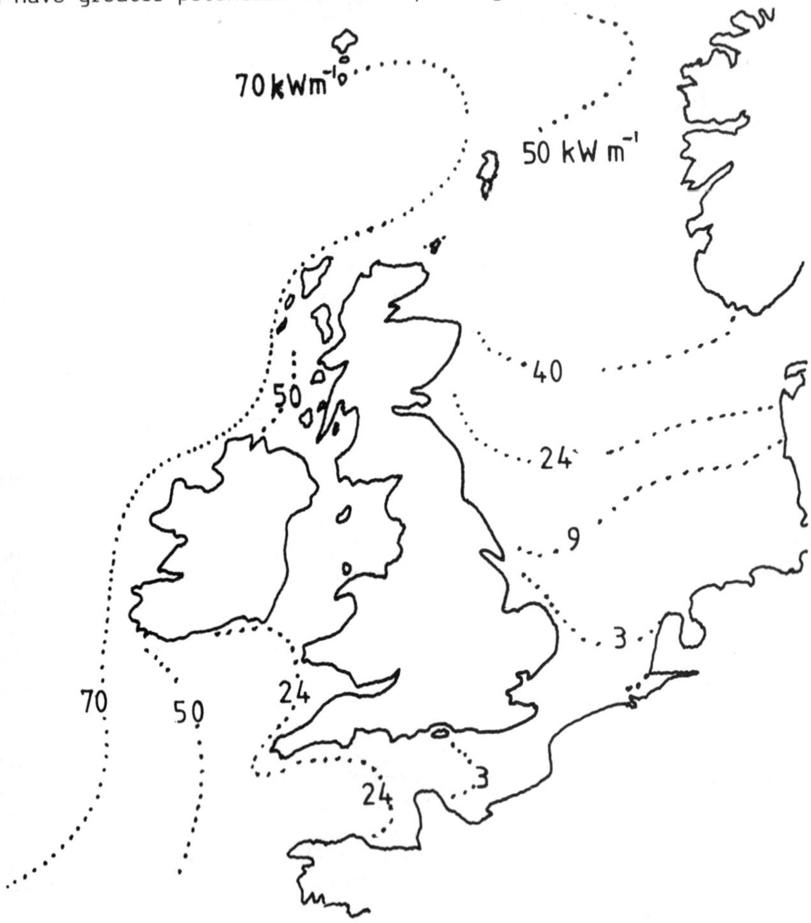

Fig.5 Wave power potential. The annual average power in the waves per unit of wavefront (kW m^{-1}).
(From Leishman J.M. and Scobie G.(1976),"the development of wave power",National Engineering Laboratory, East Kilbride, Glasgow.)

(b) Two Scottish groups are playing central roles in the research and development (Mr. Stephen Salter's group in Edinburgh University and the National Engineering Laboratory group at East Kilbride), and a reduced scale model is being tested in Loch Ness (the Clam device of Lanchester Polytechnic (Peatfield 1981)).

(c) The manufacture of large wave power devices would be most suitable for Scottish marine yards experienced in ship and oil rig construction.

Final decisions on the particular types of wave devices to be used, if any, have yet to be made. Present plans have concentrated on supplying large amounts of power, ~10,000 MW, into the U.K. national grid. It may however be more feasible to consider smaller systems supplying local needs. Thus we note that the N.E.L. has recently been supported to develop a 1 MW system to feed into the Hebridean grid to reduce dependence on diesel fuel. A similar system is being considered for Fiji in the South Pacific where intense ocean waves break onto coastal reefs. (However the waves reaching the coastal mainland of Tonga are probably far more suitable for a land based prototype device). It seems probable therefore that the first practical realisation of significant wave power use may be for island use, where the device may match community needs at a manageable and economic scale.

7. WIND

Scotland has some of the most intensive and persistent wind regimes of any generally inhabited place on Earth. The islands exposed to the prevailing westerly winds are particularly attractive for harnessing wind power, since (i) the winds are favourable, (ii) space heating is desirable for most of the year, especially since heat loss is greatly increased by the winds, (iii) oil and coal supplies from the mainland are increasingly expensive, (iv) population density is low. The only negative feature is that winds reach extreme speeds (40 to 50 ms^{-1} can be expected most years with 10 year gust maximum of ~65 ms^{-1}) and machine damage may occur.

A few traditional windmills existed into the 20th Century (Barbour 1981), small battery charging aerogenerators (e.g. the Lucas Freelight) were common in the Highlands and Islands until the 1950s when grid electricity began to be extensive, and one major U.K. development of a 50 kW machine was instigated by the Electrical Research Association on Costa Hill in Orkney from 1954 - 1955 (Golding 1976). As elsewhere in Europe and America, wind development then lapsed until recently.

After the 1973 oil disruptions various planning studies were supported by the Department of Energy to consider the economic and technical possibilities of wind power. However the programme moved in a hesitant manner without any firm commitment to build machines, despite definite programmes for construction being established in the USA, Denmark and elsewhere. All this changed dramatically in late 1981 when the Hydro Board formulated its plans for Orkney. As a result of the Board's initiatives, wind power developments are now more advanced in Scotland than elsewhere in the U.K. Despite this however the general U.K. experience is slight beside that of the USA and Denmark in particular.

The most successful trial by the Hydro Board of a Danish 22 kW machine in Orkney is noteworthy. In 7 months operation it has produced 26,000 kWh at an average power of 12 kW. Thus it can be expected that a machine in Orkney will produce at least twice the power of an identical machine in a favourable site in Denmark. The economic consequences of this improved performance are obvious.

Elsewhere in Scotland perhaps 5 American 10 kW machines have been erected. Two British commercial machines have been installed; a "7 kW" Trimble Mills machine and one P.I. Ltd. vertical axis machine. A few (possibly 30) smaller machines have been purchased or built by enthusiasts. There are definite signs that this beginning will develop rapidly, although there is no declared policy of purchasing privately generated power for the grid and financial assistance is not available from Government to purchase machines.

A Glasgow based firm, James Howden Ltd., have won a contract to build a 200 kW American machine designed by Wind Turbine Generators Inc., Buffalo, New York State. This is for the Central Electricity Generating Board of England and Wales, and will be the Board's first test machine. It will be sited at Carmarthen Bay in South Wales.

A particular encouragement is the possibility of the manufacture of small (10 kW to 50 kW) machines in the Highlands and Islands. It is obvious in Denmark that manufacturing has concentrated in Jutland where winds are most favourable and initial sales are within reach of the manufacturers for maintenance and repair. A similar strategy should be possible in Scotland. Associated with this is the opportunity to develop a wind machine test site and national research facility. In the near future (1982) a 250 kW grid linked machine will be erected in Orkney by the Hydro Board, and a 3 MW machine is planned for 1984. These machines are important, but should not detract from recognition that the major energy use is for heat and transport. Machines satisfying these latter needs may possibly be more numerous and produce a greater total power than machines erected by the Electricity Boards.

The rapidly increasing commitment to wind energy in Scotland by manufacturing companies, the N.S.H.E.B., St. Andrews University, Napier College, and Strathclyde University have encouraged plans for a Scottish Wind Energy Association.

CONCLUSIONS

We conclude by listing the positive and negative aspects affecting renewable energy supply, development and use in Scotland.

(i) Positive Aspects

(a) There is a wide variety of supply options, including some of world significance.

(b) Many island and remote areas have communities with serious energy supply difficulties due to the high price of fossil fuels. Such situations provide strong incentives to develop renewable supplies from local resources.

(c) The small scale of application in these favourable areas provide working experience for systems beneficial to many overseas countries.

(d) Scotland has a strong base of mechanical, electrical and electronic industry with experience of marine and rugged construction. Thus the manufacturing base exists for the production of devices both for home and export sales.

(e) There is considerable experience in certain biomass industries.

(f) North Sea oil revenues can sometimes be channelled for developments of local application.

(g) Specialised authorities already exist to support or finance developments where renewable supplies are first likely to be economically attractive.

(h) Renewable supply options exist to provide heat and fuel local transport.

(ii) Negative Aspects

(a) Existing energy sources of oil, coal and nuclear power are now sufficient. In particular there is considerable over-capacity of generation in the mainland electrical grid. This distracts interest and financial support from renewables.

(b) Scotland's general energy policy is embedded with England's and planned from central government. There is little opportunity therefore to match local needs to local resources. The exception is electricity supply, but even here U.K. national policies are a strong influence.

(c) There is generally a lack of interest in renewable supply options and the experiences of other countries.

(d) There are few government grants or other encouragements to establish renewable supplies. For instance, there is no positive policy for private suppliers to sell to the electrical grid.

(e) There are few energy supply requirements associated with planning. In particular there are no requirements that relate to passive solar heat gains for buildings.

(iii) Overall Assessment

The renewable energy resources of Scotland have yet to be fully assessed, but are obviously vast. When coupled with the prospects of manufacture from Scottish Industry and development from Scottish research institutions the future should be encouraging. Whether or not these potentials will be realised however is a different question.

[1] A list of individuals engaged in renewable energy supply is available separately from John Twidell, Energy Studies Unit, University of Strathclyde, Glasgow.

REFERENCES

General References

E.R.I.C. 1980 - In J.W. Twidell (Ed.), Energy for Rural and Island Communities - Proceedings of the 1980 Conference. Published by Pergamon Press, Oxford, in 1981.

E.R.I.C. 1981 - In J.W. Twidell (Ed.), Energy for Rural and Island Communities - Proceedings of the 1981 Conference. To be published by Pergamon Press in 1982.

Named References

BAILEY, P.H. and W.E. FERGUSON (1981). A solar barn for grain and hay drying experiments in Scotland. In the proceedings of the Solar World Forum 1981. Pergamon Press, Oxford, to be published.

BARBOUR, D. and J.W. TWIDELL (1981). Energy use on the Island of North Ronaldsay, Orkney. In E.R.I.C. 1980, ibid., Pergamon Press, Oxford. 39 - 51

B.H.R.A. (1981). Second International Symposium on Wave and Tidal Energy 1981. B.H.R.A., Bedford.

CALLAGHAN, T.V., G.J. LAWSON and R. SCOTT (1981). Bracken as an energy crop. In Proceedings of the World Solar Forum, Pergamon Press, Oxford, to be published.

CAVE, P.R. (1981). Tidal current power utilization using moved turbines. In E.R.I.C. 1981, ibid., Pergamon Press, Oxford, to be published.

COCKBURN, H. (1981). Utilisation of waste heat in a malt distillery. In E.R.I.C. 1981, ibid., Pergamon Press, Oxford. 21 - 26.

DAVEY, N. (1923). Studies in tidal power. Constable and Co.Ltd., London.

D.O.E.,(1981a). Renewable Energy sources in the United Kingdom. Paper for the U.N. Conference on New and Renewable sources of energy, Kenya, 1981.

D.O.E., (1981b). Digest of United Kingdom Energy Statistics 1981. Department of Energy, published by H.M.S.O. London.

ELLIOT, G. and G. ROXBURGH (1981). Wave Energy Studies at the U.K. National Engineering Laboratory. Second symposium on Wave and Tidal Energy 1981. B.H.R.A., Bedford.

ENERGY CONSCIOUS DESIGN, (1981). Passive Solar Housing in the U.K. Report to the Energy Technology Support Unit, Harwell, by D. Turrent, J. Doggart and R. Ferraro of Energy Conscious Design, London.

E.T.S.U., Energy Technology Support Unit, A.E.R.E., Harwell, Didcot. The Unit has published and sponsored a large range of reports concerning developments in energy supply and use.

E.T.S.U., (1976). Geothermal energy: the case for research in the United Kingdom. Energy paper No. 9 prepared by J.D. Garnish. H.M.S.O., London.

FRANCIS, E.E. (1981a). Small scale hydro electric development in England and Wales. Proceedings of the Third International Conference on Future Energy Concepts. I.E.E., London.

FRANCIS, E.E. (1981b). Small scale hydro electric power. In E.R.I.C. 1981, ibid., Pergamon Press, Oxford.

FRASER, A.I. (1981). The use of wood for domestic and industrial energy in rural areas of Britain and the transfer of wood energy technology to developing countries. In E.R.I.C. 1981, ibid., Pergamon Press, Oxford.

GOLDING, E.W. (1976). The generation of electricity by Wind Power. E. & F.N. Spon Ltd., London.

GRANT, A.D. (1981). Power generation from tidal flows for navigation buoys. Second International Symposium on Wave and Tidal Energy, B.H.R.A. Fluid Engineering, Bedford, 117 - 127.

H.I.D.B., (1981). Private communication based on work by the Institute for Sparsely Populated Areas, University of Aberdeen, for the Highlands and Islands Development Board.

H.M.S.O., (1943). Hydro-Electric Development (Scotland) Act. H.M.S.O., London.

H.M.S.O., (1981). Scottish Abstract of Statistics. No. 10/1981. N.B. energy data for 1979 published in 1981. H.M.S.O., Edinburgh.

LARKIN, S.B.C. (1981). Private communication, and in Larkin et al (1981). Production and distribution of agricultural wastes in the U.K. and their potential for use as an energy source. In F. Vogt (Ed.) Energy Conservation and Use of Renewable Energies in the Bio-Industries. Pergamon Press, Oxford. 335 - 351.

LEACH, G. (1979). A low energy strategy for the United Kingdom. Science Reviews Ltd., London.

MACGREGOR, K. (1981a). Solar Heat for the Highlands and Islands of Scotland. In E.R.I.C. 1980, ibid, Pergamon Press, Oxford. 185 - 190.

MACGREGOR, K. (1981b). A comparison of the climatic suitability of various locations in the European Community for Solar space heating. In Proceedings of the World Solar Forum, Pergamon Press, Oxford, to be published.

MITCHELL, C.P. (1981). A study of the feasibility of coppice as a lowland farm enterprise. In E.R.I.C. 1981, ibid, Pergamon Press, Oxford, to be published.

N.S.H.E.B., (1965). Annual Report of the North of Scotland Hydro Electricity, 1964/65. N.S.H.E.B., Edinburgh.

N.S.H.E.B., (1981). Annual Report of the North of Scotland Hydro Electricity 1980/81. N.S.H.E.B., Edinburgh.

PALZ, W. and P. CHARTIER (1980). Energy from biomass in Europe. Applied Science Publishers, Barking.

PEATFIELD, A.M. (1981). The Clam wave energy converter to provide electricity for island and coastal communities. In E.R.I.C. 1981, ibid, Pergamon Press, Oxford.

POOLEY, D. (1981). The Wind Energy Research and Development Program of the U.K. Department of Energy. Proceedings of the International Colloquium on Wind Energy, Brighton 1981. B.H.R.A. Fluid Engineering, Bedford.

PORTEUS, A. (1981). Refuse derived fuels. Applied Science Publishers, Barking.

RIDDOCH, F. and J. WILSON (1981). Solar Cells in Scotland. In E.R.I.C. 1980, ibid., Pergamon Press, Oxford.

S.S.E.B. (1980). Annual Report of the South of Scotland Electricity Board 1979/80. S.S.E.B., Glasgow.

S.R.T. (1978). Renewable Energy Sources in Highland Communities. Report of the Science, Regigion and Technology Committee, Church of Scotland, Edinburgh.

TREBLE, F.C. (1980). Photovoltaic pilot projects in the European Community. Sun at Work in Britain. 11, 47 - 51.

VAN BRAKEL,J. (1980). "The Ignis Fatuus of Biogas". Delft University Press. Delft.

APPENDIX A.PUBLIC BODIES AND INSTITUTIONS : SEE TABLE 1 FOR CODES.

ORGANISATIONS	AREAS OF WORK	DETAILS OF WORK
B.B.C., QUEEN MARGARET DRIVE, GLASGOW.	7	WIND ASSISTED TELECOMMUNICATIONS LINK.
EDINBURGH DISTRICT COUNCIL, GEORGE IV BRIDGE, EDINBURGH EH1 1UQ.	4	SUPPORTING A PASSIVE SOLAR-HEATING PROJECT ON A MULTI-STOREY BLOCK OF FLATS.
FIFE REGIONAL COUNCIL, DIR.OF ARCHITECTURAL SERVICES FIFE HOUSE 03, GLENROTHES.	4,8	8-ENERGY SAVING;USE OF PASSIVE SOLAR ENERGY.STUDENTS HOSTEL WITH DOMESTIC HOT WATER HEATED BY SOLAR PANELS.
FOUNTAIN BRIDGE HOUSING ASSOC. 161 DUNDEE STREET EDINBURGH.	4	SOLAR HEATING FOR 15 TENEMENT HOUSES
GLASGOW DISTRICT COUNCIL PARKS DEPT, 302 BUCHANAN ST, GLASGOW. 041 331 2641.	1	BIOGAS FROM GRASS CUTTINGS.
H.I.D.B. BRIDGE HOUSE, 27 BRIDGE ST, INVERNESS.	GENERAL	ENERGY POLICY FOR RURAL AREAS.
IONA COMMUNITY, REV J.CARRIE, THE MANSE, SOUTH QUEENSFERRY, EDINBURGH EH30 9HY.	4,8	A 17SQ.M. SOLAR COLLECTOR ON THE ROOF OR THE RESIDENTIAL YOUTH CENTRE AT IONA. 8-INVESTIGATING THE USE OF HEAT PUMPS ON IONA.
METEOROLOGICAL OFFICE 231 CORSTORPHINE ROAD EDINBURGH EH12 6ER 031 334 9721	4,7,2	ROUTINE METOEROLOGICAL DATA FOR SCOTLAND
NATIONAL TRUST FOR SCOTLAND, 5 CHARLOTTE SQ, EDINBURGH EH2 4DU.	7	COOPERATION WITH SHETLAND ISLANDS COUNCIL ON WIND/DIESEL SCHEME FOR FAIRISLE.
NATURE CONSERVANCY COUNCIL 12 HOPE TERRACE EDINBURGH EH9 2AS 031 447 4784	GENERAL	ENVIRONMENTAL IMPACT AND POLICY ASPECTS OF RENEWABLE ENERGY SUPPLIES.

N.S.H.E.B. 16 ROTHESAY TERRACE EDINBURGH EH3 7SE 031 225 1361	2,7	1050 MW HYDRO PLANT IN SERVICE. 22KW WIND GENERATOR IN SERVICE. 250KW AEROGENERATOR TO BE INSTALLED 1982 3MW AEROGENERATOR TO BE INSTALLED 1984 TRIMBLEMILL IN SERVICE IN THE NEAR FUTURE.
ORKNEY ISLAND COUNCIL, COUNCIL OFFICES, KIRKWALL,	7	COOPERATED IN PLANNING OF N.S.H.E.B WIND PROJECTS IN ORKNEY. INVOLVED IN WIND PROJECT ON NORTH RONALDSAY.
S.D.A. 120 BOTHWELL ST GLASGOW G2 7JP 041 248 2700	GENERAL	SUPPORT CAPABILITIES FOR GRANT AID TO TECHNOLOGY PROJECTS PLUS INVESTMENT CAPABILITY AS JOINT PARTNER IN VIABLE MANUFACTURING PROJECT.
S.S.E.B. CATHCART HOUSE SPEAN ST. GLASGOW 041 637 7177	2,7,8	8- GENERAL ENERGY USE. 123MW HYDRO-ELECTRIC PLANT. SUPPORT FOR ORKNEY WIND DEVELOPMENT
SCOTTISH DEV DEPT 43 JEFFREY ST EDINBURGH 031 556 9233.	4	JOINT R&D PROJECT TO STUDY PASSIVE AND ACTIVE SOLAR HEATING IN A SCHOOL IN IRVINE.
SCOTTISH PEAT & LAND DEVELOPMENT ASSOCIATION(SPALDA) SECR:MR.D.WATHEN, KILMARTIN FARM, GLENURQUART, INVERNESSHIRE.	1	PEAT AND BIOMASS.
SCOTISH SOLAR ENERGY GROUP, SECR:DR.J.I.B.WILSON, HERIOT-WATT UNIVERSITY, EDINBURGH.	4	THE FUNCTION OF THIS GROUP IS TO PROMOTE RESEARCH, DEVELOPMENT, AND APPLICATION OF SOLAR ENERGY IN SCOTLAND.
SCOTTISH TELECOMS BOARD	4	PHOTOVOLTAIC PANELS TO POWER RADIO TELEPHONE LINKS.
SHETLAND ISLANDS COUNCIL, TOWNHALL, LERWICK, SHETLAND.	7	INVOLVEMENT IN WIND ENERGY SCHEME FOR FAIRISLE.
SOLAR TRADE ASSOCIATION, 26 STORE ST, LONDON WC1E 7BT.	4	MOST UK COMPANIES WITH AN INTEREST IN SOLAR ENERGY ARE MEMBERS OF THIS ORGANISATION
S.R.T. PROJECT, CHURCH OF SCOTLAND HOME BOARD, 121 GEORGE ST, EDINBURGH. 031 225 5722	GENERAL	STEWARDSHIP OF RESOURCES. IMPACT OF TECHNOLOGICAL CHANGE ON PEOPLE.
WESTERN ISLES ISLAND COUNCIL, SANDWICK ROAD, STORNOWAY, ISLE OF LEWIS. 0851 3773.	ALL	INTERESTED IN SMALL SCALE APPLICATIONS TO SUIT LOCAL NEEDS. INTEREST IN PEAT.

UNIVERSITIES, SEE OTHER APPENDICES.

APPENDIX B.CONSULTANTS: SEE TABLE 1 FOR CODES.

ORGANISATIONS	AREAS OF WORK	DETAILS OF WORK
BRIAN MARKS, 17 BANCHORY AVE, INCHINNAN, RENFREW. 041 812 1238.	4,7,8	APPLICATION OF RENEWABLE ENERGY PARTICULARILY TO THE HEATING OF BUILDINGS, GLASSHOUSES, TANKS, ETC.
C.T.T.SCOTLAND KIRKWOOD HOUSE, BIGGAR, LANARKSHIRE, ML12 6PP 0899 20164	2,7	LOW HEAD TURBINES AND GENERAL ALTERNATIVE ENERGY EQUIPMENT.
CROUCH&HOGG, 18 WOODSIDE CRES, GLASGOW G3 7UU. 041 332 9755	2,4,6,7,8	CIVIL ENGINEERS FOR N.S.H.E.B.PROJECTS. DEVELOPMENT OF HYDRAULIC TRANSMISSION FOR LARGE AEROGENERATORS. 8-TOTAL ENERGY CONCEPTS, HEAT PUMPS.
ECOLOGY TECHNOLOGY, 18 ABERCROMBY PLACE, EDINBURGH.	GENERAL	ALTERNATIVE ENERGY DEVICES AND ENVIRONMENTAL IMPROVEMENT. THIRD WORLD INTEREST.
INTERNATIONAL FOREST SCIENCE CONSULTANCY, 21 BIGGAR ROAD, SILVERBURN, PENICUIK EH26 1LQ.	1	ASSESSMENT, SURVEY, CONSULTANCY FOR BIOMASS RESOURCES. CHARCOAL MANUFACTURE.
ISIS ENERGY CONSERVATION LTD 50 ST MARYS STREET EDINBURGH 031 556 9812	1	PROVISION OF A WIDE RANGE OF WOOD, PEAT, WASTE AND COAL FIRED APPLIANCES.
NAPIER COLLEGE WIND ENERGY GROUP, COLINTON ROAD, EDINBURGH EH10 5DT 031 447 7070.	2,4,7	ENERGY CENTRE. CONSULTANCY SERVICE FOR INDUSTRY AND COMMERCE. DEVELOPMENT OF SMALL VERTICAL AXIS WIND TURBINES.
NEI PEEBLES, EAST PILTON, EDINBURGH EH5 2XT 031 552 6261	2,3,5,6,7	WIDE RANGE OF HYDRO-ELECTRIC GENERATORS. COLLABORATED IN STUDIES FOR TIDAL AND WIND SCHEMES.
N.E.L. EAST KILBRIDE GLASGOW G75 0QU 035 52 20222	2,5,6,7	DESIGN OF TURBINES FOR HYDRO AND TIDAL SCHEMES. DEVELOPMENT OF OSCILLATING WATER COLUMN WAVE ENERGY DEVICE. STUDIES FOR WIND ENERGY SCHEMES.
ROXBURGH & PARTNERS MIRRWEN HOUSE MAXWELL ST, PAISLEY.	6	WAVE ENERGY SYSTEMS.

SCOTTISH MARINE BIOLOGICAL ASSOC. DUNSTAFFNAGE MARINE RESEARCH LAB, P.O. BOX 3, ARGYLL PA34 4AD 0631 62244	1,6	ASSESSMENT OF MARINE FOULING ON WAVE ENERGY DEVICES OFF THE HEBRIDES. INTEREST IN PRODUCTION OF METHANE FROM SEAWEED.
UNIV - ABERDEEN ENGINEERING DEPT, MARISCAL COLLEGE, ABERDEEN AB9 1FR. 0224 40241	8	8-STUDIES OF HEAT PUMPING AND THERMAL STORAGE IN ABERDEEN.
UNIV - HERIOT-WATT PHYSICS DEPT, RICCARTON, EDINBURGH. 031 556 9812.	4	DEVELOPMENT OF AMORPHOUS SILICON SOLAR-CELLS
DEPT OF BUILDING, CHAMBERS ST, EDINBURGH.	4	SOLAR WATER HEATING, SPACE HEATING, INSULATION, AND GENERAL CONSTRUCTION TECHNOLOGY. TESTING.
UNIV - ST.ANDREWS PHYSICS DEPARTMENT ST.ANDREWS, FIFE KY16 9SS 0334 76161.	7	DEVELOPMENT OF VERTICAL-AXIS WINDMILLS. ASSISTANCE WITH DEVELOPMENT OF WINTERBOTHAM ALTERNATOR.
WIND&WATER POWER LTD, 1 HUNTLY ST, EDINBURGH EH3 5HB. 031 556 6529.	7	WIND POWER CONSULTANTS.
Y-ARD LIMITED, CHARING CROSS TOWER, GLASGOW G2 4PP. 041 204 2737	5,6,7	AVAILABILITY/RELIABILITY STUDIES. POWER TRANSMISSION SYSTEMS. MOORING.

APPENDIX C. MANUFACTURERS AND AGENTS: SEE TABLE 1 FOR CODES.

ORGANISATIONS	AREAS OF WORK	DETAILS OF WORK
BACKWOODSMAN, BARCALDINE, CONNEL, ARGYLL.	1	MARKETTING OF WOOD-BURNING STOVES.
BONK&CO.LTD, MAIN ST, NORTH KESSOCK, INVERNESS, KESSOCK 256.	1	MARKETTING OF WOOD-BURNING STOVES.
BRITISH AEROSPACE LTD, (DYNAMICS GROUP), MANOR ROAD, HATFIELD, HERTS. HATFIELD 62300.	7	BRITISH AEROSPACE DYNAMICS GROUP IS A MEMBER OF THE WIND ENERGY GROUP, RESPONSIBLE FOR DESIGN AND CONSTRUCTION OF 250KW AND 3MW WIND TURBINE GENERATORS FOR ORKNEY.
C.T.T.SCOTLAND, KIRKWOOD HOUSE, BIGGAR, LANARKSHIRE, ML12 6PP 0899 20164	2,7	LOW HEAD TURBINES AND GENERAL ALTERNATIVE ENERGY EQUIPMENT.
HIGHLAND HEAT SAVERS, MIDFEARN, ARDGAY 505.	1	PROVISION OF WOOD-BURNING STOVES AND ENERGY SAVING DEVICES.
JAMES GREY & SON LTD., 89 GEORGE ST, EDINBURGH, 031 556 3152	4	DESIGN AND INSTALLATION OF SOLAR WATER HEATING SYSTEMS.
INTERNATIONAL FOREST SCIENCE CONSULTANCY, 21 BIGGAR RD, SILVERBURN, PENICUIK, EDINBURGH EH26 ILQ.	1	MANUFACTURERS OF CHARCOAL.
ISIS ENERGY CONSERVATION LTD, 50 ST MARYS STREET, EDINBURGH. 031 556 9812	1	PROVISION OF A WIDE RANGE OF WOOD, PEAT, WASTE AND COAL FIRED APPLIANCES.
MCKELLAR ENGINEERING LTD, GRANTON-ON-SPEY. 0479 2577	2,8	MANUFACTURERS OF LOW HEAD TURBINES. DEVELOPING WATER-POWERED HEAT PUMPS.
NEI PEEBLES, EAST PILTON, EDINBURGH EH5 2XT. 031 552 6261	2,3,5,6,7	WIDE RANGE OF HYDRO-ELECTRIC GENERATORS. HAVE COLLABORATED IN STUDIES FOR TIDAL AND WIND SCHEMES.

ROBOTHERM NATURAL ENERGY SYSTEMS, ROBOFLEX LTD, LOANINGSIDE, BALFRON, GLASGOW G63 0QF. 035 066 329	4	INSTALLERS OF SOLAR PANELS AND OTHER NATURAL ENERGY DEVICES.
SINCLAIR ENGINEERING, WOODYARD, BRIDGEND, THURSO. THURSO 2494.	7	AGENT FOR WIND GENERATORS.
SOLARWAY SCOTLAND LTD, SOLARWAY HOUSE, 13-15 TOWER STREET. LEITH, EDINBURGH EH6 7BX. 031 554 0358.	4,8	MARKET,DESIGN AND INSTALLATION OF SOLAR SYSTEMS FOR ALL TYPES OF USE. 8-SUPPLY AND INSTALLATION OF HEAT PUMPS.
WEIR PUMPS LTD, 149 NEWLANDS RD, CATHCART, GLASGOW.G44 4EX 041 639 7141.	2,3	SUPPLIER OF PACKAGED MINI/MICRO HYDRO-ELECTRIC GENERATORS.CURRENTLY INVOLVED IN SUPPLY AND DEVELOPMENT OF COMPLETE RANGE OF GEOTHERMAL PUMPING PLANT.
WIND&WATER POWER LTD, 1 HUNTLY ST. EDINBURGH EH3 5HB. 031 556 6529.	7	WIND POWER CONSULTANTS. PROPOSED MANUFACTURE OF 'OSPREY' WIND GENERATOR.

APPENDIX D. RESEARCH AND DEVELOPMENT: SEE TABLE 1 FOR CODES.

ORGANISATIONS	AREAS OF WORK	DETAILS OF WORK
FORESTRY COMMISSION, NORTHERN RESEARCH STATION, ROSLIN, MIDLOTHIAN. 031 445 2176.	1	NO SPECIFIC PROJECTS ON THE PRODUCTION OF BIOMASS FOR ENERGY, BUT NUMEROUS PROJECTS IN COMMERCIAL SILVICULTURE IN UPLANDS AND ISLANDS IN BRITAIN.
NAPIER COLLEGE DEPT.OF MECH.ENG. COLINTON ROAD EDINBURGH EH10 5DT 031 447 7070	2,4,7	DEVELOPMENT OF SMALL VERTICAL-AXIS WIND TURBINES, WORK ON WATER-POWERED HEAT PUMP, SOLAR-FINS FOR SOLAR WATER HEATERS.
NEI PEEBLES, EAST PILTON EDINBURGH EH5 2XT 031 552 6261	2,3,5,6,7	WIDE RANGE OF HYDRO-ELECTRIC GENERATORS. COLLABORATED IN STUDIES FOR TIDAL AND WIND SCHEMES.
N.E.L. EAST KILBRIDE GLASGOW G75 0QU 035 52 20222	2,5,6,7	DESIGN OF TURBINES FOR HYDRO AND TIDAL SCHEMES. DEVELOPMENT OF OSCILLATING WATER COLUMN WAVE ENERGY DEVICE. STUDIES FOR WIND ENERGY SCHEMES.
* PENTLANDS SCOTCH WHISKY RESEARCH LTD., 84 SLATEFORD RD, EDINBURGH EH11 1QU	1	ANAEROBIC DIGESTION OF DISTILLERY WASTES FOR THE PURIFICATION OF EFFLUENT AND THE PRODUCTION OF BIOGAS.
* ROWETT RESEARCH INSTITUTE, GREENBURN ROAD, BUCKSBURN, ABERDEEN, AB2 9SB. 0224 712751.	1	RESEARCH INTO PRODUCTION OF METHANE FROM AGRICULTURAL WASTES, PARTICULARILY ANIMAL EXCRETA AND CROP WASTES.
SCOTTISH INSTITUTE OF AGRICULTURAL ENGINEERING, BUSH ESTATE, PENICUIK, EDINBURGH EH26 0PH. 031 445 2147	1,4	APPLICATION OF SOLAR ENERGY TO THE DRYING AND STORAGE OF AGRICULTURAL CROPS.
SCOTTISH MARINE BIOLOGICAL ASSOC. DUNSTAFFNAGE MARINE RESEARCH LAB, P.O. BOX 3, ARGYLL PA34 4AD 0631 62244	1,6	ASSESSMENT OF MARINE FOULING ON WAVE ENERGY DEVICES OFF THE HEBRIDES. INTEREST IN PRODUCTION OF METHANE FROM SEAWEED.
UNIV.OF ABERDEEN ENGINEERING DEPT, MARISCAL COLLEGE, ABERDEEN AB9 1FR. 0224 40241	8	8-STUDIES OF HEAT PUMPING AND THERMAL STORAGE IN ABERDEEN.

(DR.P.MITCHELL), DEPT OF FORESTRY, ST.MACHAR DRIVE,	1,5	EXPERIMENTAL STUDY OF SINGLE STEM TREES AS ENERGY CROPS.ASSESSMENT OF THE RESOURCE.
UNIVERSITY OF DUNDEE DEPT.OF MECH.ENG., DUNDEE DD1 4HN.	5	TIDAL POWER STUDIES.
UNIVERSITY OF EDINBURGH, (MR.S.H.SALTER), DEPT OF MECHANICAL ENGIN. MAYFIELD RD, EDINBURGH 9. 031 667 1081	6,8	PART OF THE WAVE ENERGY GROUP,DEVELOPING FULL-SCALE DESIGNS OF WAVE ENERGY CONVERSION DEVICES. 8- FLYWHEEL STORAGE.
(MR.G.M.H.CARRIE), DEPT OF MECHANICAL ENGIN.	4	COMPUTER SIMULATION OF SOLAR HEATED DOMESTIC HOT-WATER SYSTEMS.
DEPT OF ARCHITECTURE, CHAMBERS ST, EDINBURGH. 031 667 1011	4	SIMULATION OF REGIONAL CLIMATES TO EVALUATE BUILDING ENCLOSURES,WITH REFERENCE TO THE SOLAR GAIN OF LIGHTWEIGHT STRUCTURES.
(MR.H.DICKINSON), SCHOOL OF ENGINEERING, THE KINGS BUILDINGS, EDINBURGH.	GENERAL	APPROPRIATE TECHNOLOGY,PARTICULARILY FOR THE THIRD WORLD.RURAL ELECTRIFICATION.
(DR.P.DRYBURGH), SCHOOL OF ENGINEERING.	1	PEAT USE IN SCOTLAND.
UNIVERSITY OF GLASGOW, STATISTICS DEPT., UNIVERSITY GARDENS, GLASGOW G12 8QW.	4	RESEARCH ROJECT ON STATISTICAL PROBLEMS CONNECTED WITH SOLAR ENERGY STORAGE.
HERIOT-WATT UNIVERSITY, PHYSICS DEPT, RICCARTON, EDINBURGH. 031 449 5111	4	DEVELOPMENT OF AMORPHOUS SILICON SOLAR-CELLS.
DEPT OF BUILDING, CHAMBERS ST, EDINBURGH. 031 225 8432.	4	SOLAR ASSISTED AIR HEATING SYSTEM FOR A HOUSE.
UNIVERSITY OF ST.ANDREWS PHYSICS DEPARTMENT ST.ANDREWS, FIFE KY16 9SS 0334 76161.	7	DEVELOPMENT OF VERTICAL-AXIS WINDMILLS. ASSISTANCE WITH DEVELOPMENT OF WINTERBOTHAM ALTERNATOR.
UNIVERSITY OF STRATHCLYDE, ENERGY STUDIES UNIT, 100 MONTROSE ST, GLASGOW G40LZ	1	EXPLORATION OF BIOMASS UTILIZATION IN THE THIRD WORLD.

APPLIED PHYSICS DEPT, JOHN ANDERSON BUILDING,	4,7	SOLAR CROP DRYING IN THE THIRD WORLD. WIND ENERGY SYSTEM FOR NORTH RONALDSAY.
DEPT.OF ARCHITECTURE& BUILDING SCIENCE, 131 ROTTEN ROW, GLASGOW. 041 552 4400.	4	DEVELOPMENT OF AIR-HEATING SOLAR COLLECTORS FOR SPACE AND WATER HEATING.
DEPT.OF MECHANICAL ENG.,	2,5,6	DEVELOPMENT OF HYDRO AND WIND TURBINES. ANALYSIS OF OSCILLATING WATER COLUMNS.USE OF TIDAL ENERGY SYSTEM FOR NAVIGATIONAL BUOYS.
WEIR PUMPS LTD, 149 NEWLANDS RD, CATHCART, GLASGOW G44 43X. 041 639 7141.	2,3	DEVELOPMENT OF COMPLETE RANGE OF GEOTHERMAL PUMPING PLANT.
WIND&WATER POWER LTD, 1 HUNTLY ST, EDINBURGH EH3 5HB. 031 556 6529.	7	WIND POWER CONSULTANTS.DESIGNING AND MANUFACTURING 'OSPREY'10KW AEROGENERATOR.
* NORTH OF SCOTLAND HYDROELECTRIC BOARD, 16 ROTHESAY TERRACE, EDINBURGH.	2,7	PITLOCHRY RESEARCH LABORATORY.
* ROBERT GORDON'S INSTITUTE, ABERDEEN.	4	MONITORING LOW ENERGY/SOLAR HOSTEL.

APPENDIX E. EDUCATION : SEE TABLE 1 FOR CODES.

ORGANISATIONS	AREAS OF WORK	DETAILS OF WORK
FIFE REGIONAL COUNCIL, ENERGY MANAGER, DIR.OF ARCH.SERVICES, FIFE HOUSE 03, GLENROTHES. GLENR 754411.	GENERAL	EDUCATION IN ENERGY CONSERVATION.
GLASGOW ECOLOGY PARTY, 88 MOSSGIEL RD, GLASGOW G43 2DA. 041 632 2705	GENERAL	EDUCATIONAL MATERIAL.
HERIOT-WATT UNIVERSITY PHYSICS DEPT RICCARTON, EDINBURGH. 031 449 5111	4	
DEPT.OF BUILDING	4	LABORATORY PROJECTS-SOLAR PANEL COMPARISON.
NAPIER COLLEGE DEPT.OF MECH.ENG. COLINTON ROAD EDINBURGH EH10 5DT 031 447 7070	2,4,7	B.SC ENERGY ENGINEERING. SHORT COURSES ON ENERGY TOPICS.
UNIVERSITY OF STRATHCLYDE, APPLIED PHYSICS DEPT, JOHN ANDERSON BUILDING,	4,7	COURSE ON RENEWABLE ENERGY TECHNOLOGIES.
WHITE,PETER ARCHITECT, 22 RUTLAND SQ, EDINBURGH EH1 2BB.	GENERAL	IN CONJUNSTION WITH FRIENDS OF THE EARTH RUNS AN ENERGY ADVICE SERVICE FOR BUILDING USERS.

APPENDIX F. INSTALLATIONS: SEE TABLE 1 FOR CODES.

OPERATOR	TYPE OF PLANT	DETAILS AND LOCATION
ACHLOA CENTRE, ABERFELDY.		LOCAL DEMONSTRATION OF PRIVATE HYDRO AND SMALL WIND GENERATOR.
MR.R.W.DALRYMPLE, TRANSMISSION MANAGER, B.B.C., FORT WILLIAM TRANSMISSION ST. LOWER BANAVIE, FORT WILLIAM, INVERNESS-SHIRE.	7	500 W TORNADO AEROGENERATOR AT DYCHLIMORE. BACKUP TO BE PROVIDED BY SOLAR CELLS.
CAMPHILL COMMUNITY GROUP, NEWTONDEE, ABERDEEN.	4	15 SQUARE METRE SOLAR WATER HEATING INSTALLATION ON A NEW HOSTEL.
CENTRAL REGIONAL COUNCIL, VIEWFORTH, STIRLING. FK8 23T	4	SOLAR HEATING SYSTEM AIDED BY HEAT PUMPS IN ALVA ACADEMY, CLACKMANNAN.
* MR.CHARLES CRICHTON, MONZIE CASTLE, MONZIE, CRIEFF, PERTHSHIRE.	2	OPERATES A 200KW HYDRO-ELECTRIC SCHEMEE TO PROVIDE ELECTRICITY FOR THE ESTATE AND LOCAL AREA.
EDINBURGH DIST.COUNCIL, COUNCIL OFFICES, HIGH ST, EDINBURGH.	4	A 300 SQ.M."SOLAR-SKIN" ON A MULTI-STOREY BLOCK OF FLATS AT OXGANGS, EDINBURGH.
FIFE REGIONAL COUNCIL, DIR.OF ARCHITECTURAL SERVICES FIFE HOUSE, GLENROTHES.	4	STUDENTS HOSTEL IN KIRKCALDY WITH HOT WATER HEATED BY SOLAR PANELS.
FOUNTAINBRIDGE HOUSING ASSOC. 161 DUNDEE ST, EDINBURGH.	4	SOLAR PANELS FOR BACKGROUND WATER HEATING IN AN EDINBURGH TENEMENT WITH 15 HOMES.
GLASGOW DISTRICT COUNCIL, PARKS DEPT, 302 BUCHANAN ST, GLASGOW. 041 331 2641.	1	BIOGAS FROM GRASS CUTTINGS AND OTHER WASTE MATERIALS TO PROVIDE GREENHOUSE HEAT IN BELLAHOUSTON PARK.
* IONA COMMUNITY, REV.JOHN CARRIE, THE MANSE, SOUTH QUEENSFERY, EDINBURGH EH30 9HY.	4	A 17 SQUARE METRE SOLAR COLLECTOR ON THE ROOF OF THE RESIDENTIAL YOUTH CENTRE ON IONA.
LONG JOHN DISTILLERIES LTD., 123-157 BOTHWELL ST, GLASGOW G2 7AY.	1	BIOGAS INSTALLATION AT THEIR GLENUGIE DISTILLERY, PETERHEAD.

	MILLER HOMES NORTHERN LTD., MILLER HOUSE, 18 GROATHILL AVE, EDINBURGH.	4	TWO HOUSES ON THEIR BUCKSTONE ESTATE, EDINBURGH INCORPORATE THE CALOR 31 PHASE CHANGE CHEMICAL STORAGE UNIT.
	N.S.H.B., 16 ROTHESAY TERRACE, EDINBURGH EH3 7SE	2,7	OPERATES 1750MW OF HYDRO AND PUMPED STORAGE PLANT. HAS RECENTLY INSTALLED A 22KW WIND GENERATOR IN ORKNEY.
	MR.A.M.PEATFIELD, LANCHESTER POLYTECHNIC, COVENTRY. 0203 24155.	6	"CLAM" WAVE ENERGY DEVICE AT DORES, LOCH NESS.
	SCOTTISH INSTITUTE OF AGRICULTURAL ENGN. BUSH ESTATE, PENICUIK, EDINBURGH. EH26 0PH 031 445 2147	1,4	SOLAR CROP DRYING BARN.
*	SOLARWAY SCOTLAND LTD., SOLARWAY HOUSE, 13-15 TOWER ST, LEITH, EDINBURGH.	4	THIS COMPANY HAVE INSTALLED OVER 1500 SOLAR SYSTEMS IN THE LAST FIVE YEARS. CUSTOMERS INCLUDE MULTI-OWNERSHIP LODGES & HOTEL AT LOCH RANNOCH, THE MACROBERT TRUST & KYLE & CARRICK DISTRICT COUNCIL.
	S.S.E.B., CATHCART HOUSE, GLASGOW S4. 041 637 7177.	2	TWO HYDRO SCHEMES ON THE FALLS OF CLYDE AND IN GALLOWAY, PRODUCING 134.4MW. SOLAR/HEAT PUMP SWIMMING POOL, ALVA, CLACKMANNANSHIRE.
	STANLEY.P.MORRISON LTD, 13 ROYAL CRES, GLASGOW G3 7SU.	1	USE OF DISTILLERY WASTE AT GLENGARIOCH DISTILLERY.
	STRATHCLYDE REG.COUNCIL, EDUCATION STRATHCLYDE HOUSE, 20 INDIA ST, GLASGOW G2.	4	PASSIVE SOLAR HEATING AND HEAT RECOVERY IN FENCEDYKE SCHOOL, IRVINE.
	TOMATIN DISTILLERS CO.LTD. 34 DOVER STREET, LONDON W1X4HX	1	USE OF DISTILLERY WASTE.
	TRIMBLE WINDMILLS, 55 EAST PARADE, HARROWGATE, N.YORKSHIRE HG1 5LQ	7	TWO WINDMILLS PRODUCED BY THIS COMPANY HAVE HAVE BEEN INSTALLED IN SCOTLAND; ONE IN BARRA, THE OTHER IN THURSO.
*	HERIOT-WATT UNIVERSITY, (MR.F.W.FLEMING), DEPT.OF BUILDING, CHAMBERS ST, EDINBURGH. 031 225 8432	4	SOLAR-ASSISTED AIR HEATING SYSTEM IN A HOUSE. LINLITHGOW

WIND & WATER POWER LTD, 1 HUNTLY ST, EDINBURGH EH3 5HB. 031 556 6529.	7	THIS COMPANY HAVBE INSTALLED A NUMBER OF SMALL WIND MACHINES.
* CUNNINGHAM, MR. R. 94, MAXHOLM ROAD, KILMARNOCK, KA1 4JQ.	4	LOW ENERGY HOUSE WITH ACTIVE SOLAR HEATING.
* INT. FOREST SCIENCE CONSULTANCY, PENICUIK, EDINBURGH.		CHARCOAL MANUFACTURING.
* SCOTTISH TELECOMMUNICATIONS BOARD.		PHOTOVOLTAIC POWER FOR TELECOMMUNICATION LINK ON SOAY.
* TURNBURRY TRUST, LOCH RANNOCH.	4	HOUSING SCHEME WITH ACTIVE SOLAR HEATING.

Initial Conclusions Regarding the Usage of Renewable Energy Sources on Tresco, Isles of Scilly

T. L. Shaw and I. D. Mays

Sir Robert McAlpine & Sons Limited, UK

1. Introduction

Despite the current apparent surplus of conventional energy sources due to low economic activity, interest in the renewable sources is steadily increasing. The principal reason for this is the continuing uncertainty about future oil supplies and the persistent, substantial increases in the price which has to be paid for this dwindling resource. As long as it remains uncertain to what extent nuclear power can take over the role of oil, so allowing our dependence on this precious fuel to fall to a more realistic level, we must strive to develop options capable of filling the gap left by reduced oil production.

The form taken by this gap must be clearly defined in order to identify the rate at which the renewables should preferably be exploited, the form in which this energy is needed and the cost which the consumer can be expected to pay. It is therefore a question of looking at markets and it seems reasonable to conclude that none is more urgent than that of the remote communities, for whom the versatility of oil makes it such a potentially ideal energy source were it not for the excessive cost penalty it then imposes due to the extra transport requirement and isolation from the economic benefits of, for example, the bulk generation of electricity from a mix of fuels.

However, remote communities are usually well endowed with higher than average densities of renewable energy, especially from wind and also from the sea by those with a coastline. Space for the necessary installations is much less a problem, not least because the amounts of energy needed are relatively small.

Sir Robert M^c^Alpine & Sons Limited have for some time been involved in renewable energy projects. Principal amongst these are wind, waves and tides.

In this paper, we outline some of the work in renewables technology by M^c^Alpine and the results of a study recently made for the Island of Tresco, Isles of Scilly (Fig.1.) in which we assessed the present and likely future energy supplies of the island up to the turn of the Century. These are compared with the contributions which the renewable sources available to that island could sensibly make towards reducing its need for fossil fuels. By virtue of its location, Tresco is in a position to take energy from the winds, waves, tides, direct solar and, using heat pumps, indirectly from solar by exploiting either the sea or its small lakes.

2. Current Energy Supplies

Energy usage on Tresco may be divided into four main categories:-

Space heating
Water heating
Cooking
Lighting, refrigeration and other minor domestic and agricultural electrical loads.

A large number of dwellings use locally available wood (of which there is an abundance) for space and water heating; the Abbey, the Hotel, the New Inn and two larger cottages use oil for this purpose. Coal is used for space and water heating in some cottages and for cooking where 'Rayburn' stoves are utilized. No electric immersion heaters are used.

The majority of cooking on the island is powered by Calor gas. This minimises the load on the diesel-driven electricity generating equipment, which supplies lighting, refrigerators, minor domestic appliances and an agricultural cold store.

The quantities of imported energy supplies and their costs are described below:-

(a) Coal

Coal is brought to the island in 20kg bags and is of 3 types, house coal, 'Maxiheat' and 'Sunbrite'. The average landed 1980 price for this is about £2.50 per bag including 25p per bag transportation cost. The current annual consumption is 5400 bags giving a total cost of £13,500.

(b) Bottled Gas

Calor gas is brought to the island in 3 different sizes of bottles, 32lb, 42lb and 104lb. The current annual consumption of these is 250, 70 and 150 bottles respectively. The cost of this is £7,144 but this figure excludes usage by the hotel and so a more realistic total figure might be about £10,000 p.a.

(c) Heating Oil

The annual consumption of heating oil is 25,000 gallons at 0.66p per gallon, an annual cost of £16,500.

(d) Fuel Oil

Fuel oil constitutes the largest component of the energy imports at 31,000 gallons. The rate is similar to that of gas oil giving an annual cost of £20,460.

The total annual cost of energy, excluding local distribution, indigenous supplies and maintenance of plant is therefore about £60,000

Usage

Due to the nature of the island's industry the seasonal energy consumption pattern is rather unusual. The major industry is summer tourism which is also a comparatively large consumer of energy. Other industries such as flower and potato growing have a relatively low energy demand.

The residential population of Tresco has a core of 140 people who live on the island all year round. During the tourist season, however, this figure nearly triples to about 400. An increase of a further 20% is possible by 1990 giving a potential summer total of about 525 people.

Figure 2 shows in arbitrary units the month by month variation of energy cost for each of the component fuels throughout the year. The balance of use between the fuels and recent unit prices were used to construct this diagram, but changes in units costs and absolute and differential changes make the overall pattern of more significance than the details at the time of preparation. The monthly distribution has been estimated for gas, gas oil and coal throughout the year to give the annual totals. Most interestingly the monthly cost of fuel oil consumption is roughly constant, where one might have expected a summer peak associated with the increase in population and use of refrigeration.

Commentary

In attempting to explain this it is useful to describe the electricity generating plant. Three generators exist. Two of these are Lister-Blackstone EV2 diesel generating sets, each rated at 56kW and both approximately 20 years old. The third is a Mirlees system rated at 70kW but is rather older (1943) and is more expensive to maintain. The 2 x EV2 are used primarily, as these jointly are able to cope with the maximum demand of 100kW. The Mirlees is used during periods of maintenance or breakdown, the latter of which is apparently rare.

Although the winter demand for electricity is likely to be below the summer demand, the use of some electric fires may then take the load above that which can be supplied by one EV2. The need to use two sets, not at full load, i.e. less efficiently, will mean a fuel requirement approaching that during the summer when the diesels are working efficiently closer to full load.

The use of coal will have a winter peak for space heating but there will be a lower usage during the summer for domestic water heating as shown in Fig.2. Similarly, gas oil for heating will have higher use in the winter, but a considerable amount is also needed in summer to provide hot water for the increased residential population. The hotel and the New Inn are closed from the end of October until the beginning of March, and this is reflected in the anticipated consumption. Central heating would be used at these centres during the colder periods of their seasons. The use of cooking gas is higher during the summer season when visitors to the island are present, as indicated in Fig. 2.

The total energy cost has a winter peak but this is not as pronounced as it might have been had the island not been a centre of summer tourism. An important point to note is that increases in the number of summer visitors envisaged by 1990 will bring summer energy consumption up to a level such that monthly energy cost will be fairly constant throughout the year, assuming patterns of use and relative fuel costs are unchanged.

3. Possible functions for new sources

The various sources of renewable energy available to the island could not totally displace the conventional sources unless either very major capital expenditure or a much less secure quality of supply is acceptable. The partial displacement of existing sources is assumed here, in keeping with the contributions which each new source could sensibly make. Each sector of consumer demand will not necessarily be penetrated equally beneficially by the renewables.

(a) Solar Power

This source is best directed towards the supply of warm water. The month-by-month requirement for this has not been assessed though it will tend to follow the seasonal variation in residential population.

The contribution made by solar power is somewhat intermittent but this would not make it unacceptable as a basic source of water heating. Gas oil will be needed to top up as necessary. If half the summer heating load foreseen for 1990 is taken by solar, the saving in oil costs will be about £4,000/year (6% total bill, at 1980 prices).

The total area of solar panels involved would be about 220 or 225 or 56 conventionally sized flat panel units. Other more efficient devices, such as passive tube, are now on the market or being developed and are well worth consideration.

Steady progress is being made with the technology of solar collector design and costs are tending to come down as more units are installed. Although on present costs their value is at best marginal, the annual value of the savings they offer will increase. Against this must be balanced the relatively short life thought to apply to at least some makes of unit on the market. Some collectors being developed are less dependent on direct sunlight and will give a more constant output. All factors considered, there is probably a good case for installing a limited number of collectors of the most suitable design available to give direct experience of their use and value.

(b) Wind Power (See also 4.1)

The annual wind energy source on Tresco is far in excess of its total energy demand and the technology exists to harness it. However, there may be aesthetic and amenity considerations in siting large machines, also the energy these would give appears from the wind records available for the Isles of Scilly to be about five times higher in winter than in summer. Although this bias does not match the unusually consistent demand pattern for electricity alone, the seasonal match is better for electricity and heating combined (Fig. 3). A wind turbine could make a major penetration of winter demand for heating and for electricity.

Wind data are available from Bishop Rock Lighthouse and St. Mary's Airport. It is probable that both records suggest lower speeds than occur over the hills of Tresco. On-site measurements are needed to show the energy available to the island from this source. it is probable, however, that the output of a 25m, 130kW, machine, standing 25m off the ground, would vary thoughout the year as shown in Fig.3. If used for electricity and heating via heat pumps within the demand pattern foreseen for 1990, this machine could supply a monthly average maximum of about 95% and a monthly average minimum of about 30% total load, the average being of the order of 60%.

It should be noted however that due to the high installed capacity of a wind turbine relative to a diesel generator, careful further study is necessary regarding the optimum interface, mode of generation and storage media. Because the resource lacks firmness, it may still be necessary to have available sufficient diesel plant to meet the peak load. The windmill is essentially a fuel saver. However, by lessening the load on diesel generators it will also prolong their life and probably reduce the need for maintenance, that is as long as the wind source is adequately persistent at any speed. Further studies based on relevant wind data are needed to quantify this. Firmness could be improved by using the tidal scheme (d) for storage of energy.

It is normally assumed that windmills should be used to generate electricity though there is also a case for directly coupling their mechanical output to the shaft of the compressor of a heat pump (see (c) below). A single unit could meet both functions. For topographic and logic reasons, however, it is more likely that a small 'lowland' windmill would better suit the relatively low power which might be used to drive heat pumps drawing upon the thermal energy resource of a lake or the sea. The major source of wind is at exposed upland sites, from which the transmitted electricity could still in part be used to drive motors linked to heat pumps sited as suggested above.

(c) Heat Pumps

This is a device for extracting heat from a relatively low grade but abundant source and intensifying it for supply at higher temperature but smaller rate. A well designed heat pump operating over the temperature step-up range of about 10 to 60°C would deliver at least three times as much thermal energy as the electrical or mechanical energy needed to drive its compressor. A continuous input gives a continuous output, which is an unusual characteristic of a heating source (oil and gas boilers operate intermittently). The supply temperature of 55-60°C should therefore be sufficient to meet most if not all of the demands served by an oil or gas-fired heating system set to run at 70-75°C. Surveys of heating system efficiency and losses from buildings will show whether or not the heat pump alternative will perform the same function without additional expense on insulation (see (g) below).

Heat pumps from a size of a few kW to a few MW are in satisfactory operation. Although improvements in efficiency can be expected, the principles involved have been well demonstrated over many years. More than 2 million units are in service in U.S.A. and there are more than 100,000 in Germany. The U.K. market has been slow to develop but is now picking up. Most units use air as the heat source because it is universally available. Water is better for several reasons not least because its annual temperature variation is smaller and the minimum temperature of flowing water is above zero. Tresco has the added advantage of access to both relatively shallow warm lake and sea water, either of which could be utilised by a heat pump.

The hot water output of this device must be piped to and circulated round the building in question. A long run of supply pipe will incur substantial cost and some heat losses. The device should therefore preferably be located close to the consumer. If a piped radiator system has also to be installed this will add significantly to capital costs.

Tresco Abbey, the Island Hotel and the New Inn are well sited to be fed by heat pumps, drawing upon Abbey Lake (TA) and the sea (IH and NI). Links to adjacent buildings could also be considered. Fig.3 suggests the penetration which installations at these three locations could make into the heating load at present met by gas oil. Because the principal benefit would be realised in the winter, the units would not necessarily increase the annual peak demand for electricity above the maximum level foreseen for 1990 (July/August). It is also worth noting that, if necessary, use of heat pumps could be restricted to daily periods of low electricity demand, e.g. at night and during the middle of the day. However, larger heat pumps (or oil fired supplementation) might then be needed to make good the fall in total heat supply as units would be operating for only a percentage of the time.

(d) Tidal Power

The tide range off the Isles of Scilly varies from a maximum at springs of about 6m to a minimum at neaps of about 4m. The full spring-neap-spring cycle takes the normal 2 week period (approximately) hence the source is otherwise available with little variation throughout the year.

A principal cost elements of small tidal power schemes is that of isolating an adequate area of water from the sea and constructing the power house for the turbo-generators. The topography of Tresco already provides a lake (Great Pool) of appropriate area to supply a significant proportion of the island's electrical needs, separated from the sea by a low bank less than than 100m wide. The water in the lake is said to be brackish, indeed it may once have been a bay linked directly to the sea but now isolated from it by the accretion of sedimentary material.

The output from this scheme would be increased if a small amount of pumping could be carried out when the tide is high. This may be possible at night. Excess output from, say, a windmill could be stored in this way. The efficiency of this type of scheme goes down as more pumping is done. It is better to raise the level of a larger area through a small height than a small area by a lot. This option might require that some of the reed beds and the organic deposits which now occupy the margins of much of Great Pool should be removed, to increase the 'live' water volume of the lake.

Our initial estimates suggest that the average output of the scheme throughout the lower half of each tide cycle would be about 70 kW with up to 40% more on spring tides and up to 40% less on neaps. The average pumping load would be about 25kW and the level of Great Pool would rise (and fall) by about 0.3m from its present level.

A detailed survey of the site would be needed before firm numbers for this option can be provided.

(e) Wave Power

The resource available off the north coast of Tresco and the west coast of the Scillies generally is enormous compared with local energy requirements. At the present time the technology to harness it economically has not been demonstrated in the sea though work in hand on the development of several possible devices may prompt this step within the next 3-4 years. The size of demonstration unit anticipated at present will have a rated capacity of about 2MW. Present estimates suggest that as an energy source in its own right it will be relatively expensive and could only be justified if Government supported it on an experimental basis, which is what it would principally be, at least initially.

The size of unit to suit Tresco's needs would be much smaller and correspondingly more expensive per kWh generated. Wave power should not therefore be regarded as a viable source for Tresco alone at the present time.

(f) Conservation

About 40% of Tresco's annual energy requirement is for heating (Fig. 2). General experience shows that this can be substantially and quickly reduced by various means and that these can be implemented quickly and economically using local labour. It has not so far been possible to do more than generally assess the order of improvement which these measures could make if applied consistently to Tresco's various buildings, but it is readily clear that there is substantial scope for benefit.
A 20% reduction in total energy demand for the same level of comfort ought to be attainable. Various ways of contributing to this should be explored.

4. McAlpine Activites in Renewable Energy

4.1 Wind

Over the last six years there has been a upsurge in interest in wind energy in the wake of the oil price rises of the early 1970's. Many countries have national wind energy programmes. These programme have been set-up to investigate the potential for obtaining a proportion of the countries energy needs from the wind, and in some cases to bring about the exploitation of that potential.

In the U.K. there are two industrial consortia having development programmes for large, megawatt sized, wind turbines with support funding from the Department of Energy. One consortium the Wind Energy Group, is led by Taylor Woodrow Construction Ltd and is concentrating on a horizontal axis design. The other is led by Sir Robert McAlpine and Sons Ltd and is developing a British concept, proposed by Dr Peter Musgrove of Reading University, the Variable Geometry Vertical Axis Wind Turbine. This has numerous advantages over the more conventional horizontal axis machines being developed around the world.

The turbines main characteristics are as follows:-

1) It is omnidirectional and so does not need to be turned to face the wind.

2) The output shaft is vertical allowing major transmission components to be situated at ground level.

3) Its performance is as good as and possibily superior to that of horizontal axis machines as it does not suffer the power loss associated with the latter that occurs when the wind veers off the axis of the rotor.

4) The blade shape is simple, straight, untapered and untwisted.

5) The power output of the turbine is controlable by variation of geometry. This allows the machine to operate always with the blades in the unstalled region, thus avoiding the undesirable effects of cyclic stalling.

6) The turbine has a high shut down speed, thus minimising energy loss due to shut downs in high wind speeds.

7) With its lower rotational speed, the dynamic requirements on the tower are less severe.

8) The blades are not subject to reversing stresses due to gravity as they rotate. It is possible, therefore that this type of machine can be built to larger sizes than horizontal axis machines, an important point when considering offshore siting.

The industrial consortium "Vertical Axis Wind Turbines" led by McAlpines also comprises NEI Cranes, Engineering and Power Development Consultants and Aircraft Designs (Bembridge) Ltd. A phased development programme has been instigated and the 2nd phase, the design of a 25 metre diameter, prototype, intermediate size machine, has just been completed (figure 4). It is hoped that this machine will be constructed in 1982/83.

4.2 Waves

Research into wave power only started seriously in 1975/76, but progress has been rapid thanks to extensive support for the study of concepts and design requirements by the U.K. Department of Energy. It is now timely to review progress and prospects for the advance of the more promising techniques through to large scale demonstration in the sea and to the supply of energy to consumers.

The Bristol Cylinder which was originally developed at Bristol University and is now the subject of extensive physical model, computer and design studies led by McAlpine, is generally regarded as one of the most promising wave energy devices so far identified.

Its operating principles were describes in the 1980 Proceedings of this Conference, and have generally changed little in basic detail. However, exhaustive studies have advanced our understanding of both its performance and its optimum specification to a stage where it is possible to judge how best it should be designed and how it will perform over a wide range of sea conditions. Reports are now in preparation covering all aspects of the device, for presentation to the UK Department of Energy at the end of 1981.

The programme for the further development of this and other wave energy devices will be decided on the basis of these reports.

The full wave energy programme has confirmed the substantial scale on which this source could contribute to the nation's requirements. Although more obviously a source of electricity, wave energy could also be used for the production of hydrogen and ammonia. This could be done close to the energy source, and the product shipped by tanker as required. In this way the capacity of the tanker to act as a store or buffer to accumulate the product at its variable rate of supply according to incident wave energy density is important.

The same, of course, is true for offshore wind energy devices, or indeed for the intermittent component of any energy source which may be more difficult and less economic to integrate into the daily pattern of an electrical demand fed mainly by large thermal power stations.

The relative novelty of wave energy means that many components have yet to be tested at the scale, in the numbers and exposed in the way which will be required. It is now timely for this work to be done. There is every reason to believe that conventional, proven engineering can form the heart of this type of project, using materials with a well established record of use in a marine environment. The times required for testing need not be very long and two years should be ample for most of the work. Accelerated methods of study are possible, as is also the direct use of marine experience with structures of similar type elsewhere.

The Bristol Cylinder now comprises standard engineering components, installed and maintained by proven offshore underwater methods and with vessels to a common standard. This is seen as a vital requirement to advance what is, to the energy world a new technolgy. Refinements can come later, and there is every reason to expect that they will, without much delay. New, faster and cheaper designs and procedures follow almost as a matter of course in any project of this nature. To ignore that inevitability at the outset of a project when estimating quantity production costs is to threaten the development before it is properly born. However, the converse of proper cost appraisal prior to injecting risk capital is also vital.

From the thorough, comprehensive studies made of all principal aspects of the Bristol Cylinder during the last 2½ years, the case for taking its development to large scale component testing and demonstration module design has been shown to be strong.

5. Conclusions and Recommendations

It is clear from the foregoing that the renewable sources of energy, in addition to conservation measures, could make a significant impact on the fossil fuel consumption of Tresco.

This preliminary study has identified the more promising options. These will need further detailed study before implementation. Clearly the options with the lowest capital cost and highest economic benefits should be employed first. The order of priority of implementation is given below:-

1) Conservation.
2) Heat pumps.
3) Solar panels - summer water heating.
4) Increased planting of trees and use of wood.
5) Tidal scheme.
6) Wind power. (Current costs of medium sized wind turbines are high as these are largely development machines. Costs will drop with increased production. Small wind turbines could however be used with marginal economics in the current situation).
7) Wave Power (further development needed before suitable devices are available commercially).

TRESCO, ISLES OF SCILLY.

FIG. 1.

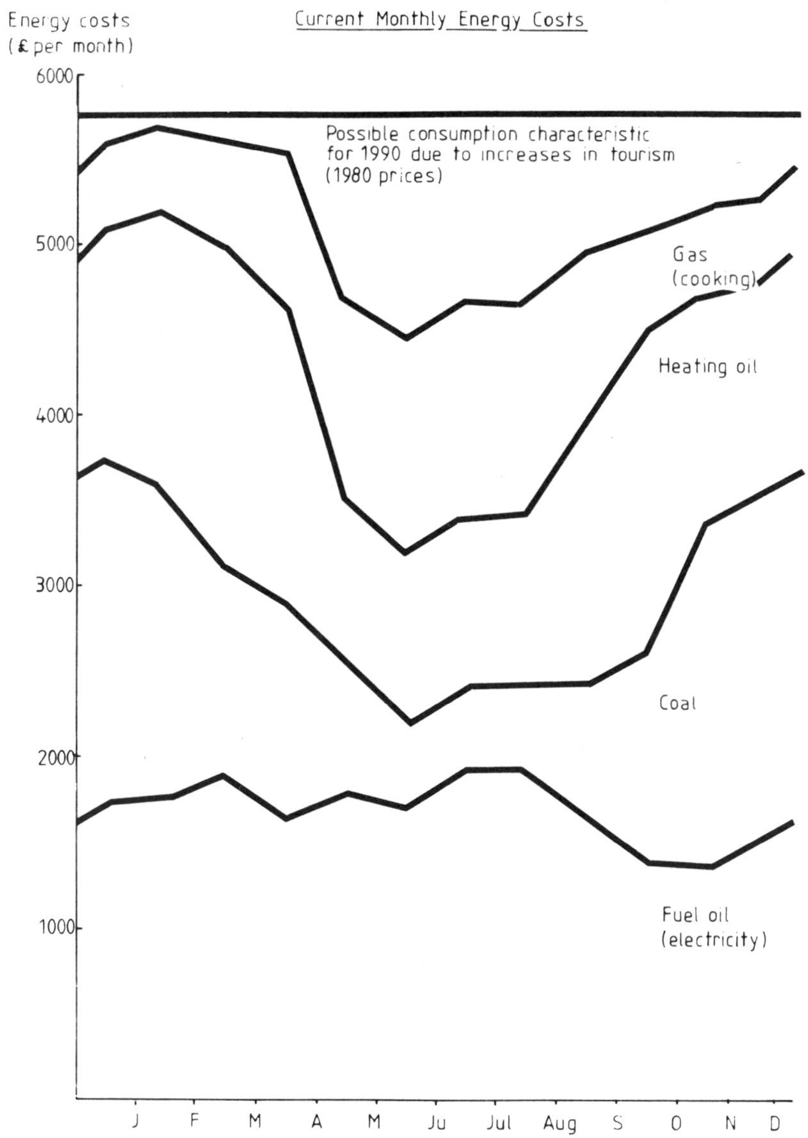
Energy costs
(£ per month)
Current Monthly Energy Costs
6000
5000
4000
3000
2000
1000
Possible consumption characteristic
for 1990 due to increases in tourism
(1980 prices)
Gas
(cooking)
Heating oil
Coal
Fuel oil
(electricity)
J
F
M
A
M
Ju
Jul
Aug
S
O
N
D

FIG 2

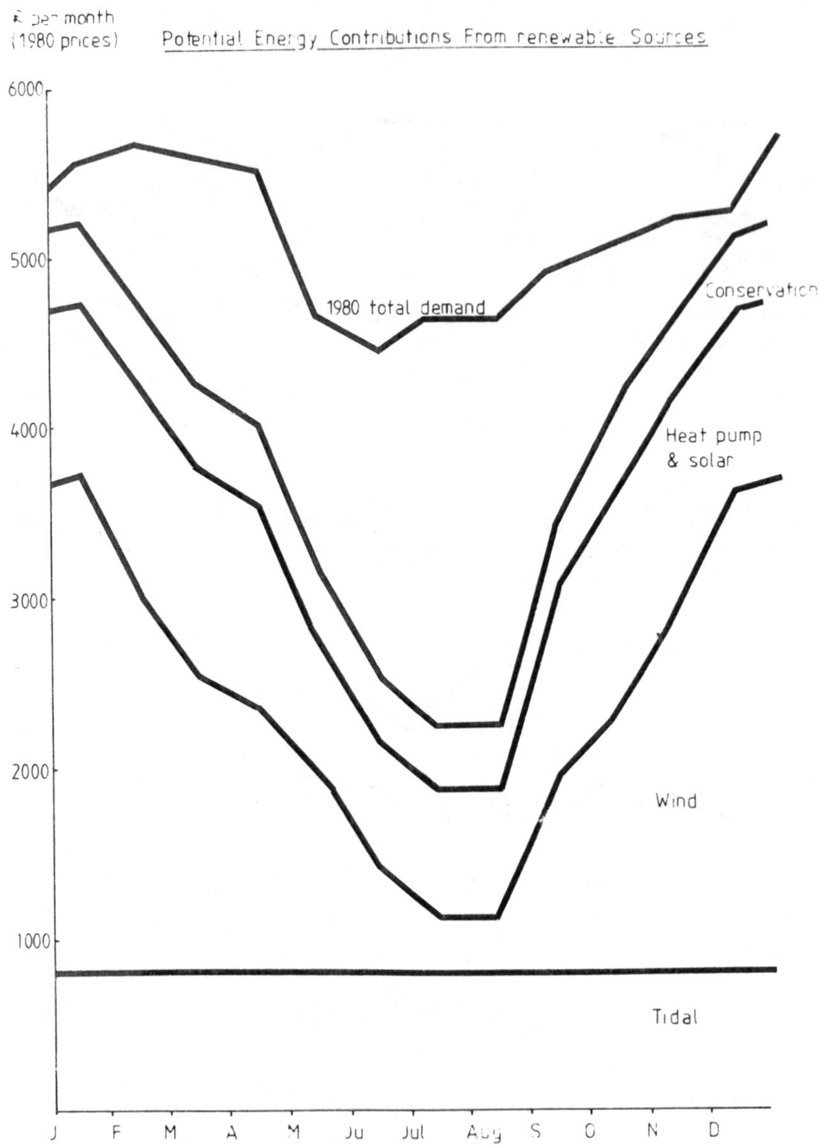

£ per month
(1980 prices)
Potential Energy Contributions From renewable Sources
6000
5000
4000
3000
2000
1000
1980 total demand
Conservation
Heat pump
& solar
Wind
Tidal
J
F
M
A
M
Ju
Jul
Aug
S
O
N
D

FIG 3

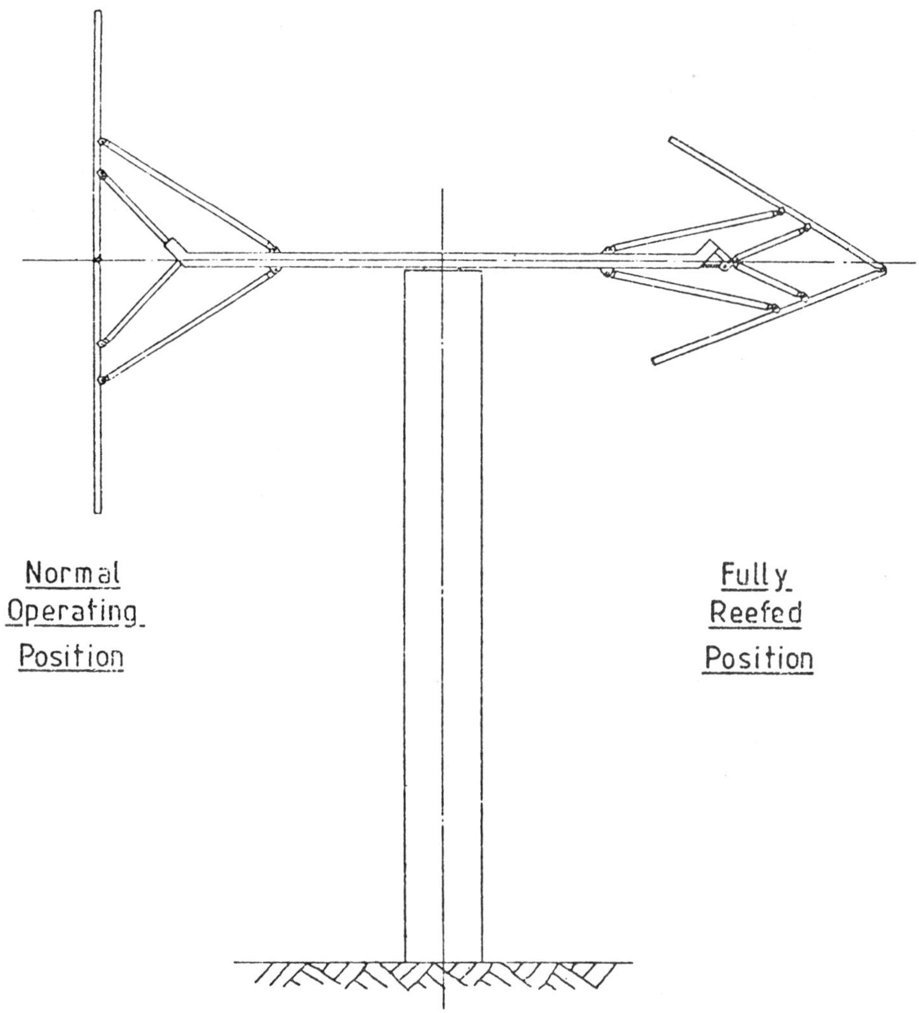

VERTICAL AXIS WIND TURBINE DIAGRAM SHOWING ROTOR REEFING MECHANISM

FIG.4

Second Law of Thermodynamics as a Planning Tool for Rural Development[1]

F. Butera

Istituto di Fisica Tecnica, Viale delle Scienze, 90125 Palermo, Italy

ABSTRACT

Population migration from poorer rural areas to richer industrial ones has been a very common phenomenon in Italy. These depleted rural areas are characterized by a slow pace of development or socio-economic decline. Territorial areas of this kind, defined as Limited Resources Areas (LRA), and their possible development, have been studied in the last two years in Italy, focusing attention on the energy parameter, the role of which is taken into account from the very early stages of the planning process.
Methodologies used and most important results are briefly outlined, as well as the need of a reliable index for energy projects evaluation. It is proposed to derive such an index from the second law of thermodynamics, focusing the attention on entropy flows. The advantage of such an approach - that allows planners to take into account both the energy and the environmental issues - are discussed within the framework of the generic issue of development planning. Examples of applications in the island of Pantelleria are given.

KEYWORDS

Development; energy; entropy, exergy; information; planning.

INTRODUCTION

It has been experienced that the growing cost of energy sources poses serious additional restrictions to socio-economic development of less developed areas. As a consequence the gap between developed and developing countries/areas in the world is widening. It is also recognized that this trend may be reversed only if development plans include rationalization of the territorial energy system and a large use of renewable energy resources.
On this ground a research project has been launched by National Research Council

---

(1) Work financed by CNR-Progetto Finalizzato Energetica.

(CNR) of Italy, whose main objectives may be outlined as follows :
1) Evaluate the impact of the introduction of the design parameter "rational use of energy" in the planning and the development of a Limited Resources Area (LRA). LRA is defined as a mainly rural area at present subject to a process of social and economic decline.
2) Promote, by means of the use of the energy parameter, a new development approach of LRAs, emphasizing the use of local resources in order to obtain a balanced and time-lasting socio-economic growth.
3) Produce a methodology to be used by planners for the analysis of the energy system of LRAs.
4) Identify the criteria that the planners should use in order to evaluate the impact that the LRAs energy system analysis may have on their development guidelines.
CNR research is based on case studies. Six LRAs were chosen and interdisciplinary research groups were set up for each of them.
One of those areas is the island of Pantelleria (83 Km^2; 8,000 inhabitants), in the middle of Mediterranean sea, used - in this paper - as a reference for second law analysis.

ENERGY READING OF THE TERRITORY

Energy, up to now, has been treated as a function on only two variables : quantity and cost. If one looks at the territorial energy system, instead, it appears that energy is function of many variables : quantity, quality, spatial distribution and time dynamics of both demand and supply.
In order to analyse all these factors, a sub-methodology has been developed and introduced in the usual planning methodology : the "energy reading of the territory" (Butera, 1981).
The energy reading of the territory consists of the collection and suitable aggregation of all the energy-relevant data, in order to determine the territorial distribution of (i) all the potential and actual energy sources (renewable or not) both in quantity and in quality and (ii) energy end uses.
For the LRAs studied, both solar radiation maps have been produced, (taking into account the shadows cast by mountains and/or hills), and wind maps (wind speed and direction being strongly affected by orography), by means of ad hoc developed computer models. By overlapping such maps with others, where information about physical characteristics of the terrain is giveng (slope, orientation, soil composition and stability, etc) the best areas for any purpose may be selected (Adamoli, 1981).
Hydro-potential studies, including historical surveys on hydro-power exploitation in the past, have been carried out.
Biomass, wind, geothermal and wave energy potential have been also examined according to the characteristics of the area.
A detailed energy demand analysis has been carried out in each area.
The energy demand analysis also included the collection of detailed energy-related data on the existing building stock, the domestic energy consumption appliances used and their usage pattern (Adamoli, 1981; Matteoli, 1981).
Once the step "energy reading of the territory" is accomplished, it is possible to identify the most promising energy projects, i.e. projects of rationalization of the energy system or involving renewable energy sources utilization. At this stage a question mark appears on the table of the planner : according to what

sort of planning strategy energy projects should be chosen and ranked?
For many planners this question may not arise : they follow the bare monetary approach, therefore the selection parameters is the time of return of the investment. Other planners, more concerned, will use a cost-benefit analysis, trying to include in the evaluation also social and environmental issues. Unfortunately most of these issues cannot be monetized and the result is not satisfactory.
As an example, lets take Pantelleria case-study.
The energy reading of the territory showed (Rodonò, 1980) :
1) 49% of the fuel imported is used for electricity production and 36.7% is used for transportation
2) 47% of electricity consumption in the domestic sector is for water heating
3) energy demand for space heating is very low, because of the short and mild winter, but it is all electric heating
4) potential for solar, wind and geothermal energy is very high :
- 4.8 Kwh/m^2 day on horizontal surface for solar energy
- 1.7 Kw per m^2 of blade-swept area obtainable by wind energy
- 200 TJ/year (thermal) for geothermal energy.

The economy of the island is mainly based on agriculture (vineyard and capers) and tourism. The main constraints to development are the lack of water and the unreliability of the harbour. Because of the absence of fresh water resources, irrigation is not possible. Per capita water consumption has been estimated 25 litres/day; 40% of this amount is rain water collected in individual reservoirs and 60% is shipped from the mainland.
Several short and long term energy projects may be identified :
- water desalination with several options : solar heat, geothermal heat, diesel-electric generators heat waste, wind powered electro-dialysis
- electricity production : by solar energy (photovoltaic), by wind energy, by geothermal energy
- hot water production by solar energy
- greenhouses for agriculture
- existing building insulation.

Some of these energy projects are being studied, evaluating also their social and economic impact (Matteoli, 1981).
The most suitable projects will be identified and priorities will be stated; but will it be the right answer to the problem of a long lasting development leading to an improved quality of life? Unfortunately none of the methodologies presently available is able to cope with the complexity of the problem; only a few parameters at time can be optimized, and little or no projection of the future consequences of the projects on the whole system is possible.

A COMPREHENSIVE APPROACH

In order to try to solve such an intricate question it is here proposed a comprehensive approach based on second law of thermodynamics, i.e. on the entropy analysis of energy systems.
An open system (chemical components reacting in a container; a biological system; an ecological system; a city, etc) is able to mantain (or even increase) its internal organization because is fed by a flow of "organization" coming from the external environment, i.e. a flow of low entropy.
The variation of entropy dS during a time interval dt may be written (Nicolis,1977):

$$dS = d_iS + d_eS \qquad 1)$$

where d_iS is the internal entropy production, due to irreversible processes inside the system, always ≥ 0; and d_eS is the flow of entropy due to the exchanges with the surroundings. For an isolate system d_eS is zero and therefore $dS = d_iS \geq 0$: entropy cannot decrease. In an open system, although d_iS is never negative, the flux term d_eS has no definite sign. As a result, during evolution a system may reach a state where entropy is smaller than at the start, or mantain a steady state.

If we apply these concepts to a biological system, d_iS is the entropy production due to the internal metabolism, i.e. to all those irreversible processes necessary to mantain the system organized, alive; d_eS is the food, the oxygen, the water. The same concepts may be applied to the territorial energy system, where d_iS is the metabolic entropy production due to all the energy conversion processes; d_eS is the flow of fuel or electricity or negative entropy embodied in the materials crossing the border of the territorial system.

Rationalization of the energy system must then mean reduction of the "metabolic" entropy production, and this approach - as it will be shown later - gives an answer to the problems and uncertanties expressed previously about energy projects. The negative entropy flow d_eS may be expressed as $d_eS = d_eS_{nr} + d_eS_r$, where d_eS_{nr} is the non-renewable negative entropy flow, i.e. net flow of oil, gas, coal, electricity (if produced by non-renewable energy sources) and materials; d_eS_r is the renewable negative entropy flow, i.e. solar radiation, wind, hydropower, biomass, etc.

The renewable negative entropy (negentropy) flow d_eS_r crosses the borders of the territorial system either if we use it or we don't. In both cases this high quality energy is degraded into low quality. There is then a sort of "base" entropy production to which is added the degradation of d_eS_{nr}. In a mature ecosystem $d_eS = d_eS_r$, and the entropy production d_iS is only a part of the total degradation of d_eS_r. In our territorial systems d_iS is almost entirely the result of the degradation of d_eS_{nr} :we are therefore are responsible for an "excess entropy" production. This is a very important point and it is strictly linked to the problem of environmental pollution and its cost for human society. The excess entropy is "noise" for the natural ecosystem. Ecological systems are drowned in a flood of noise : the flow of high and low entropy we pour in the environment. In order to repair the demage, the current practice is to "filter" this noise with anti-pollution devices; for producing, installing and letting them work an additional negentropy flow is activated and the overall entropy production is furthemore increased. In fact organization in a living system can be mantained only at the expenses of other organization.

ENTROPY ANALYSIS OF ENERGY PROJECTS

A few more concepts must be introduced in order to apply the general principles described in the previous paragraph. Taking into account the second law of thermodynamics, energy may be expressed in the form (Szargut, 1980) : $E = EX + AN$, where E is energy; EX is exergy, i.e. that part of any form of energy that can be converted into shaft work; AN is anergy, i.e. the complementary part of energy that

cannot be converted into exergy at all.
The laws of thermodynamics are then reduced to this simple statement : the sum of exergy and anergy remains constant in every process. Real processes convert some exergy (high quality energy) into anergy (useless energy). Exergy is exempt from the law of conservation, so that its balance must be closed up by the anergy production occurring inside the considered system.
Exergy calculated for a hypothetical reversible process coincides with what many authors call available energy or available work.
Anergy is also called lost work or exergy loss, and is related to entropy production : $AN = T_o \Delta S$, where : T_o is the temperature (°K) of the surrounding environment and ΔS is the entropy increase in the system.
A very important parameter that can be derived is the exergetic efficiency

$$\eta = \text{useful exergetic effect/consumption of the feeding exergy} \qquad 2)$$

The denominator of 2) is the sum of four terms : a) exergy of the energy source; b) exergy consumed for the production of the materials constituting the system; c) exergy consumed for building the system; d) exergy consumed for the maintenance of the system. Therefore in a real system $\eta = 1$ is an upper impossible value.
The higher is the value of η, the lower is the metabolic entropy production $d_i S$ of the system. A plot of η versus the size of a system shows always a maximum.
It may be objected that exergetic analysis does not take into any account economics. Actually, planning with the second law of thermodynamics implies planning with long term economics, reminding how economics is defined : "the study of the allocation of scarce means among competing ends". All the scarce means we deal with may be lead back to only one scarce mean : the negentropy flow.
The relationship among development, growth, scarcity, quality of life, economics, planning and entropy have been analysed by many authors (Georgescu-Roegen, 1971; Daly, 1979; Lustig, 1979), all of them pointing out the need of this kind of nonmonetary approach to economics.
As an example, one of the proposed Pantelleria's energy projects has been examined: flat plate collectors for domestic hot water production.
Calculations have been carried out for a commercial system to be installed in a multi-storey building with 10 apartments. The performance of the system has been evaluated by means of a simplified method (Butera, 1978). The produced exergy is calculated by means of the second law efficiency for solar energy systems, according to Kreider (1978).
The exergy spent for producing the system has been estimated from energy data (Gugliermetti, 1980), related to commercially available systems. No allowance has been made for installation and maintenance exergy and for electricity needed for water circulation; for this reason system's life has been considered 10 years. Results (Table 1) show that the optimum is for a solar fraction of 40%.
In the last two columns of Table 1 two important parameters have been included. The first is the "reproduction" capability of the system, i.e. the ratio of the exergy produced by the system to the exergy spent to produce it.
For solar fractions above 80% the exergy used to build the system is more than the exergy obtained, i.e. the entropy produced to build the system's components is greater than the solar negentropy gained. The second parameter is the redundance $R = 1 - \eta$ of the system. Redundance is an index used in information theory and indicates how badly the potential information of the source is used in transmission messages. Some redundance is anyway necessary because this gives flexibility to

the system : it allows the correct interpretation of the message even if some noise is disturbing it.
The same concept has been used in ecology (Odum, 1970). An ecological system shows always a certain amount of redundance to protect itself from noise, i.e. impredictable variation of the flow of energy, matter, living organism across it.
In other words, a redundance equal to zero means that the system is perfectly organized, that all the potential information is used, that the entropy production is the minimum possible, but also that the system is very fragile; the opposite when R = 1. A system with high redundance requires large flows of negentropy to self-sustain.
Table 1 shows that the minimum redundance of the solar energy system analysed is 0.58. Associated to this value there is the fragility of the system : a 60% probability of hot water "black out". Complete reliability of the system is associated to a redundance of 0.82. A different system, made of more appropriate components, may show a lower redundance. Exergetic analysis allows a better choice of materials. When hot water is produced by means of an electric resistance, the exergetic redundance of the system is nearly 1 (R = 0.94) : the electric system is less "informed". There is another conceptual difference between the two systems : the exergetic redundance of the solar energy system is "structural" redundance, due to the exergy spent for producing the system. Otherwise the exergetic redundance of the electric device is a "flowing" redundance, due to the waste of a flow of non-renewable high quality energy. This result may be generalized : the shift from non-renewable to renewable sources of energy requires the constitution of a "capital" of structural exergetic redundance.

TABLE 1 Exergetic Analysis of a Solar Energy System

% solar	coll. m^2	produced exergy(GJ/yr)	exergy to build the system(GJ/yr)	reproduction capability	η	R
20	5	1.98	1.35	1.46	0.42	0.58
40	10	3.96	2.57	1.54	0.42	0.58
60	16	5.94	4.32	1.36	0.40	0.44
80	28	7.92	7.56	1.05	0.30	0.70
100	60	9.9	16.2	0.6	0.18	0.82

CONCLUSIONS

It has been shown that the energy reading of the territory is only a first step in the more complex procedure that should be followed in order to give more comprehensive answers to the intricate problem involving the relationship among development, energy and environment.
Exergetic analysis (or entropy analysis) is the following step, more appropriate and rich of potential success.
The main conclusions that may be drawn, as a consequence of the entropy approach applied to territorial planning are the following :
a) Exergetic analysis shows us that for energy projects absolute optimum solutions can be found, that take into account both energy and environment. These absolute

optima are not "static" optima : they change as the organization of the system changes, being linked to the exergetic cost of production of components.
Optima are independent of the current economic analysis, i.e. are avoided the uncertainties deriving from the arbitrary assumptions generally made about interest rates, future cost of energy, monetary value of abstract benefits etc. Moreover, exergetic analysis allows the identification of the optimum size of energy converters : the debate on large scale or small scale projects may be settled on a scientific base.

b) Institutions have a powerful tool to decide wich project should be funded or incentivated and to which amount.
The aim of the planner becomes just one : to minimize metabolic entropy production d_iS and to substitute non-renewable negentropy flow d_eS_{nr} with renewable negentropy flow d_eS_r.

c) In order to reduce the flowing and structural redundance it is necessary a large input of information (a cogeneration system is more complex of the sum of an electricity generation plant and a heating system, and it may be - if exergetically optimized - less redundant).
The information flow in the system is not free : it requires negentropy (Brillonin, 1962). The information input is therefore combined with exergy input. This information-exergy is stored in the system and allows a reduction of entropy production, i.e. of the non renewabale negentropy flow (Butera, 1979; 1980). This is the kind of substitution to aim to, not the simplistic and useless substitution of an expensive or non renewable energy source with a cheaper or renewable energy source.

d) The greater is the entropy of the energy source, the greater must be the information-exergy stored in the energy conversion system. Introduction of renewable energy sources in the territorial energy system requires a large investment of both exergy and technological information, because the entropy of renewable energy sources is higher than the entropy of fossil fuels.
It is necessary to rearrange the territorial energy system in such a way that non-renewable exergy resources (oil, gas, coal) are used to create new stocks of exergy, instead of using them as flows. Non-renewable exergy sources should be extracted, coupled with information and stored as structure of the energy system. As a consequence large financial investments are necessary, larger than those recognized as necessary in the pre-energy crisis development philosophy.
The large input of information required implies a very fast growth of the education level of both population and institutions.

e) In order to reduce the financial effort necessary for creating an informed and flexible territorial energy system, initially the structural redundance of the system could be kept low, using - instead - the flexibility (i.e. redundance) of population and individuals.But, again, this can be obtained only if cultural level of population and consciousness of the problem is high.

f) The previous remarks throw us in the core of the problem : scientifically and technically everything may appear quite clear. What about the social issue? What kind of social system is behind the entropy analysis of development?
No definite answer may be given to this problem. We do not have other models of systems working with a philosophy similar to the one outlined, except ecosystems; the approach followed in this paper is still very far from being fully examined in all its implications. It is more a theoretical framework than a methodology : more research and debate is therefore necessary.

It appears quite clearly, anyway, that planning territorial energy systems - especially when renewable energy sources are involved - requires a new planning philosophy. This is particularly urgent for rural developing areas, in order to achieve a real and long-lasting development : their stock of information-exergy is far lower than that of developed areas, and it must be built up properly.

REFERENCES

Adamoli, M, E. Manzini and C. Ratto (1981). Uomini, energia e territorio. Ricerca su una vallata ligure : la Fontanabuona. Proc. of Quarto Seminario RERE/PFE/CNR PEG, Milano, pp.295-309.

Brillonin, L. (1962). Science and Information Theory. 2nd ed. Academic Press. N.Y.

Butera, F., G. Cannistraro and G. Silvestrini (1978). Solar energy for industrial processes : a simplified design method. Revue Internationale d'Heliotechnique, 1er semestre, 8-14.

Butera, F. (1979). Quale energia per quale società. Mazzotta, Milano.

Butera, F. (1980). Informazione ed approvvigionamento energetico dei sistemi territoriali. HTE, N° 3, 60-67.

Butera, F. (1981). The energy parameter role in the development of Limited Resources Areas. Proc. of First International Energy Agency Conference on New Energy Conservations Technologies and their Commercialization. Berlin.

Daly, H.E. (1979). Entropy, Growth and the Political Economy of Scarcity. In V.K. Smith (Ed.), Scarcity and growth reconsidered, Johns Hopkins University Press, Baltimore and London. pp. 67-94.

Georgescu-Roegen, N. (1971). The Entropy Law and the Economic Process. Harward University Press, Cambridge, Ma.

Gugliermetti, F. and A. Porru (1980). Impianti solari a pannelli piani : energia di fabbricazione ed energia prodotta. In Il Nodo Energia, E.A. Fiera di Bologna, Bologna. pp. 309-315.

Kreider J.F. (1978). Solar energy and the second law of thermodynamics. Proc. of International Solar Energy Congress, New Delhi.

Matteoli, L. (1981). Studio per l'integrazione energetica e ambientale dell'isola di Pantelleria. Proc. of Quarto Seminario RERE/PFE/CNR. PEG, Milano. pp.311-320.

Nicolis, G and I. Prigogine (1977). Self Organization in Non-equilibrium Systems : from Dissipative Structures to Order trhough Fluctuations. Wiley-Interscience, London.

Odum, H.T. (1970). Environment, Power and Society. Wiley-Interscience, New York.

Rodonò, G. and R. Volpes (1980). Indagine sperimentale sul sistema energetico dell'isola di Pantelleria. Quad. Ist. Fisica Tecnica dell'Univ. di Palermo, n.43.

Szargut J. (1980). International progress in second law analysis. Energy, 5, 709-718.

The Potential and Need for Energy Co-operatives

P. J. Meynell

British Anaerobic and Biomass Association Ltd., The White House, Little Bedwyn, Marlborough, Wiltshire, UK

ABSTRACT

This paper presents the idea of energy co-operatives as a means for introducing energy saving technologies into social situations. Because of economies of scale etc, many technologies lend themselves to co-operative action, as opposed to individual action. However, if co-operative action is to be effective, the motivation must exist in all the members and it must be seen to be satisfying a definite felt need. Nevertheless there are considerable advantages to group action which is seen as the only way in which some technologies can be used by the poorest sectors of developed and developing countries.

KEYWORDS

Energy conservation; new and renewable energy sources; co-operatives; group action; community energy supplies; uptake of energy technologies.

INTRODUCTION

With the considerable interest in energy conservation and new and renewable energy sources, there has been a considerable emphasis on the technologies involved and their improvement. This is very necessary, for without further development many of the technologies would remain nonviable. However, the social aspects are often neglected and it is these which may well define the viability in a given situation. This paper attempts to put the case for co-operative and group action for adoption and adaptation of some energy saving technologies to the varying social needs and conditions throughout the world. The research for this paper was carried out for the proposal for a joint FAO/ICA (Food and Agriculture Organisation/International Co-operative Alliance) programme for energy use and conservation in agricultural co-operatives and other rural groups. It was prepared in collaboration with the Plunkett Foundation for Co-operative Studies for implementation after the UN Conference on New and Renewable Energy Sources. (Meynell and McKone, 1981).

WHAT IS MEANT BY CO-OPERATIVE ACTION?

There is nothing magical about co-operative action; a co-operative is merely an association of people i.e. the members, who have agreed to join together to achieve a certain purpose. In most instances these purposes are financial or for the provision of services to increase the efficiency of the member's own businesses etc. The only difference between co-operative and private enterprise is essentially the degree of control held by the members over the management and allocation of the benefits. There are many forms of co-operative or group action ranging from the large formalised co-operatives to the more informal, smaller groups, communities or pre-co-operatives. Similarly the range of activities is very wide, from the industrial, and retailing co-operatives through to agricultural supply and marketing co-operatives, and rural development groups set up to stimulate some economic activity in deprived rural areas. It is the latter two areas which will be considered more fully in this paper, because they relate more directly to energy for rural and island communities. Nevertheless in existing co-operatives, as with any enterprise, the search for ways of reducing the cost of inputs such as fuel is of vital importance to continued economic viability.

Agriculture is especially interesting as far as energy is concerned, because it not only consumes energy but also holds the potential for being the largest producer of alternative forms of energy especially biomass. In many countries of the world, both developed and less developed, agricultural co-operatives form by far the largest single sector within agriculture. Thus there are three distinct energy functions for the co-operative. The first is to reduce conventional energy usage and hence costs, by a variety of means such as insulation, better maintenance and operation of machinery or by substitution of conventional fuels with alternative energy sources e.g. wind/water pumping. Thus the prime activity of the co-operative is not changed, but the inputs to production have been altered; alternative energy sources are produced and used on site.

Secondly there is the energy co-operative in the strictest sense, the enterprise which produces a fuel for outside sale as one of the principal economic activities. Although perhaps with limited possibilities in the U.K., the overseas potential for energy co-operatives producing, say power alcohol from biomass, is great.

Thirdly and perhaps most appropriately for community use, there is the production by the group of energy supplies - electricity and/or fuel - for the domestic use of members. The prime examples of this is the community fuel wood lot, most successfully exploited in Korea, China and India (Noronha, 1980) (FAO, 1978), and the village biogas plant e.g. the different approaches found in China and India.

WHY CO-OPERATIVE ACTION?

Leaving on one side those energy technologies which any enterprise should consider as a means of reducing input costs, be they co-operative or not, there are certain technologies which lend themselves to group action. Thus, any dairy should consider installing a heat exchanger to transfer the heat of the fresh milk to a smaller quantity of water for washing down etc, but the use of sewage and animal wastes to produce biogas could be organised co-operatively.

There is no doubt that co-operative action is hard work and can be very inefficient, but it is argued that for many groups and communities, it may be the only way in which they can take advantage of economies of scale <u>and</u> make use of the technologies. There is very little incentive to do something co-operatively if it can be done easily by individuals. The motivation and conviction that group action will achieve a certain felt need is necessary in order for that action to succeed - for all members to pull together. The advantages need to be obvious in order to overcome the problems which will certainly arise.

Table 1 shows some of these benefits which may be applicable to group action and

Table 2 shows how these benefits can be achieved - the advantages of group action over individual action. Thus a co-operative may add a new product to its activities, and generate more income; with changes in price structures the production of power alcohol or vegetable oils to replace diesel oil may become more profitable to the group than food production. It may wish to generate more income by adding value to an existing product or by reducing wastage in that product; vegetable processing and crop drying may achieve both these ends. In these cases the individual might not be able to afford the processing or drying plant and thus turn to the co-operative to provide a joint facility. One example comes from France where a co-operative of 300 farmers have built a grain drying facility using straw as the main fuel.

Similarly the need for economies of scale both in purchase of capital equipment and in the operation using communal resources, may encourage co-operative ventures. The production of biogas at a centralised plant fed by animal wastes brought in by the members forms the basis for the village energy supply in several Indian villages. Similar ideas have been considered for biogas production from seaweed in Scottish island communities. Any scheme in which capital equipment needs to be bought as replacement for regular supplies of fossil fuels e.g. solar water pumps, wind pumps, animal draught power etc, lends itself to co-operative action if the purchase price is outside the individual's means, and if the smallest viable scale provides more than the individual would require.

The other area in which co-operatives have traditionally been most effective is in the provision of services which it would be costly or difficult for the individual to obtain. These may incidentally provide another income generating activity for the community and so be useful socially. Maintenance and servicing of equipment for energy efficient running, and training and advisory services for energy saving are examples. The appointment of co-operative advisers to perform energy budgets on farms and advise accordingly has already proved its effectiveness in Ireland. (Co-op Ireland April 1981). Similarly a livestock co-operative might organise a maintenance team for the biogas plants owned by the members.

The other area of co-operative activity has been in purchasing and marketing of inputs and products. There are a number of co-operative organisations which buy fossil fuels in bulk and so reduce input costs for members. Indeed in Canada co-operatives have been encouraged to move into the oil and gas exploration business, but this is for largely political reasons. In an isolated area of the U.K. where the local petrol stations were increasingly threatened by the distributors who wanted to discontinue supplying such small outlets, the local co-operative made an agreement with the oil company to take over the supplies, knowing the vital role played by these garages in the rural community. (Farmers Weekly 10.4.1981).

Marketing is often an activity neglected by farmers more interested in the production side. Co-operative marketing removes this burden and makes it more efficient. In terms of energy co-operatives, marketing can be envisaged where there are a number of small producers of charcoal, for instance, who might be unable to reach the more profitable town market because of distance; they thus would have to rely on the middle man, unless co-operative marketing could arrange the packaging, transport and storage of the charcoal for them. In India, for example, wood collectors are very marginalised people and a marketing group would begin to help them out of the trap between the middlemen pressing for low prices on the one hand and harassment by the Forestry Service on the other.

Co-operative action even at this level does provide the group with a voice and this can be very important politically in representing the interests of the group. At higher levels, the political muscle is increased and indeed in Japan the co-operative apex organisation has obtained government subsidies for the construction of small scale hydropower plants generating between 150-10,000 kilowatts for local communities.

Table 3 shows some of the advantages for development agencies in dealing with groups for the transfer of energy saving technologies. This is not considered further in this paper, but is included for the sake of completeness of Table 4 which represents a subjective assessment of the suitability of many energy saving techniques for co-operative and group action.

These, then are some of the advantages offered by group action. The illustrations given show the potential, and highlight the fact that usually it is the co-operative activity which provides the means for using the technology not vice versa. In many situations the uptake of technology is slow and difficult - an example of this being the use of improved woodstoves - partly because of the social conditions. An existing co-operative provides the social institution in which proven technologies can be fitted.

Many of the new technologies require an integration and overall management of resources which would be impossible to achieve on a small scale. They may be achievable using a number of distinct operations each dependent upon the other and this may only work in the context of co-operative enterprise. The concept of integrated farming, usually incorporating some form of anaerobic digestion of animal waste at its core, is inherently co-operative. A banana co-operative in Columbia is starting a cattle business fed on waste fruit and grass. The cattle waste is digested to produce biogas to light the cattle sheds; worms are grown on the digested waste which is used as a fertiliser for the grass and fish ponds and the worms are fed to geese so that from the banana business a whole series of new products are derived.

HOW TO ACHIEVE CO-OPERATIVE ACTION FOR ENERGY?

With an existing co-operative the problem of achieving some action on energy saving should not be too difficult since the framework already exists. Any activity which has the potential for reducing costs should be examined carefully in the light of the situation. The problem is that with fossil fuels the choice of energy sources is relatively small and well documented. Now with rising fossil fuel costs, the choice of alternative fuels is much wider with varying degrees of suitability and stages of development - a much more difficult decision. There is a real need for expert advice for co-operatives to sort out what is and is not feasible.

However, it is the community aspect which offers the greater challenge, because often the main common interest which people in a community have is the fact that they all live there. The more isolated communities have been amongst the first to realise their dependence upon expensive and unreliable fuel supplies and that they must organise themselves to make use of local resources. Not many have got further than discussions amongst themselves about the possibilities. In very recent years such community discussions have been held in isolated communities of Norway and Scotland, and this must be the way to start.

Notwithstanding the concern which these communities and others are showing, it is probably fair to say that fossil fuel prices and availability are still the most cost effective and reliable energy sources and that until this changes, alternative methods will remain discussion points only. In developing countries the position may be very different, especially since the commonest fuel is still wood and charcoal. The urgency for community fuel wood lots and other means of easing the pressure upon an already deforested countryside will grow.

This highlights the necessity in any community project, that for success it must be seen to satisfy a felt need of the community and to be high up in the priorities of these needs.

A good example of this is the solar water still installed about 10 years ago for an island community off Haiti, in which the need for a reliable supply of drinking water became increasingly felt as a result of climatic changes due to deforestation. The whole community became involved, including the co-operative, church and local power holders, so that with technical help from a US University they provided themselves with an adequate and safe water supply. (Alward 1980).

The likelihood is that the need for fuel will rise steadily up the list of priorities; until recently the abundance of fuel has not warranted much community action - energy has not been in the past a great motivation for communities. This is changing and it is apparent that in the words of an FAO paper on Energy in Agriculture and Rural Development (FAO 1980) "a good share of the action for using improved renewable energy sources will depend upon communal action and relevant aspects of this must be incorporated into programmes in support of participatory, local community organisations".

REFERENCES

Alward, R. (1980). In Ceres (FAO) Dec. 1980.
Co-op Ireland, April 1981. ICOS Energy Conservation Workshop.
FAO (1978). Forestry Paper No. 7. Forestry for local community development. FAO. Rome.
FAO (1980). Energy in Agriculture and Rural Development. COAG/81/9 FAO Rome.
Farmers Weekly 10.4.81.
Meynell, P. J. and McKone, C. (1981). Proposal for a joint FAO/ICA Programme for energy use and conservation in agricultural co-operatives and other rural groups. Plunkett Foundation, Oxford.
Noronha, R. (1980) Village woodlots: are they a solution? Paper published by National Academy of Sciences, Washington D.C.

Energy Demand Analysis as a Planning Instrument in Limited Resources Areas: the Enza River Valley Case

W. Ganapini

CNEN-FARE, CSN-Casaccia, Rome, Italy

ABSTRACT

Energy demand analysis has been applied to the Enza River Valley, a representative limited resources area of the Appennines. An investigation about the effects on its energy and natural resources balance deriving from the prevalent development model has been carried out. We showed that it would be necessary to apply to appropriate technologies in order to exploit correctly human, natural and energy local resources.

KEYWORDS

Limited resources area; energy analysis; land use policies; appropriate technologies; energy saving .

INTRODUCTION

The aim of this research, started in 1978, has been to single out and study operative instruments in order to change the underdevelopment situation of limited resources areas (LRA).
How to reach the goal of an "appropriate" development, characterised by a rational use of local resources, for these areas ?
The rational recovery of LRA and the optimum use of local energy resources, mainly of the renewable, are affected by the availability of appropriate natural, social, economic political and cultural conditions.
On the other hand, it is necessary to apply to appropriate technologies, which should have a relatively work's high intensity, in order to determine an industrial decentralization into small units widespread over the area, to exploit human, natural and energy local

resources, to save energy and materials and to respect natural, social and cultural pre-existences(Ganapini, 1980).
The research task has been divided in two parts : the first one, finalized to the definition of a resources geography, the second one concerning the elaboration of a development plan and of the operative instruments of intervention.
Schematically, resources have been divided into physical, anthropic and energy resources and their investigation has been carried out as follows :

1. Physical geography of LRA

1.1 Morphological analysis (altimetries, slopes, catchment-basins, solar radiation);

1.2 Geological analysis (lithology, permeability, erosion);

1.3 Soil potential use analysis.

2. Anthropic geography of LRA

2.1 Actual soil use analysis;

2.2 Pollution sources analysis (localization, quality and quantity of pollutant emissions);

2.3 Hierarchy of settlemens (rank, size);

2.4 Demographic analysis (random sampling and process computing of structural data concerning the population);

2.5 Employment balance.

3. Energy resources geography

3.1 Actual energy demand and balance in LRA (Basic methodology to enphasize energy saving opportunities);

3.2 Local energy resources availability (Basic methodology for a preliminary definition of energy choises for the development plan).

Besides, we have to develop historical research on human settlements in order to define the essential links between population and territory, between resources and technologies.

THE ENZA RIVER VALLEY

Analysing the Appennines as a typical italian LRA, we chose Enza River Valley as a representative test-area. The Enza River Valley is the natural boundary between Reggio Emilia and Parma provinces.
It originates at 1425 m. above sea-level (Monte Palerà), runs through 112 km. and has an average slope of about 7 %.

Its basin measures 670 sq. km.; the mountainous territory (above 1000 m.) represents about 18 % of the whole basin, the hilly country (over 500 m.) is 37 %, the lowest hilly surface (200 m. - 500 m.) is 31 %, the plain is 14 %.
From a geolithological point of view, the basin is characterized by horizons belonging to the quaternary.
From an agronomic point of view, there is a medium mixture soil, with a good hydric receptivity.
Concerning the atmospheric precipitations, we found 800-850 mm. H_2O/y as average value.
Seasonal average is the following :

spring	190-200	mm.	H_2O
summer	150	mm.	H_2O
autumn	260	mm.	H_2O
winter	190-200	mm.	H_2O.

The examined catchment-basin suffers from soil erosion deriving from considerable slopes of hills and tributaries and from poor vegetation covering.
The existing coppices do not contribute to soil erosion control and water detention: moreover prevalent tillage methods have damaged a so weak balance.
The upper valley, which fell into decay since the Second World War, shows serious alterations : many landslides, for instance, have destroyed Succiso village.
In the hilly territory, we found a high pollution degree, restulting from uncontrolled landfilling of urban solid waste, sewer water and intensive animal breeding (pigs, calves, ecc.) slurries.
The population of this area is 35.000 people, with an extremely decentralized kind of settlements.
On an average, there are 3.2 people in each family and 1.6 of them has a job; 1.5 people is over 60 years old.
Concerning working activities of the head of family, we have 1 % of managers, 8 % of employees, 5 % of craftsmen, 5 % of tradesmen, 34 % of workers, 18 % of farmers, 29 % of retired persons.
In the case of the whole active population, we found 1 % of managers, 14 % of employees, 6 % of craftsmen, 9 % of tradesmen, 50 % of workers, 20 % of farmers.

METHODOLOGICAL ASPECTS OF PLANNING PROPOSAL

The data related to the points 1.2.3. (Introduction) have been translated into thematic charts (ratio 1 : 50.000); crossing them, we obtained an ecological and energy reading for the examined valley. From this analysis we single out the critical areas, for which renewable sources (sun, wind, biomass) are available and where there is, at the same time, a lack of conventional sources (electricity, fuel,

etc.) or an improper use of them.
In this way, we defined a development plan finalized to an energy conservative model, to be pursued whether through short-term interventions (modifications of actual production cycles and of prevalent technologies) or through middle-and long-term choices (suggestion of new land use policies, new industrial and agricultural activities, etc.).
Planning, in our opinion, is not only a "method for a rational use of resources", but also a social instrument to achieve both civil and democratic growth. From this point of view, we have considered rural areas not as residual in comparison to the urban ones : the agricultural soil, in our planning model, has been treated as a finished resource. Agriculture has not to be considered only from a short-term economic point of view, according with the prevalent development model; for a new, long-term, approach, it is necessary to measure physical resources availability, to protect these resources from non agricultural uses, to assess human work needs required to exploit these resources and to evaluate financial costs of this policy.
Commercial and public administration activities have to be rationalized in order to increase their economic and social productivity, maintaining their actual employment level; therefore, it is necessary that public government gives a wide range of professional choices to young people, by improving technical training.
On the other hand, industry has to secure not only turnover's covering, but also new employment occasions, based on local resources (forestry, animal productions, etc.) exploitation and finalized to minimize commuting phenomena.
In this way it would be possible to absorb a young manpower.
Following these general lines, we have considered energy demand analysis one of the most important instruments for planning.

ENERGY DEMAND ANALYSIS

The main goal of our investigation was a correct recording of all data concerning final thermal and electric energy uses in the LRA (Artioli, Ganapini, Visentini; 1980).
We excluded from the analysis field all the energy consumptions related to commercial activity, public transport and other public services,due to the difficulty of finding reliable information sources.
In order to characterize a representative sample, we divided the Enza River Valley into three altimetric areas : inside each area we chose one municipality.
From the registry-office lists of each municipality, we extracted a random sample of 150 families, according to their domicile (village, hamlet, single house).
Recording of data has been based on personal interviews made at home.

In this way we evaluated home and farm energy consumptions.
The 50 % of the housing estate has been built before 1945; about 99 % of it is grid connected, but only 10 % is connected with a methane-pipe.
Regarding heating systems, 98 % of the families applies to single equipments (stoves), with an average utilization period of 150 d/y.
Table 1 synthesizes the annual energy consumption for a representative family of the valley.

TABLE 1 Annual energy consumption for a representative family of the Enza River Valley

Energy source	Annual consumption
Naphta	2.600 kg.
Kerosene	400 kg.
Wood	2.900 kg.
Electricity	1.150 kWh

Annual heating and lighting costs amount to about 700 U.S.dollars (1980). Because of the commuting flow from villages to the plain, each family possesses, on average, 1.4 car, running 15.000 km/y, with an annual gasoline cost of about 2.500 U.S. dollars (1980).
Recording of industrial energy consumptions was made by sending a questionnaire to 124 factories of the Enza Valley, 95 of them (76.6%) are manufactories, 23 (18.6%) food industries, 6 (4.8%) building contractors. 45 factories (36.3%) employ 2 workers, 36 (29%) employ 3÷10 workers, 43 (34.7%) have over 10 workers.
Table 2 shows the different energy sources utilized by these factories.

TABLE 2 Energy sources utilized by factories (%)

Energy source	Food industries	Manufactories	Building contractors
Electricity	100.0	100.0	100.0
Fuel oil	43.5	33.7	16.7
Gas oil	69.6	69.5	66.7
Wood	4.3	14.7	16.7
Gasoline	21.7	22.1	50.0
L.P.G.	8.7	7.4	-
Methane	-	21.1	-

Annual energy consumption for industrial activities are showed in Table 3.

TABLE 3 Industrial energy uses analysis (*)

Final use	Food industries	Manufactories	Building contractors
<100°C	FO 108.2 GO 150.7 W 17.0 LPG 21.8	EE 7.4 FO 501.3 GO 794.5 W 80.4 LPG 15.0 M 1629.5	EE 18.7 FO 3.0 GO 50.0 W 3.0
100°-250°C	FO 38.8	EE 16.6 FO 1064.1 GO 475.0 G 1.0 LPG 0.2 M 24.3	G 9.1
> 250°C	EE 1.5	FO 2257.7 GO 3.0 LPG 611.8 M 2219.6	
Lighting	EE 27.1	EE 309.8 GO 1.1	EE 18.2
Motive-power	EE 1055.2	EE 813.8 GO 0.5 G 2.9	EE 455.9
External transport	FO 122.7 G 2.9	GO 207.6 G 35.5 LPG 5.8 M 11.0	GO 1.2 G 4.4
Internal transport	FO 96.7 G 0.3	GO 60.0 G 4.2	GO 5.0

(*) Electric energy = EE (10^3 kWh/year)
Fuel oil = FO (tons /year)
Gas oil = GO (10^3 l /year)
Wood = W (tons /year)
Gasoline = G (10^3 l /year)
LPG = LPG (tons /year)
Methane = M (10^3 m^3 /year)

The above mentioned data show a strong similarity with the industrial energy demand in developed areas.
The obvious result is that industrial activities settled in Enza River Valley are not in harmony with the environmental and social peculiarities of this LRA. For this reason, it is necessary to implement new industrial activities : pastures and forests have to be recovered to produce animal fodder and wood, that can be utilized to feed stoves or as raw material for chemical processes; agricultural and animal byproducts and wastes represent an important fertilizer and energy resource (Ganapini, 1978).
On the other hand, we found that the family's electric energy consumptions and transport costs are increasing; the first ones due to an improved life style, the second depending on commuting flow from the LRA to the plain.
There are many differences between the families of villages, hamlets and single houses; for this reason, it appears necessary to apply to a wide range of technical solutions in order to reach the environmental protection and energy saving goals (Butera, 1979).

CONCLUSIONS

Energy demand analysis as applied to LRA allows to enphasize the effects af an unbalanced development model. A new planning approach is required, to prevent irreversible decay of LRA.
First of all, it is necessary a complete knowledge of physical, anthropic and energy resources in the area. Translating these data on thematic charts and crossing them, it would be possible to single out the optimum land use policies.
Population and local public authorities have to be involved in this process; in italian situation, besides, it is necessary to fill a legal gap regarding natural and energy resources management.
Appropriate technologies have to be developed, in order to exploit, in a way consistent with the LRA peculiarities, local human and natural resources.
Energy represents, in this context, an important parameter to consider, but not the only parameter, because energy is a service (Colombo, Lanzavecchia, Zorzoli; 1980).
New energy services, applyng to saving technologies and to renewable sources, could accelerate the achievement of above described aims.

REFERENCES

Artioli, G.P., Ganapini, W., and G. Visintini (1980). "Ecological and energy analysis of the upper Enza River Valley", Final report, ANR-PFE, Rome.
Butera, F. (1979). "Quale energia per quale società", Mazzotta Ed., Milan.
Colombo, U., Lanzavecchia, G., Zorzoli, G.B. (1980). "La speranza tecnologica", ETAS-Libri Ed., Milan.
Ganapini, W. (1978). "La risorsa rifiuti", ETAS-Libri Ed., Milan.
Ganapini, W. (1980). "Oltre l'ecologia", ETAS-Libri Ed., Milan.

The Aberfeldy Project

H. Liddell

Achloa Centre, Aberfeldy, Scotland, UK

INTRODUCTION

Aberfeldy is situated in the geographical centre of Scotland, and is the main town for the Upper Tay Valley. In 1974 a group of prominent citizens in this agriculture-cum-tourist area decided, in the wake of the OPEC price rises and the collapse of the livestock market, that they should abandon faith in successive forms of central government as a basis for solving the long-term problems of rural areas.

In 1981 the context has not changed - the pattern has entrenched itself and the group has seen no necessity to change its original manifesto. In fact, it has strenghtened its resolve by moving from mere discussion and small bursts of limited objective activities to establishing a permanent limited company, with charitable status - the Breadalbane Institute - in order to sponsor more direct action.

PHASE ONE - THE LOCAL INITIATIVE.

Quotations from two members of the original self-reliance group will suffice to explain the reasons for its foundation.

> Dr. Walter Yellowlies: "Throughout Scotland the grocers' vans are plying their trade, unloading white bread, margarine, tinned vegetables, tinned soup, tinned meat and processed foods at the cottage doors. Behind the vans come the doctors' cars bearing penicillin, sulphonamide, iron and vitamin pills, while the people break out in boils, ulcers, festering tonsils and rotten teeth, or in increasing numbers are consumed by cancer. All this in a land where piles of stones and crumbling cottages are a monument in a deserted countryside to a people who once had the vigour to grow all their own foods and the raw material for most of their clothing".

> John Campbell-Smith: "Milk was once provided by dairy herds in the glens, but these farmers switched over to beef production as a result of government incentives. Milk is now imported to every farm in bottles from Perth. There used to be an abattoir in the town, but this was closed, cattle now travel to Perth to be slaughtered, then travel back to Aberfeldy shops (a round trip of 70 miles)".

In September 1974 the Group produced the following manifesto: "We .... conceive it as our duty to discuss, and if possible to reach the following objectives:

1. to urge, by every means possible, government policies which will enable our agriculture and horticulture to make Britain self-supporting in essential foods and in the requirement for their production.

2. to grow in the land of the Upper Tay Valley as much food as possible in as much variety as possible for consumption by those living in the valley, and for sale of surpluses for any who want to buy.

3. to achieve these objectives without impairing soil fertility for the future, without polluting our surface waters, and without extravagant use of fossil fuel energy.

4. to spread the knowledge of the relationship between soil fertility, food and health.

In the three or four months of Summer, areas like Aberfeldy are a hive of activity - the lambing season followed by the harvest, coincide with the peak of tourist pressure. But throughout the remainder of the year - and particularly in the Winter - there is time during the longer evenings to reflect on the nature of rural life. Every few weeks through the Winters of '74-'76, the Aberfeldy Group met to discuss the possibilities for a permanently sustainable future for the area. The dozen or so local farmers, businessmen, professionals and councillors were reported by the national press as 'doom-mongers' in a period when such 'oddities' were being sought out. The same people are now being re-investigated by the same journalists - six or seven years later, as a ray of optimism in a world of gloom.

The Group's first period of activity generated little more than a local gardening competition and an expertise as interviewees. But the sense of frustration at not being able to achieve more tangible results was itself educational. Two things were needed - hard information and a means of buying somebody's time to work full-time on the project.

PHASE TWO - THE "ABERFELDY PROJECT".

The process of gathering hard information was carried out by the Alternative Technology Group at Hull School of Architecture, who offered their services in 1976 and produced a series of reports until the end of 1977 - covering a basic analysis of the area and looking specifically at the potential for self-sufficiency in food and energy in the valley. The Aberfeldy Group felt that their studies gave chapter and verse to what they already thought intuitively, and had gleaned from their own rapid analysis.

These studies were based on fairly straightforward questions. How much food does a person need for a healthy diet, and in what variety? How much land would be required to feed the whole population of the area? How much energy is used at present? How much of this could be produced from natural sources of energy in the area? The information-gathering involved a fair degree of original research and produced some surprising results - especially in the extent to which both the total output and the variety of produce had declined since 1800 and also the large energy potential from local forests.

There are currently two or three fully organic farms, a few operational small-scale hydro installations (one at 150 kW), sporadic solar collectors, a health food section in one of the grocer's shops and so on. However, until recently, there has been no real attempt to coordinate the Group's ideas in the area under a set of consistent and deliberate policies. This is where the need for full-time involvement comes in and has been the starting point for phase three, which has been getting under way since the beginning of 1980.

PHASE THREE - VALLEY INTERPRETATION.

Patrick Geddes the famous biologist/town planner/holistic thinker, who was born in Perth, used a model showing various modes whereby things change. It incorporates a fairly complicated set of interrelationships but can be put simply as in diagram 1 where progression is anti-clockwise.

'Acts' are those things perceived as the current 'reality'. 'Facts' are the statistical/analytical interpretation of that reality. 'Thoughts' are the creative elements that emerge from the 'facts'. 'Deeds' are the implementation of policies, strategies and actions resulting from 'thoughts'. This in term leads to a new perceived 'reality' (or 'Acts') and so on.

Diagram 2 shows how a mixture of new and existing organisations have been linked to act as instruments of change in the Upper Tay Valley. The linkages between the different organisations are partly formal and partly informal, with the latter being by far the most important.

To demonstrate how these relationships work consider the example of our proposal for an interpretive facility (originally called an 'Eco-park' for the want of a better name). The Achloa Centre, a limited company providing educational and research services, put forward a discussion document to the Breadalbane Institute. The document suggested that a number of diverse sites and activities, embracing aspects of the Institute's aspiration, could be brought together under a loose federation. This would give one type of picture of the valley showing the possibilities, by examples, for various sorts of self-reliance. Sites include organic farms, hydro-power installations, a small-scale forestry machinery factory, a water-powered oatmeal mill, a fish farm, etc.. The idea is to have a central facility in Aberfeldy from which visitors can go to about fifteen demonstrations of various aspects of self-reliance,through a past, present and future perspective.

The Breadalbane Institute successfully sought funding from Perth and Kinross District Council for a grant to carry out a feasibility study covering the financial and physical potential for the idea. The study was then commissioned from Finalyson and Campbell, a local architectural and energy consultancy practice. In doing this study Finlayson and Campbell has been able to call on a number of planning and architectural studies done by students at the Achloa Centre. The web of local cooperation and contact runs through all modes of the Geddes 4 point diagram.

However, words are an inadequate means of communicating the texture and feeling of the kind of activity that is starting to emerge in the Aberfeldy area. It is hoped that the appended collage of photographs can convey a little more of this.

The process of change in the valley is very slow - but there are considerable advantages to this, in that it becomes an oasis of labour intensity in a desert of unemployment.

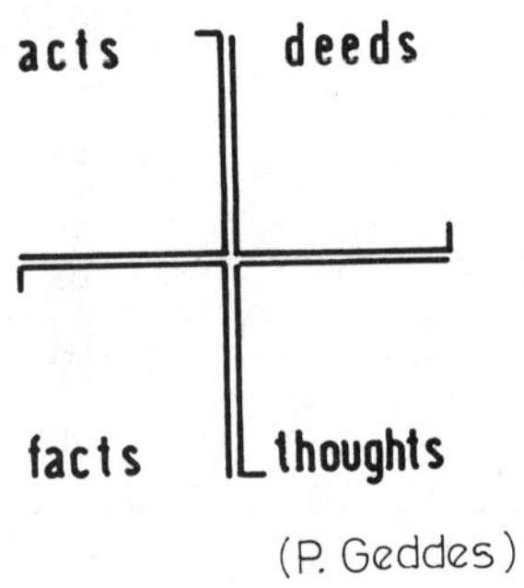

diag. 1. Theory

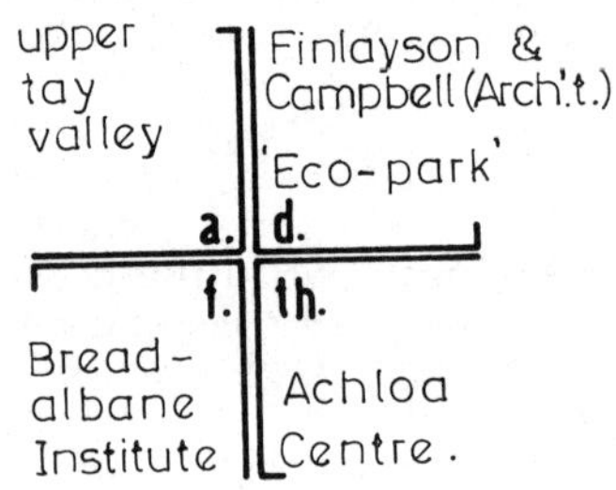

diag. 2. Practice

Aberfeldy Project Images

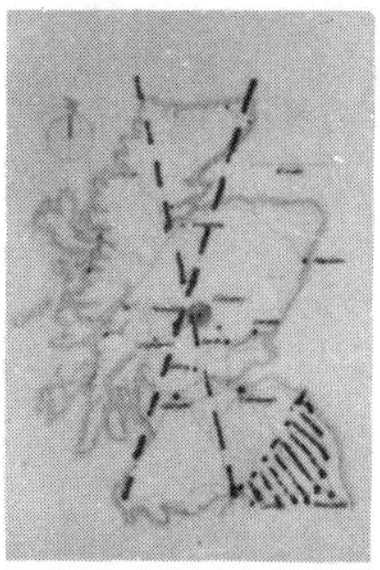

Aberfeldy
heart of Scotland

view looking North

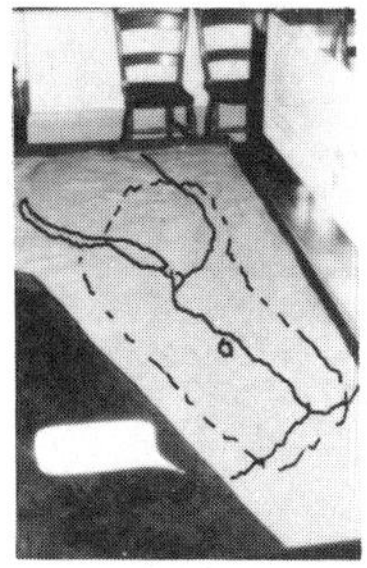

study area

derelict village

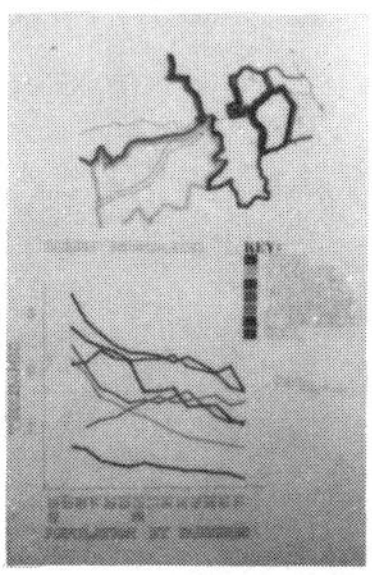

population decline

habitable derelict castle

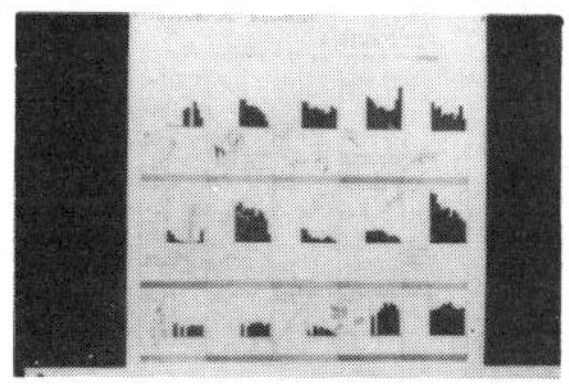

agricultural reduction

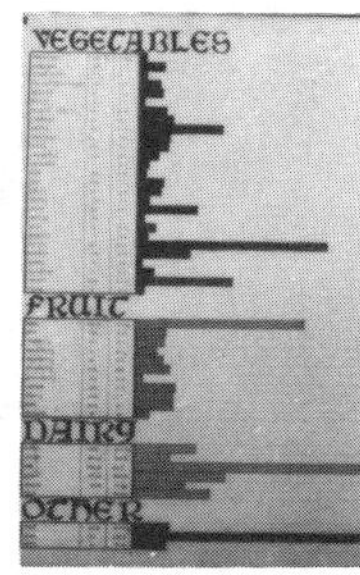

diet 1 acre/person

solar hotel Rannoch

wind potential

traditional working
hydro power

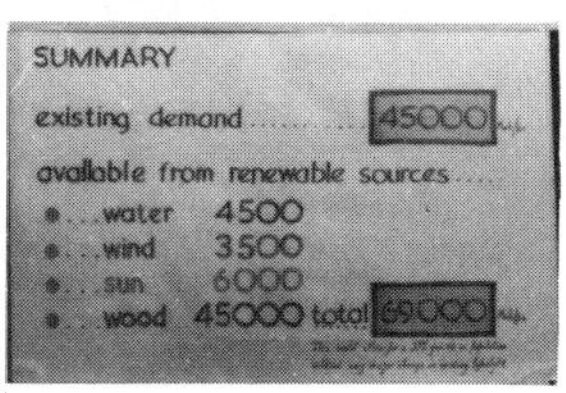

100% self sufficient
in bio-fuel

Achloa Centre

demolish

model

design

technical drawings

← same students & residents → (Scottish–Norwegian)

roof structure

topping out

slating

interior – before

–and–after

finished house

courses:

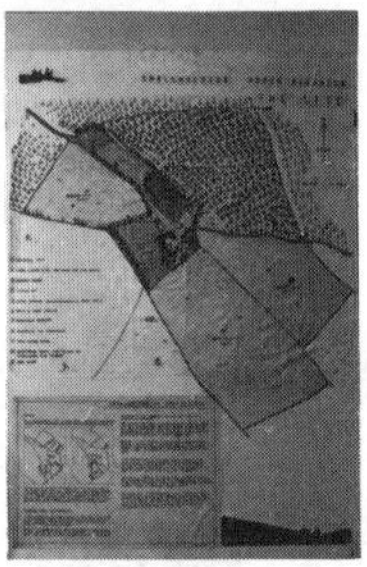

– farm plans

fish farm proposal

and hydro

on farm estate

local concrete plant

mill briefing

– exhibition centre and trades exchange

local planner at course review Oslo

A6.6

Energy and Planning in Limited Resources Areas: the Fontanabuona Case+

E. Manzini* and C. F. Ratto**

*Dipartimento di Programmazione e Produzione Edilizia, Politecnico di Milano, Via Bonardi 3, 20133 Milano, Italy
**Istituto di Scienze Fisiche and Unità di Ricerca di Genova del GNSM-CNR, Viale Benedetto XV,5, 16132 Genova, Italy

ABSTRACT

The Fontanabuona, a Valley of about 200 sq.kms, near the town of Genova, in Italy, has been studied during two years by a multi-discipline staff.
This research has been performed within the framework of an ensemble of studies, financed by the Italian National Research Committee (CNR), aiming at the definition of a methodology allowing the analysis of the energy system and the evaluation of its role in the planning procedure relative to limited resources areas.
The study has covered three main lines of research:
i) analysis of natural parameters with special attention to renewable energy sources;
ii) analysis of parameters characterising the human community with special attention to energy consumption and end uses;
iii) analysis of the relationship between energy system and planning tools.
Finally, efforts have been concentrated on setting up a few "projects"; the most important amongst them are relative to :
i) hydroelectric production on a small scale;
ii) retrofit of traditional housing system and integration between new and traditional settlements;
iii) cultivation and rational use of wood.

KEYWORDS

Development; energy; planning; case study.

+ Work financed by CNR, Progetto Finalizzato Energetica, Sottoprogetto RERE, Tema N, under contract N° 80.00936.92.

INTRODUCTION

The present paper regards a research performed by a multi-discipline group, composed by engineers, physicists, geologists, chemists, architects, agronomy experts, economists and historians. Such research work began about two years ago; we are now near to the end of the methodological phase, phase which has been financed by the Italian National Research Committee (CNR).

The main goals of this research are:

i) the setting up of tools allowing the inventory of the natural resources and their mapping;

ii) the setting up of tools allowing the analysis of the structure of energy consumptions and end-uses;

iii) the definition of development objectives which could contribute to invert the present tendency toward social and economic weakening;

iv) the identification of energy producing or conserving technologies which, within the framework defined in point iii, could be significant for the whole area;

v) the setting up of a planning procedure being able, from its very first steps, to take into account the energy parameter.

Our research considers a specific area (the Fontanabuona Valley, in the Regione Liguria, in Italy) but aims at setting up a methodology suitable for application to other areas with analogous characteristics of social and economic weakness.

DESCRIPTION OF THE CASE STUDY

The relation between the Fontanabuona Valley and the Ligurian coast is a typical example of unequal development of neighbouring areas, one of which is "strong" and the other one is "weak". Such a situation is very common in Liguria, as well as in many other Regions, not only in Southern but also in Northern Italy. In Liguria this unbalanced development is the consequence, at the same time, of the break-down of the economy of the inland, once based on agriculture and forestry, and of the attraction by the industrial and commercial development of the coastal areas.

Aside to this trend, new housing and industrial settlements have been established, during the last years, in some marginal areas. These new settlements are not the fruit of a change of the trend, but rather the consequence of the land saturation in urban areas. Hence such phenomena do not help the economic development of weak areas, for usually the best lands are occupied, which should instead be used for more productive activities.

The Fontanabuona Valley is about 30 kms long, with a total surface of nearly 200 square kms. This Valley, which is nearly parallel to the coast, is limited by mountains whose tops reach 1350 ms above the sea level, at North, about 800 ms a.s.l. at West and South: the bottom of the Valley is at nearly 100 ms a.s.l.

The average temperatures are about 21° in Summer and 6° in Winter; the

TABLE I Annual Energy Consumptions per capita (10^6 kcal)

SECTOR	Non-electric			Electric		
	Fontanabuona	Liguria	Italy	Fontanabuona	Liguria	Italy
AGRICULTURAL	0.10	0.31	0.33	0.01	0.01	0.03
INDUSTRIAL	1.30	6.57	5.43	0.66	0.89	1.17
SERVICES	0.19	0.48	0.35	0.11	0.15	0.15
RESIDENTIAL	3.21	5.25	4.88	0.33	0.51	0.52
TRANSPORTS	1.67	3.34	3.29	——	0.26	0.07
Total	6.47	15.95	14.28	1.11	1.82	1.94

TABLE II Annual Energy Consumptions at 1990, according to different scenarios: see the text (10^9 kcal)

SECTOR	1980	A		B		C	
		N.ES.	ES.	N.ES.	ES.	N.ES.	ES.
AGRICULTURAL	1.3	1.0	1.0	8.2	7.4	8.2	7.4
INDUSTRIAL	23.5	23.7	23.2	28.9	26.9	28.9	26.9
SERVICES	3.6	3.7	3.5	4.9	4.5	4.9	4.5
RESIDENTIAL	42.5	51.9	41.8	59.5	48.0	73.0	59.4
TRANSPORTS	20.0	19.7	19.7	21.4	21.4	30.6	30.6
Total	90.9	100.0	89.2	122.9	110.1	145.6	128.8

snow is rare but rain more abundant than on the coast.
The population amounts to about 12000 people, from the 26000 people living in the Valley in 1881; presently the population is concentrated mainly in the bottom of the Valley and, in higher altitude villages, shows up an age composition typical of marginal areas.
The agriculture is very poor, mainly based on part-time activities; the land has been, for centuries, fractionated in very small estates. At present times the chestnut and oak woods cover about 70 sq.kms, the grazings about 40 sq.kms; land used for other agricultural uses amounts to less than 10 sq.kms.
The main industry of the Valley is that more or less directly related to the extraction of slate (in Italian the name of the blackboard is "lavagna", i.e. the name of the river flowing in the bottom of Fontanabuona Valley).
Small industries have been developed only recently, mainly in the bottom of the Valley, but with a loose relation with other activities of the area.
Our analysis of the natural, economic and social characteristics of the Valley has brought us to identify the following significant or critical points:

i) a fairly favourable situation from the point of view of the climate (good insolation on the wider side of the Valley, side which is also protected from cold wind) and of the quantity of water (both for agricultural uses and energy production);
ii) a good importance of the woods, although they are now almost abandoned;
iii) a limited extension of flat lands and a possible competition, on them, between housing and agricultural uses;
iv) a complex demographic dynamics which has produced a decrease of the population, but, at the same time, a migration within the area, from the hills to the bottom of the Valley;
v) the presence of historical settlements on the South-facing side of the Valley, at a mid-altitude;
vi) a weak economic structure, but in which the slate industry has been able to produce other related activities and shows up a good level of local initiative;
vii) energy global consumptions per capita which are less than one half of the values relative to Italy or Liguria; a distribution of energy consumptions among different sources and end-uses distorted with respect to the Italian and the Ligurian situations (see table I; see also the contributions by Butera and by Ganapini to this Conference).

DEVELOPMENT LINES AND SUGGESTED SOLUTIONS

From our analysis of the Fontanabuona Valley, we have drawn the conclusion that any hypothesis of balanced development of the area has to be based on the following three major objectives:

i) maintaining and enhancing forestry and cattle breeding at higher elevations; maintaining and enhancing of intensive agriculture in less sloping lands;

ii) maintaining and enhancing of mid-altitude settlements, both by retrofitting existing houses and by constructing new-ones;

iii) maintaining of activities related to slate extraction and enhancing of industries and artisan-activities related to the transformation of the products of the Valley.

These objectives imply the following priorities in the domain of energy production and conservation:

i) minimizing energy consumptions in housing, both by energy conserving interventions in existing buildings and by solar and/or energy conserving solutions in constructing new buildings;

ii) maximum use of local resources;

iii) integration between different sectors (e.g. industry and agriculture, agriculture and housing....) and use of energy-cascading or total-energy systems.

Within this framework, we are presently studying a few technological solutions and/or land-use proposals; because of the methodological character of the research, we are considering only those solutions which can concern a large fraction of the area and/or can affect the planning procedure.

These solutions and proposals, which we call "projects", can be sketchy summarized as follows:

i) Housing. This project has two main aspects: identification of retrofit solutions for the traditional houses (mainly those at medium altitudes on the Southern side of the Valley); identification, for new buildings, of those (active and passive) solar solutions which can fit to the local climate;

ii) Agriculture, forestry and cattle breeding. Beçause of the weakness of this sector, the core of this project consists in a study regarding a possible productive reorganization of the woods, and in the identification of appropriate techniques allowing rational harvesting, transformation and use (partly for energy production) of the wood;

iii) Industry. Two different situations can be considered: a concentration of polluting industries in an industrial area and a diffuse scattering of "clean" activities within the residential areas. With respect to the industrial area an integrated energy system can be studied (energy cascading, total energy, integrated systems allowing effluents elimination or reuse);

iv) Renewable energies. The main project within this point consists in the study of the possibility of reopening small hydroelectric plants, shut by the National Electric Agency (ENEL). We have evaluated that the hydroelectric potentiality of the Valley amounts at nearly 10^7 kWh/year, which is a good figure, since in 1980, in Fontanabuona Valley, there was a consumption of $1.3 \cdot 10^7$ kWh of electric energy. Our first rough economic analysis indicates that the most interesting way to produce this energy consists in producing electric power for about six hours per day, during the peak-consumption times, during about 300 days per year (this means 3-4 plants with a power of about 1 MW each). In this case, and with a decentralized management of the plants, the costs can be comparable with those faced by the National Electric Agency in plants considered economic.

Other projects which will be, more sketchy, considered, regard the management of wastes and the transport patterns.

RELATIONSHIPS BETWEEN ENERGY SYSTEM, PLANNING TOOLS AND TERRITORIAL DEVELOPMENT

From the examples given before it should be clear that no one of the "projects" raises only technological problems, for it has many social, economic and territorial implications.

Indeed, even the identification of the projects themselves is the fruit of a number of preparatory tools, which we think should in some way enter a correct planning procedure being aimed at taking into account the energy parameter.

Such tools can be divided within three main categories which will be analyzed in the following sections.

Evaluation of Physical and Environmental Resources and their Mapping

The evaluation of natural resources should enter every correct traditional land-use planning procedure. In fact our approach has introduced two innovating points of view, i.e. the importance given to local resources (wood, water, sun, wind, wastes) and, hence, the importance given to the tools necessary for their quantifying and mapping.

These tools are:

i) The geo-morphologic analysis. Besides a more or less traditional mapping of the present land-use, we have performed two less conventional geological analyses; one concerning the height of the detrita coverage of the soil (this information is useful to evaluate the agricultural potentialities of the soil) and onesingling out the soil erosions and the land-slides. Both these analyses have been the basis for the elaboration of a map indicating the areas in which productive agriculture is judged possible, the areas in which grazing is suggested, the areas in which woods are judged necessary for stabilizing the soil and so on.

ii) The evaluation of the hydroelectric potentialities. The traditional procedure used by hydro-engineers has been modified in order to be appropriate to the dimensions and characteristics of an area in which only small power plants can be installed.

iii) The evaluation of the solar radiation incident on an area with complex orography. Original computer programs have been settled out, which allow the construction of maps describing both morphologic parameters (elevations, slopes, orientations of the terrain) and solar irradiations on surfaces with any slope and orientation, in any time-period of the year, taking into account the shadows cast by hills and mountains surrounding the area being considered.

iv) The evaluation of the relative intensities of the wind blowing on an area with complex orography. A computer program has been written which allows the construction of maps describing intensity and direction of wind, at different elevations above the terrain, given direction and intensity of wind at high elevations.

v) The construction of thematic maps. Since computer programs described in points iii) and iv) use the same digital description of the terrain, we have settled out a computer program allowing the construction of maps relative to any combination of parameters describing morphology of the terrain, insolations and wind intensities and/or directions. It is clear that this tool, which allows the selection of best areas for any purpose, can be a very crucial tool for any planning procedure which aims at taking into account natural potentialities.

Evaluation of Energy Consumption and End-Uses in Present Situation and in Different Scenarios

To reconstruct the structure of the energy system in a marginal area is not an easy task; one difficulty arises from the fact that such an area is too small for having statistical records of energy consumptions and too large to perform direct surveys (for instance with the method of interviews).

Furthermore the situation is very differentiated according to the energy source; for instance, while ENEL keeps records of electric consumptions for every municipality (Fontanabuona Valley is divided in ten municipalities) divided according to a few main uses (domestic, public illumination, power, etc.) for the other energy sources no precise data are available.

Therefore we have decided to use all the figures we could find, relative either to the Fontanabuona, to the Regione Liguria or even Italy altogether. In this manner we have been able to estimate for the Fontanabuona, with different levels of accuracy, the energy consumptions per capita relative to the different sources and/or end-uses, in the different sectors (housing, agriculture, industry, transports).

Another important step in our study of the energy system consisted in the elaboration of possible scenarios of social and economic situation of the Fontanabuona Valley in year 1990.

The scenarios we drew are the following ones:

i) Inertial evolution (A). This scenario corresponds to the hypothesis that present trends continue during the next ten years, in the absence of any mayor public or private interventions on the situation;

ii) Social and economic strengthening (B). This scenario is based on the hypothesis that agricultural activities receive a new momentum, that industrial activities connected with the transformation of local resources take pace and, finally, that the slate industry follows its present development trend;

iii) Housing expansion (C). This scenario is based on the hypothesis that the existing plan of opening a tunnel connecting the Valley to the coast is realized: this should provoke a development, in the

Valley, of housing at the service of the coast. For the moment, the other hypothesis on which this scenario is based have been taken equal to those of scenario B.

For every scenario we considered two different cases:
i) no energy-saving policy (N.ES)
ii) energy-saving measures are taken (ES).

The energy consumptions relative to the six scenarios were evaluated, starting from the coefficients giving the energy consumptions per inhabitant or per industry being evaluated in the present situation, modified according to the development hypotheses made (see Table II).

Evaluation of Economic and Institutional Frameworks

Since the Fontanabuona Valley is a weak area near a strong one, two opposite solutions to the energy problem can, in principle, be considered:
i) all the energy being required can be imported from the outside of the Valley; remark that in Fontanabuona an efficient high- and average-voltage electricity distribution system already exists, which we evaluated sufficient also for consumptions at 1990;
ii) the local energy system can be completely redrawn, in order to allow the maximum exploitation of local resources.

The second road has the advantage of making the energy production--transportation-management sector not a sector purely at the service of the other ones, but a productive sector itself, hence one which could act as a catalyst of the development of the area.

The two roads, as well as any project or technological solution, have to be compared on the basis of a cost-benefit analysis. This analysis cannot be a traditional one, since it has to take into account both the national level and the local one. This is not an easy task, which is presently faced by a group of economists (Mebane and others, 1980) who recently began to collaborate with our research group.

To assess the possibilities of realizing the projects being proposed, one needs also to know the institutional framework with which one has to interact.

In our case the levels involved should in principle be: i) the Italian Central Government; ii) the Government of Regione Liguria (Italy is divided into 20 Regions, each of which has a degree of authonomy in planning); iii) the "Provincia di Genova" (one of the four sections of the Regione Liguria); iv) the "Comunità Montana della Fontanabuona" (an administrative authority relative to an area a little larger than the Fontanabuona Valley itself) and, at last, v) the ten "Comuni" (municipalities) in which the Valley is divided.

We have evaluated that levels ii), v) and, mainly, iv) are the most important for the realization of our projects.

We have also analyzed the existing relationships between these levels and the relationships which should exist in order to allow an efficient role of these structures in land-use and local resources exploitation planning.

In synthesis, we have found a confirmation of the fact that "the bottle-neck is the lack of institutional and social structures capable of managing" a correct land-use planning procedure and an energy system based on renewable and local resources (Butera, 1981b).

CONCLUSIONS

We hope that these sketchy considerations have been able to give an idea of our approach to the study of the Fontanabuona Valley and of its energy system and potentialities. We are convinced that every planning procedure aiming at correctly taking into account the local resources is necessarily very complex, but we hope that our work will be able to be a first step, among others (Butera, 1981 a, 1981 b), in providing methodologies and tools for such planning.

REFERENCES

Adamoli, M., E. Manzini and C.F. Ratto (1980). Uomini, energia e territorio. Ricerca su una vallata ligure: la Fontanabuona. In Politica edilizia e gestione del territorio; il nodo energia. SAIE, Bologna. pp. 67-69.

Adamoli, M., E. Manzini and C.F. Ratto (1981). Uomini, energia e territorio. Ricerca su una vallata ligure: la Fontanabuona. In Sottoprogetto RERE. Atti del quarto seminario informativo. Edizioni PEG, Milano. pp. 295-305.

Butera, F. (1981a). Tema N: analisi del contributo che lo studio del sistema energetico può dare alla pianificazione delle ARL. In Sottoprogetto RERE. Atto del quarto seminario informativo. Edizioni PEG, Milano. pp. 287-294.

Butera, F. (1981b). The energy parameter role in the development of limited resources areas. In Proc. IEA. Conference on new energy conservation technologies and their commercialization. Berlin (to be published).

Cozzi, G., P. Genco and F. Isotta (1978). Il sistema energetico regionale: il caso della Liguria. Franco Angeli Editore, Roma.

ENI (1978). Direzione e programmazione. Gli usi finali dell'energia in Italia. Roma.

Mebane, W., G. Panella and A. Pecchio (1980). Metodologie economiche per la valutazione degli strumenti dello sviluppo delle ARL. Interim report to CNR-PFE.

TOPIC B

Wind Power

North of Scotland Hydro-Electric Board Experience with 22 kW Aerogenerator

I. T. Pope

North of Scotland Hydro-Electric Board, 16 Rothesay Terrace,
Edinburgh, Scotland, UK

ABSTRACT

The paper outlines the thinking behind the North of Scotland Hydro-Electric Board's decision to install a small aerogenerator on Orkney, and on the reasons for then choosing the 22 kW Windmatic machine for this application. The construction of this machine is briefly described, together with its operating regime. Site experience is then discussed from delivery and erection through the first six months of the machine's life. The simple data logging system designed for the aerogenerator is described. The problems which have been met in the first six months are mentioned, together with steps taken to overcome them. Estimates are made of the potential generation from the machine once these difficulties are solved but the point is made that it is still too early to draw conclusions about maintenance requirements or economic life.

KEYWORDS

Aerogenerator; Wind Power; Data Logging; Small Systems.

INTRODUCTION

Why should an electricity utility such as the NSHEB be interested in a 22 kW aerogenerator in the first place? After all 22 kW is rather small even by the scale of the Board's isolated island systems. Well, the short answer to this question is that, firstly, Scotland is a particularly windy place and some of the Board's consumers can be expected to think of erecting an aerogenerator themselves. Given that this is likely to happen the Board thought that the right answer was to try out a suitable small commercial machine itself. It would then be able to obtain experience of the peculiar problems of operating small aerogenerators and be in a position to advise any of our consumers who were interested. We would also be able ourselves to confirm our theoretical calculations as to what does happen when a small aerogenerator is connected into the rural 11 kV distribution network. As a second point the Board is seriously interested in wind power as a means of reducing fuel costs in isolated systems - you will hear more of this in the next paper - and the 22 kW machine gives us a first and useful contact with the technology of aerogeneration.

Surveying the market in 1979 - 1980 it seemed to us that the Danish Windmatic machine came nearest to meeting our requirements. A number had been made and sold and were operating successfully. Also the machine was essentially simple and robust: for example, features such as feathering mechanisms and generator slip-rings were not featured in the design. Although wind speeds in Orkney are higher than those in Denmark the manufacturer felt confident that the machine could be modified to suit Scottish conditions. The modifications which were in practice made consisted of:

1 Using a stronger design of tower.

2 Incorporating a wind pressure operated braking device which had already been fitted to two of the Windmatic machines but was not a standard feature.

3 Specifying a 415 V (nominal) stator voltage, the Danish standard being 380 V.

The data logging system was also considered necessary on this trial installation. The Board's Research Laboratory designed and built the suitable simple system.

Following agreement to purchase the machine was erected at Berriedale Farm on South Ronaldsay (Orkney) in December 1980. The Windmatic range comprises machines from 10 m to 14 m diameter; the output range is from 22 kW to 55 kW, and a slightly larger machine of 17 m diameter is under development. The machine in Orkney is the smallest in that range, having 10 m diameter blades and a nominal output of 22 kW.

SPECIFICATION OF MACHINE

The usual tower consists of three 6 m sections giving a hub height of 18.5 m which is normally sufficient to lift the turbine clear of obstructions to a wind. Where strong wind may be expected it is possible to erect the 22 kW machine on a shortened 12 m tower made up from the top two sections of the standard tower. This reduces the cost of the tower and foundation and this option was chosen by the Board for the installation in Orkney: the site is open and unobstructed.

The machines have a three bladed, fixed pitch, upwind turbine which is mounted on a substantial hollow steel shaft supported in heavy duty bearings and inclined upwards at an angle of 10°. The blades of the turbine are bolted onto the hub and additional support is provided by tensioned guy wires stretched between the outer anchor points to prevent torsional vibration of individual blades. Provision is made in the assembly for the control line which operates the air brakes on the wings.

The blades of the air turbine are formed in glass reinforced plastic and bonded to a laminated timber beam which extends from the hub to a point at approximately 80% of the blade radius. The timber is coated throughout in glass reinforced plastic to protect it from the elements. The twin fan tails are a traditional feature of Danish windmills affording an equal response to a veering or a backing wind. These blades mounted on a common shaft provide the power to yaw the machine.

The nacelle enclosing the gearbox, generator and mechanical control gear is a substantial steel fabrication hot dip galvanised to prevent corrosion. Weather protection is provided by a moulded glass reinforced plastic cover attached by hinges and a folding strut. A general indication of the machinery arrangement is shown as Fig. 1.

The standard tower is of square section tapering from the base to the top plate. The upper two sections are unit fabrications formed from substantial round steel corner members braced with round lattice members all fully welded and hot dip galvanised. As already mentioned, the lowest section has not been necessary in this application. The tower is bolted to a mass concrete foundation block by 12 high strength bolts cast in position while accurately located by a template.

OPERATION

To gain government approval in Denmark a windmill must be provided with a number of safety devices intended to ensure a satisfactory and safe shutdown in the event of a fault and to ensure that the machine cannot continue to generate into a disconnected line. This aerogenerator is provided with air brakes and with a disc brake both of which are automatically applied should there be a failure of control power or control air pressure. If there is a mechanical failure of the power train the rising speed of the main shaft would be detected by a tacho relay initiating shutdown.

The supply conditions at the machine are monitored by a series of relays ensuring that the three phase voltages are present and that the frequency is correct to within 1 Hz. A short time delay ensures that a momentary excursion of the supply caused by local load switching does not shut down the machine. Shutdown can be initiated also by a manually reset vibration relay fitted in the nacelle or by a limit switch operated by a mechanical counter if the nacelle turns more than five turns in one direction from the start point. This latter device prevents excessive twist in the main cables. There is also an overtemperature sensor in the generator winding which will cause a shutdown until the machine has cooled to a safe temperature.

In its standard form the machine normally remains off line after a shutdown until it has been inspected. On a remote site much useful generation will be lost if the machine shuts down following a brief supply interruption and the Orkney machine is therefore to be modified to allow auto restart following a supply reconnection provided that all the monitored safety circuits are healthy.

The aerogenerator is aerodynamically self starting. As the speed approaches running speed a speed sensing relay prepares the control relay which compares the generator shaft speed with the supply frequency. As soon as the speed is just above synchronous speed the first power contactor is energised followed after one second by the main contactor. Two contactors are provided because the induction generator is dual wound as a means of minimising magnetising inrush current. The machine is now generating and the power output will rise with the wind speed. At about 13 m per second the machine will have reached its rated power and any further increase in power will be sensed by an overcurrent relay which will raise the spoiler flaps to limit the power. Should there be too much power the generator will overheat or the slip speed increase to a point where a shutdown trip will be initiated. When the wind speed falls the spoiler retracts and the blades fly clear until there is no effective output at which point the generator speed will fall below synchronous speed and be immediately disconnected by the windmill control relay until the wind strength is again sufficient for generation when the cycle will repeat.

SITE EXPERIENCE

The foundations were prepared in November 1980 and the mass concrete block poured after suspending the foundation bolts from the levelled base template into the excavation. Provision was made in casting the foundation to carry the power and control cables through the concrete in a duct to a trench and fence to a monitoring Portacabin sited within the fenced enclosure. The complete aerogenerator, including blades and tower, was transported from Newcastle upon Tyne by road and ferry direct to the site. A single 40 ft articulated lorry sufficed. The parts were unloaded by mobile crane and laid out on the site to facilitate erection on the 15 December. In favourable conditions it is possible to erect and commission the machine in one working day but on Orkney the process took longer and the winds were too strong to permit the final lift until the late evening of the 17 December. Commissioning was delayed by a cable problem until 22 December, when the machine first generated power.

Since this was the first machine of its type to operate in this country the Board arranged to install a data logging system at site to record as much significant data as possible on the behaviour of the machine and its control system, together with a meteorological log of prevailing conditions. In order to produce the data in manageable quantities the data logging system was arranged to sample the wind speed and direction every 10 seconds and "bin" the data against Beaufort 0-15 and 16 compass points. The store was read onto tape every 30 minutes and recorded along with spot readings of wind speed and direction against a logged time. By this means the daily log can be reduced to A4 page format, and a sample is shown as Figure 2.

OPERATIONAL EXPERIENCE

It is very interesting to observe that, despite the fact this site is very open and free from obstructions, the wind speeds in practically every half hour sample vary over a range of about 2:1 representing a variation in energy flow of 8:1. This variation is quite distinct and separate from normal day to day variation in wind speeds. Observing the machine in operation it was seen that as the wind speed approached the rated wind speed the spoilers became active and on occasions were observed to hunt quite rapidly - several times per minute. However, as the wind speed became yet stronger the operation of the spoilers became less frequent and they eventually retracted with the generator producing a substantially constant output of some 25-27 kW. This indicates that the turbine blades are a good compromise for operation in both the "flying" and "stalled" modes. The machine has stayed on line quite happily in some severe winds without shutting down. It has survived gusts of up to 47 metres/second and continued to generate full output when the wind was gusting up to 39 metres/second.

After six months operation the machine has generated 20,000 kWh of power. As we gain more experience with the machine and overcome the initial operating problems it seems likely that output will increase to about 60,000 kWh per annum and this represents a capacity factor of about 30%. The machine will be producing some output of course whenever the wind speed exceeds 8m/sec and this happens for about 80% of the total time.

A point to note is that during this same six months period the reactive power absorbed by the machine amounted to 23,000 kV Arh. This means that system losses have been increased by the current in the 11 kV feeder circuit. No environmental problems have been encountered except for low intensity rhythmic whistling noise which originally occurred when the wind speed was around 6 m per second in dry weather but the critical speed has now risen to about 10 m per second. It is thought to be due to a failure of one of the spoilers to seat properly but we have been unable to confirm this.

On one occasion the machine was stationary during a spell of wet weather. It was found that rain water entering at the hub had run down the spoiler control wire channels and collected in a hollow section near the tip of one of the rings. This caused a marked reduction in performance. In accordance with the manufacturer's recommendations drain holes have now been drilled at the wing tips and this has solved the problem.

The data logger has proved to be of considerable value in detecting variations in performance and operating conditions and since this is an experiment installation on a comparatively remote site it has been decided that as well as being recorded on cassette data will in future be transmitted to the Board's Research Laboratory at Pitlochry.

It is rather too early yet to predict maintenance costs and the economic life of the machine is another unknown. Kirkwall Airport in Orkney recorded a gust of 61 m/sec in 1969 and the ability to survive intact gusts of at least this speed is essential. However, the machine has certainly given us a feel for the practicalities of aerogeneration and forms a first step towards the larger machines described in the next paper.

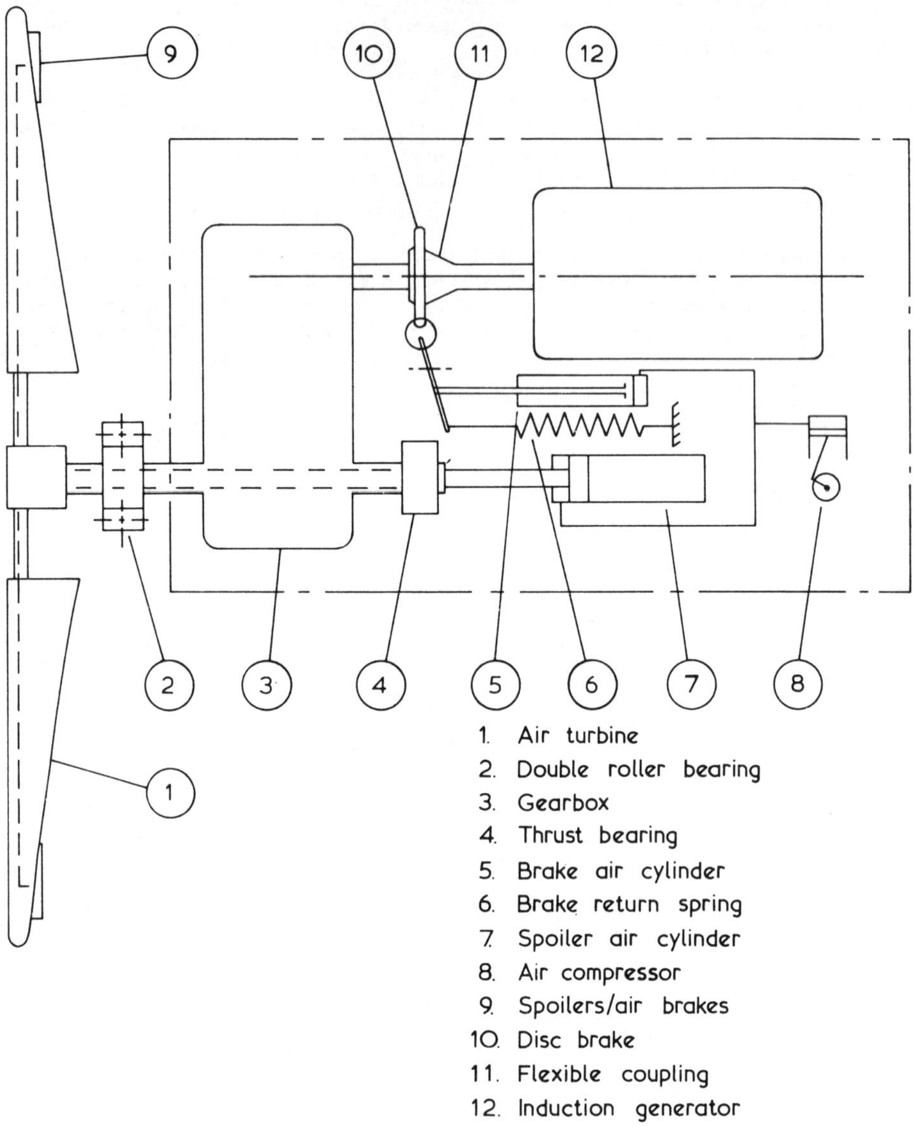

DIAGRAMMATIC MACHINERY ARRANGEMENT

Fig. 1

TYPICAL HALF HOURLY READINGS FROM DATA LOGGER

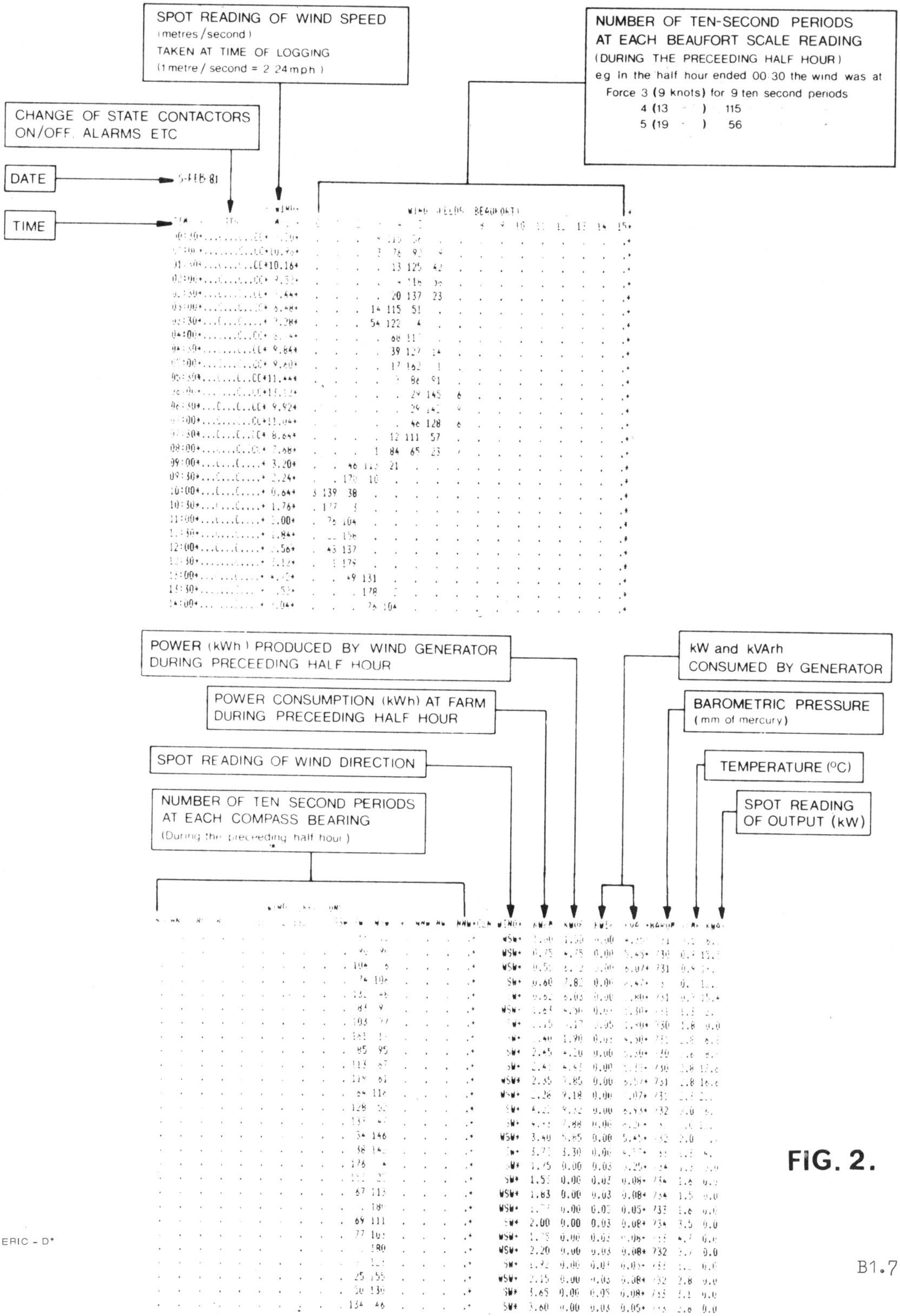

FIG. 2.

ERIC - D*

A Programme to Integrate Horizontal Axis Wind Turbines into the Orkney Distribution System

D. Lindley* and W. G. Stevenson**

*Taylor Woodrow Construction Ltd., Southall, Middlesex, UK
**North of Scotland Hydro-Electric Board, Edinburgh, Scotland, UK

ABSTRACT

A project is described which involves the design, construction, commissioning and monitoring of two large wind turbine generators on mainland Orkney. The first of these is 20 m diameter and has a rated output of 250 kW. The second is a 60 m diameter machine with a rated output of 3000 kW. The smaller machine is being built as a prototype for the larger machine and as a basis for the development of commercial medium size machines. Both machines are of the two-bladed, upwind, horizontal axis type and designed to be dynamically stiff. Both will deliver power into the Orkney distribution system.

KEYWORDS

The Island System; siting; specification; components; monitoring; economics.

INTRODUCTION

The Secretary of State for Energy announced in January this year the decision to erect a 60 m diameter 3 MW wind turbine generator on Orkney Mainland. This announcement followed a four-year programme of design, research and development by a consortium of British companies as part of the United Kingdom Department of Energy's programme of wind energy research which itself began in 1976.

In recent years the North of Scotland Hydro-Electric Board has recognised an urgent need to implement wind energy conversion systems on three island systems currently using diesel generated electricity. These are the island groups of Orkney, Shetland and the Western Isles. Electricity consumption is growing in these islands and diesel costs and hence generation costs have been rising sharply. Of these island sites, Orkney was preferred as the site for the first demonstration as a body of wind data was available that suggested the availability of a large number of sites that were amongst the best in the United Kingdom (see references 1 to 5). Orkney was also the most conveniently located for the first such demonstration having good access by sea and air.

During 1980 the Wind Energy Group went ahead with the design of a 20 m diameter wind turbine. The purpose of this machine is to act as a development prototype for the 60 m wind turbine. Experience of designing, building and operating this machine will be of substantial benefit in the development of the 60 m machine.

As the diesel system on Orkney represents a set of operational criteria typical of many hundreds of similar systems around the world the project provides the opportunity to demonstrate the credibility of United Kingdom expertise in a revival of an old technology. The machines will be of a design that can be modified for connection to "strong" UK mainland grids, at medium to high wind speed sites both onshore and offshore.

THE ISLAND SYSTEM

The approximately 70 separate islands in the achipelago forming the islands of Orkney have a land area of 97,489 hectares (379.4 sq miles) with a population of 18,139 (1979 census).

FIG.1 THE ORKNEY ELECTRICAL GRID

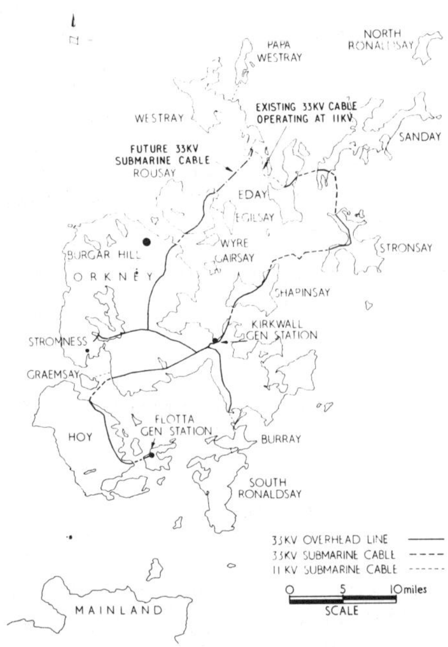

The main islands are shown in Figure 1 which also shows the location of the site chosen in relation to the existing grid. Kirkwall, the county two of Orkney, lies on the largest island called Mainland which in 1980 had 7211 occupied domestic dwellings. The total number of consumers on the island system is 8414. Though the demand for electricity has increased constantly, the percentage increase has fluctuated due to variation over the years in the severity of the winter weather. Between 1971 and 1978, the annual increase in units sold averaged 11%. Over the last three years there has been a marked drop due to the national economy drive on the use of electricity and the self-imposed economy brought on by increased tariffs.

The units generated in 1980/81 amounted to 74.7×10^6 kWh and on a consumer basis this amounts to an average 8878 units per year.

Kirkwall Power Station generates electricity at 11 kV which is transformed at the Power Station to 33 kV for the main distribution system. At main sub-stations it is transformed again to 11 kV to feed the rural networks, being transformed to 415 volts, three phase or 240 volts single phase at the consumer premises.

The increase in electricity consumption has followed a similar pattern in the Shetland and Western Isles where the effective average growth rate for the period 1970/71 to 1980/81 has been 12.7% and 7.4% respectively. The units generated in 1980/81 in these island systems amounted to 137×10^6 kWh for Shetland and 110×10^6 for the Western Isles.

SITING THE TURBINES

Mainland Orkney has many potential sites on numerous low hills rising 200 to 260 metres. Most of these were visited and assessed against a set of criteria which included consideration of road access, proximity to the 33 kV Grid, interference with TV or other communications, height, shape and exposure of hill, and environmental acceptability. 15 sites were examined in detail against this set of criteria and as a result Burgar Hill was selected. The site has a potential for the location of two further 3 MW turbines in addition to the 250 kW and 3 MW turbine now planned. Statutory consents have been given for the project and subject to contractual arrangements being satisfactorily concluded the anticipated completion for the 250 kW unit is 1982 and for the 3 MW unit 1984.

WIND SPEED

On the basis of extrapolations of existing data taken at sites on the island, the long term annual mean wind speed has been estimated for Burgar Hill as 10.5 m/s at 16.3 m (hub height of the 20 m turbine) and 12.6 m/s at 45 m (hub height of the 60 m turbine). It should be stressed that such estimates are made in the presence of many uncertaintites and measurements at the actual site are therefore essential.

TURBINE SPECIFICATIONS

The design of the machines is influenced by several considerations, the main ones of which are:

1 Burgar Hill is likely to be a very high wind speed site. The machines are, therefore, designed for survival gust speeds of 70 m/s when stationary and 40 m/s when operating.

2 Because these machines represent the first such demonstration in the UK and because of the high wind speeds, they are conservatively designed to be dynamically stiff.

3 Although the Orkney supply grid is strong as far as the 20 m machine is concerned, there is still a requirement for high quality power. This leads to the adoption of the following three features:

(a) Power trimming rotor

(b) Soft drive train

(c) Synchronous generator

These considerations have led to the design shown schematically in Figure 2.

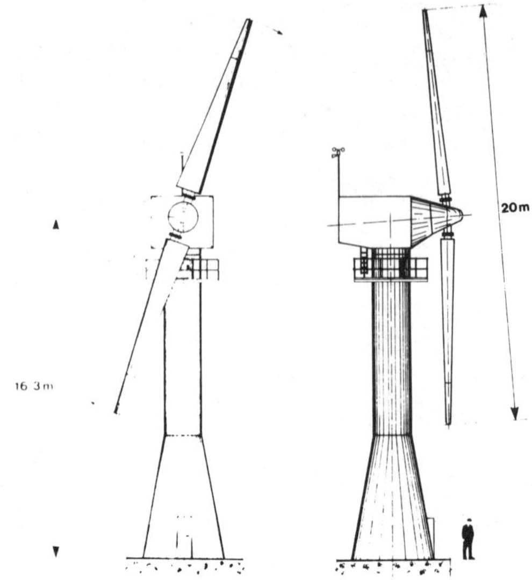

FIG. 2 ORKNEY 20m W.T.G SPECIFICATION

DIAMETER	20m
RATED POWER	250kW
RATED WIND SPEED	17m/sec
ROTATIONAL SPEED	44–88 r.p.m
BLADES	Fixed pitch. NACA 44xx series
CONTROL	Variable pitch tips
TRANSMISSION	2stage shaft mounted
GENERATOR	Synchronous. 440v. 3phase
ORIENTATION	Servo drive
TOWER	1·82m dia. steel
CONTROLLER	Microprocessor
ANNUAL ENERGY	700,000kWh at 10m/sec site

20 m Diameter Turbine Specification

The performance goal of the 20 m diameter turbine is for a rated power of 250 kW at 17 m/s and rotational speed of 88 rpm. It will begin to operate at a wind speed of 8 m/s and shut down when wind speeds exceed 27 m/s. Based upon the estimate that the annual mean wind speed at hub height will be 10.5 m/s, the annual energy output is estimated to be 700,000 kWh, giving it a capacity factor of 0.32.

60 m Diameter Turbine Specification

The 60 m turbine performance goal is for a rated output of 3 MW at 17 m/s and rotational speed of 30 rpm. Its cut-in and cut-out speeds will be similar to those of the 20 m machine. Based on the estimate that the annual mean wind speed at hub height will be 12.6 m/s its annual energy output is estimated to be about 10.5 GWh, giving it a capacity factor of 0.25.

COMPONENT DESCRIPTIONS

Both machines are to have a synchronous generator and variable pitch rotor blade tips. The main features of the 20 m machine have been given in detail by Armstrong, Ketley and Cooper (6), and are outlined in brief below. Those for the 60 m machine are subject to revision as the design contract proceeds and cannot be specified in detail at this stage.

Rotor

The two bladed rotor is located upwind of the tower. The outer 20% of each blade can be feathered through 90° to provide power trimming and aerodynamic braking of the rotor. The structural element of each blade is a fabricated steel spar. This is enclosed in a GRP shell of aerofoil section with appropriate twist and taper, the space between the spar and shell being filled with foam. The tips are continuously moved via a bell crank and linkage system by a hydraulic actuator located in the nacelle.

Transmission

The rotor shaft is carried by two roller bearings mounted in a fabricated steel housing. The shaft carries at its aft end a two stage in-line gearbox. This is allowed to rotate against the action of springs and dampers. The mechanical compliance introduced in this way allows torque spikes from the rotor, caused by gusts, to be attenuated before reaching the generator and hence the grid.

The swinging gearbox removes the high frequency torque spikes whilst the low frequency surges are removed in combination with the movement of the blade tips which spill power. The power quality is thus maintained within limits acceptable to the grid system. A more detailed description of how this is accomplished has been given by Garrad (7).

The high speed gearbox output is connected to the generator by a one-way coupling. This allows the rotor to freewheel below synchronous speed, and prevents power being drawn from the grid during wind lulls. It also is helpful in the synchronisation procedure.

A disc brake is installed on the low speed shaft, and is capable of stopping the rotor from an overspeed condition in high winds with the tips unfeathered.

Nacelle

The power train components are mounted on a stiff steel frame which is in turn mounted on the tower by a standard crane slewing ring. The yaw drive is provided by a highly geared electric motor driving a pinion meshing with the outer gear of the slewing ring. A yaw brake is permanently applied to prevent dynamic instabilities and backlash in the gear teeth.

Tower

This is formed from a cylindrical steel tube mounted below the level of the rotor disc on a conical concrete frustum. An internal ladder leads to an external platform below the level of the nacelle; a ladder from this platform provides access to the interior of the nacelle.

Mode of Operation

The overall system is shown schematically in Figure 3.

Fig.3 ORKNEY 20m WTG INSTALLATION

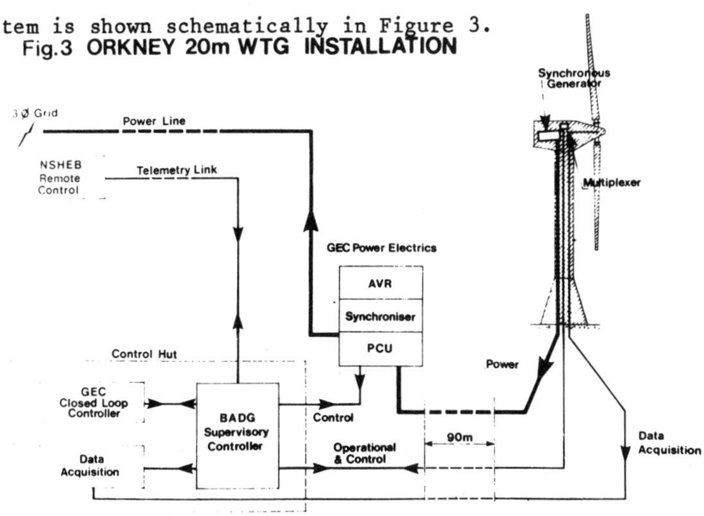

There are two modes of operation:

(a) Direct connected mode, in which the generator once having been driven to synchronous speed by the rotor is synchronised with the grid and delivers power as the rotor turns at a constant 88 rpm, and yields a wind speed versus power characteristic of a form shown in Figure 4.

Fig.4 20m WTG POWER CHARACTERISTIC

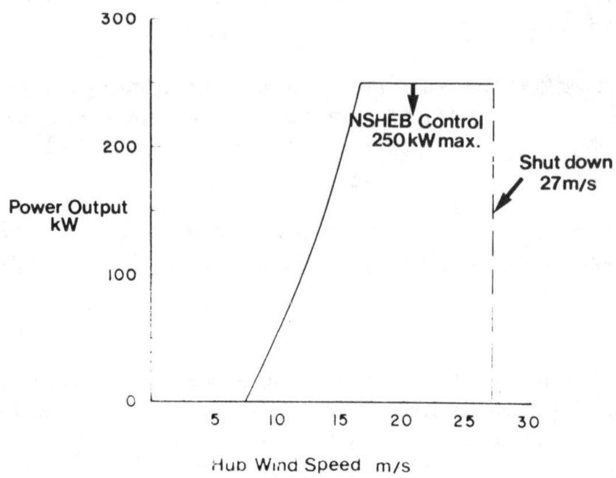

(b) Power conditioned mode (PCM) in which the rotor is allowed to vary in speed between 44 and 88 rpm, so as to follow the characteristic shown in Figure 5.

Fig.5 20m WTG SPEED CHARACTERISTICS

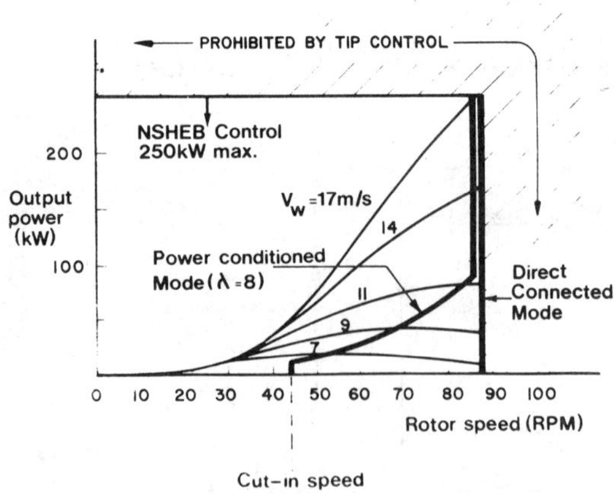

Greater detail of the mode of operation and control systems have been given by Armstrong, Ketley and Cooper (6).

MONITORING

The performance, measurement and health monitoring of the system will be accomplished via a data acquisition system designed to fulfil these functions for both 20 m and 60 m diameter turbines. The overall objectives of the monitoring programme are to:

(a) Confirm the integrity of the design.

(b) Assess the performance of the system

(c) Validate the analytical models used in the design of both machines.

Because the 20 m machine is scheduled to be commissioned 24 months at least before the 60 m machine reaches the same stage, the data acquired from the 20 m machine will provide an input to the engineering of the 60 m design.

Economics

In a twelve-month period of the 1980/81 financial year, 13,000 tonnes of heavy fuel oil and 630 tonnes of light oil were used at the Kirkwall power station. The average price of the heavy fuel oil was £113 per metric tonne at that time. The total fuel bill of £1.548 million was therefore equivalent to a fuel cost per unit of 2.69 pence, and a total generation cost of 2.94 pence per kWh. (Excluding capital interest, depreciation, staffing, etc.)

At the time of going to press the figures for 1980/81 were not available but for the 1979/80 financial year the total loss in supplying Orkney with electricity amounted to £730,000. The total loss of supplying the Orkney, Shetland and Western Isles amounted to £3.55 million in the same period, and is anticipated to be of the order of £8 million for 1980/81.

The current cost of heavy fuel oil is £129 per tonne yielding a present cost of fuel per unit generated of 3.1 pence and a total generation cost of about 3.3 pence per kWh. (Excluding capital interest, depreciation, staffing, etc.)

When the non-recurring cost portion (these include design, dynamic analysis, experimental rigs and testing, and some data acquisition hardware) of the budget cost of £5.6 million for the 60 m diameter turbine is taken into account, the remaining balance of capital costs, and including a construction contingency and project management allowance, is equivalent to a unit cost of approximately 3.2 pence/kWh. This calculation is based on a 15-year life, a 1% O and M charge, 4% fuel cost inflation rate and 5% discount rate. If one assumes a 20-year or 25-year life, the unit cost falls to approximately 2.5 pence and 2.1 pence respectively.

Production versions of this machine could naturally be manufactured, installed and commissioned at a lower cost thereby lowering even further the unit cost of electricity generated.

The Participants

The recently formed Wind Energy Group comprising Taylor Woodrow Construction Ltd, British Aerospace Public Ltd Company and GEC Power Engineering Ltd will design, construct, install and commission both machines for the North of Scotland Hydro-Electric Board (NSHEB). The UK Department of Energy will contribute £4.6 million towards the 60 m diameter machine whilst NSHEB will provide the balance of costs amounting to £1 million.

The group, NSHEB and the Department of Industry will fund the design and construction of the 20 m diameter turbine. The group will be the main contractor for the anemometry and monitoring phase for both machines which will be funded by the Department of Energy.

CONCLUSIONS

The Orkney Project offers a unique opportunity for UK industry and electrical utilities to gain operational experience of two of the world's largest wind turbines.

A demonstration of UK industrial credibility in this technology could provide access to a very large market for wind turbine generators, both at home and overseas. Potential purchasers and manufacturers of such equipment need to be convinced that the technology is mature enough to justify investment in it. It is for this reason that Governments in the USA, Sweden, Germany, Denmark, Holland, Canada and the UK are supporting the development and demonstration phase of wind turbine technology.

ACKNOWLEDGEMENTS

The authors wish to acknowledge the support given to this Project by the:

UK Department of Energy;

The North of Scotland Hydro-Electric Board;

and the Boards of:

Taylor Woodrow Construction Limited;

British Aerospace Public Limited Company;

GEC Power Engineering.

REFERENCES

1. Plant J.A., and Dunsire, A., "The Climate of Orkney" Climatological Memorandum, No 71, Meteorological Office, Edinburgh, 1974.

2. Golding, F.W., "The generation of electricity by wind power" Published by E. and F.N. Spon Ltd., London, 1955.

3. Tagg, J.R., "Wind power related to the generation of electricity by wind power", ERA Technical Report C-T115, 1957.

4. Golding, F.W., and Stodhart, A.H., "The selection and characteristics of wind power sites", ERA Technical Report C-T108, 1952.

5. Lindley, D., and Stevenson, W., "The horizontal axis wind turbine project on Orkney", presented at the 3rd British Wind Energy Association Conference at Cranfield 9-10 April 1981 - to be published by Multiscience.

6. Armstrong, J.R.C., Ketley, G.R., and Cooper, B.J., "The 20 m diameter wind turbine for Orkney". - IBID -.

7. Garrad, A.D., "Wind Turbine Transmission systems", - IBID -.

Danish Experience of Small Wind Powered Generators

P. Lundsager

Risø National Laboratory Test Site for small windmills, Roskilde DK-4000, Denmark

(This summary and comment written by Dr. John Twidell)

ABSTRACT

Summary of the Danish programme for the development of small wind powered generators (10 kW to 100 kW) and the Wind Test Site at the Risø National Laboratory.

KEYWORDS

Denmark; Wind; Generator; Test; Manufacturer.

INTRODUCTION

Mr. Per Lundsager presented a verbal report to the Conference on the experience and progress of small wind powered generators in Denmark. This summary and comment is written to present the outline details of the most successful Danish programme.

DANISH GOVERNMENT STRATEGY

The strategy was formulated around 1975 to reduce dependence on imported oil. Small wind generators up to a capacity of about 50 kW were to be encouraged for household, farm and institutional use. Large machines, say 250 kW to 3MW, would be developed mainly for electric utility use.

The small machines would be used in either of two modes: (1) as stand-alone machines producing variable voltage and frequency power for use in resistive heating loads for space and water heating, or (2) as grid-connected machines with the controlled electrical power being used either locally or exported for sale to the grid.

The small machine development would be encouraged by a State research group at the Risø National Laboratory, Roskilde. This laboratory was originally established for nuclear power research and development, but has since diversified. The wind power group tests machines for manufacturers over 3 month periods and advises on technical improvements. The service is provided free of charge on the understanding that the information is made available to the public. Regular test reports and summaries of developments are published, and a good relationship has been established with the manufacturers and the public. Test results and research conclusions are published in Danish, with a few general studies in English.

The Risø Laboratory group has played an essential part in establishing safety codes and registered modes of operation, particularly for grid linked machines. By setting a professional and fully technical approach from the onset, the Laboratory has obviously helped to establish wind-power engineering as a respected and specialist discipline in Denmark.

The Danish Government also established clear financial inducements to encourage manufacturers and purchasers of approved machines:-

(i) Value added tax (presently 20% in Denmark) would be refunded on all sales of small machines.

(ii) A substantial subsidy (originally 30% and now 20%) would be given to purchasers towards the cost of the machine. The total amount of these subsidies would be limited each year. In practice about 250 machines per year have been subsidised.

(iii) The electricity utilities agreed to purchase electricity from privately owned grid-linked wind generators at one half the unit sale price. Thus the owner of a wind generator may sell electricity to the local grid at 2 pence per kWh when the purchase price is 4p per kWh (1981 prices given as an example). Obviously this arrangement was only possible when the machine had been inspected by the utility for safe operation.

MANUFACTURERS

There are now about 20 manufacturers of small wind power generators in Denmark. Together they have manufactured and sold in Denmark about 700 machines of capacity between 10 kW and 50 kW. The 1981 production is expected to be at least 300 machines sold in Denmark and possibly several hundred more exported.

Only one manufacturer makes machines exclusively for heat via a supply of variable voltage and frequency electricity. There have been about 300 sales of these machines. All other manufacturers have made induction generation machines for grid connection. These could be adapted for self-excitation as stand-alone machines, but in practice few had been so adapted.

All but two manufacturers had began as relatively small firms with experience in, for instance, agricultural engineering, or heating and plumbing. The remaining two manufacturers were sections of large engineering companies. Most manufacturers have expanded their business considerably in the past four years, with several beginning line production of machines and employing qualified wind engineers.

USERS

The main reason for purchasing a wind generator has been to save some expense of heating oil in private households, farms or public institutions such as schools. The output from independent stand-alone machines is usually fed to separate resistive heating immersers in the central heating boiler and the hot water supply tank. Grid linked machines have their output fed to the local supply line via import/export metering. Thus by generating electricity the users save on electricity purchases and can always have the opportunity of selling excess power as export to the grid.

A common capacity of wind generator for a stand-alone machine is 10 kW. Grid linked machines of 22 kW maximum capacity have been the norm, but there are signs that this average size will increase towards 50 kW. The cost before subsidy and without V.A.T. of a 22 kW capacity machine is about £14,000. Site works are extra.

CONCLUSION

The manufacture and public acceptance of small wind generators in Denmark is now assured. The total market for small machines in Denmark is at least 20,000 (i.e. about 1 machine per 100 households or institutional buildings), and may be considerably more. In addition there is considerable export potential. Thus the present sale of a total of about 750 machines in Denmark has hardly begun to make an impression on the market. Manufacturers are all optimistic about future business once they pass through the early stages of development and organised production.

Details of Danish wind machines are available from the Risø Laboratory, from the manufacturers directly, from the Danish Trade offices, and from specialist magazines published in Denmark.

A Small-scale Wind Power Installation in Shetland

F. C. Evans and H. Winterbotham

ABSTRACT

The design and installation of a small-scale wind-powered generator on the island of Papa Stour are described. The windmill is a vertical-axis design with self-acting variable-pitch blades. The alternator is a large-diameter air-gap machine using permanent magnets. Emphasis is on the practical problems of construction, transport and installation, but a short technical description of each machine is included.

KEYWORDS

Wind-power, Vertical-axis, Air-gap, Alternator.

In 1977 the authors agreed to cooperate in the development of two novel designs which together should make up an aerogenerator of considerable promise: the Evans windmill and the Winterbotham alternator. On Saturday 18th July 1981 the erection party arrived in Lerwick, located the rig which had arrived a few days before, and saw it loaded onto the Spes Clara, a fishing boat chartered by the Shetland Islands Council, for transport to the island of Papa Stour. By the same evening it had been unloaded on the island, taken over to the selected site by one of the island tractors, and placed in position. The plan was to erect the aerogenerator and leave it working for a year.

This project was made possible by a grant from the Shetland Islands Council, which has a keen interest in windpower and has done a great deal to foster the smaller island communities. We had agreed to the proposition with some misgivings; the site was remote and the windmill not fully tested. However, the attractions of a strong wind regime in which to complete development were obvious, and so we agreed. The local Council arranged for a resident, Mr Kevin Coleman, to undertake supervision and to read the instruments, in return for the use of the heating produced.

In this Paper we give our first impressions of the island and its problems, and a brief description of the aerogenerator, supplementing the Summary. We then outline the progress made up to the date of writing (the end of July 1981) and some of the lessons learned.

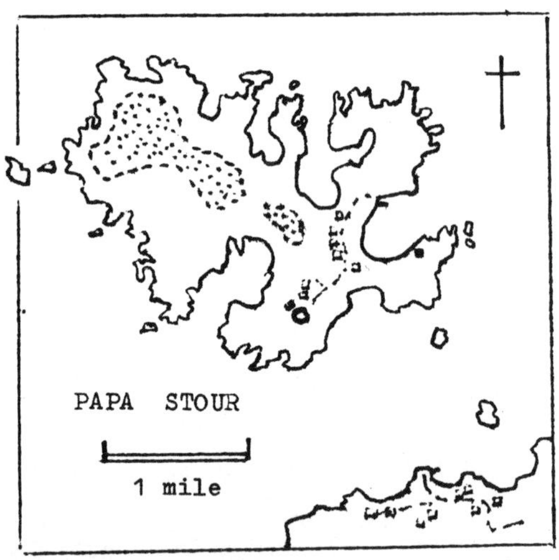

Fig.1. Location of aerogenerator O

Papa Stour is a small island off the west coast of the mainland of Shetland. A map appears at Figure 1. The channel is narrow and shallow but can be extremely rough and the island may be cut off for weeks at a time in winter. The south-eastern part, facing the mainland, has a good depth of soil and is quite productive. The remainder has been stripped of its peaty soil cover in past ages and now the surface consists of rocks and sparse microflora. The land surface is generally low. Deeply indented bays with sandy beaches are interspersed with rocky headlands and isolated rock stacks. The highest point is at the north-western end, where the Trig point height is 87 m. From this point a ridge runs south-west towards the inhabited part. This area, shown dotted on the map, would be very suitable for a "wind farm". The island is within the wind belt and a 5 - 6 m/sec wind blows almost continually. Very strong winds and gales are normal in winter. One object of the project is to study the wind regime.

Most of the occupied houses are widely separated in crofter style and lie within reach of the one tarred road, which connects the pier in the north with the School and the old Manse (the location of the aerogenerator) in the south, a distance of a mile or so. There are now some 40 people living there, including some old folk and five children. In the old days of the herring fishing there were ten times as many. Many long-abandoned ruins, and the traces of garden and track boundaries still visible in the green turf, give a sad impression; but the houses that remain are well found and some fields are under cultivation. Sheep graze on the fields in considerable numbers.

The community is supported by the families which settled on the island in the early 70s, when the original population had dwindled to a handful. A great deal of hard work has to be done in unloading coal and oil for heating and running the small diesel generators providing "mains" electricity, in loading sheep for export to the mainland, and the like. The number of able-bodied men is barely adequate for such tasks and the community is only viable because of the strong spirit of cooperation and mutual help which prevails.

Communications with the mainland have now improved. There is an island-based fishing boat which takes the mail and light goods to and fro three times a week. The Spes Clara calls from two to four times a year with heavier goods. There is an efficient radio telephone service which connects with the national STD system. There is also a good piped water supply pumped from a fresh water loch. There is no electricity supply. Most houses have their own or a shared diesel generator, of 1 to 3 kVA capacity. Opinions vary on the desirability of a central generator. Some would prefer to run their own, which they feel are at least under their own control. Independence and self-reliance are marked characteristics.

From our limited acquaintance we would guess that the Papa Stour community needs a few more families to be really viable. There is also a lack of earning capacity, to offset the rising prices of coal and oil. Wind power could be useful in providing warmth in winter and lessening fuel consumption. A more ambitious plan would be to develop the northern end of the island as a "wind farm", probably employing a number of small generators for reliability. In addition to solving the heating problem this would provide employment for a small number of technical and maintenance staff and thus increase the number of residents.

A general idea of the aerogenerator can be gained from Fig.2. It is located at the circle marked in Fig.1 and its immediate surroundings are shown in Fig.3. It is designed to be transportable and can be erected in a couple of days by two men. A rig of this size presents many engineering problems. The base pyramid contains the alternator, a winch for erection, a band brake and a 2:1 V-belt drive. The emphasis is on rigidity. The mast and arms must resist bending and vibration, and the centrifugal forces generated by the blades. The problems here are similar to those met in marine installations. The blades and their couplings must combine strength with lightness and ease of control. Aircraft design methods are appropriate.

The aerogenerator is designed to run at variable speeds and to produce alternating current for heating. It is not at present thought that a control unit will be

Fig.2.

required, as the windmill has excellent self-starting characteristics (of which, more later). An instrument board in the house provides readings of wind speed, alternator R.P.M., and power. There is a meter for recording accumulated kilowatt-hours. Two 12 volt accumulators charged by the alternator provide power for the meter board electronics. The output is absorbed in three 3 kW immersion heaters in parallel in a 45 gallon tank, connected to the domestic supply. The power cable connecting the generator to the house is 60 m long and has a loop resistance of 0.2 ohms, low engough to be negligible compared with the load resistance of 6.4 ohms. This cable is ploughed in just below the turf and on the underlying rock.

Installation proceeded rapidly in the first few days. It was found desirable to hang a temporary counterweight on the derrick to ease the strain on the winch whilst the blades were assembled. The difficulties began when one of the blades failed to take the clamping stress at the control wire fastening, during assembly. Investigation showed faulty manufacture. A new form of attachement was devised for all three blades and new parts were ordered by telephone from St Andrews University. Meanwhile a temporary ring-bolt was welded up from waste material and fitted to the faulty blade.

The assembled windmill, without the alternator being connected, was unbraked on Thursday morning in a 6 m/sec wind. It accelerated smoothly to about 60 R.P.M., ran speed-limited for about five minutes, and then the improvised bolt snapped. The wing concerned swung outwards and the rig instantly stopped, a good demonstration of the "fail-safe" action if a control wire parts.

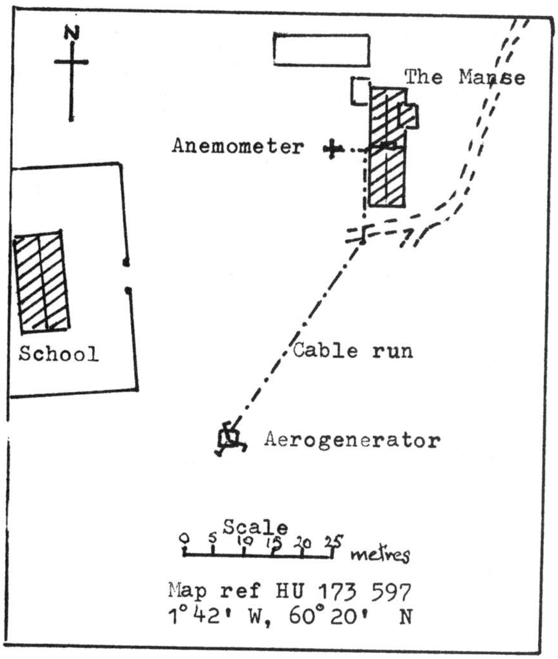

Fig.3.

The new components duly arrived, and on Tuesday night, 10 days after our arrival, the rig was again complete and was left with the brake hard on overnight. The alternator was again not connected. The brake had demonstrated its ability to hold against a 45 kg pull at a distance of 3 m. But this was not enough. During the night the wind increased to Force 6 and the rig started against the brake. It must have run for some time as the brake area was extensively scorched. Finally the brake failed, its lining burn out. The rig ran up to a speed estimated as 120 R.P.M. and possibly more, and cracked a main blade support. This time the control wire fastenings held, perhaps unfortunately.

This episode, though a remarkable demonstration of self-starting and driving home a number of useful lessons, marked the end of what could be done locally. The party returned on Saturday August 1st, two weeks after arrival, having made the rig safe and waterproof.

It is planned to return as soon as the broken parts have been studied and design improvements made, to continue the development. Our misgivings about moving to a remote place to learn about windmill construction were probably justified. To be prudent, one should never go far from workshops with an unproven design. However, so much the worse for prudence. In no other way could we have learned so much so quickly.

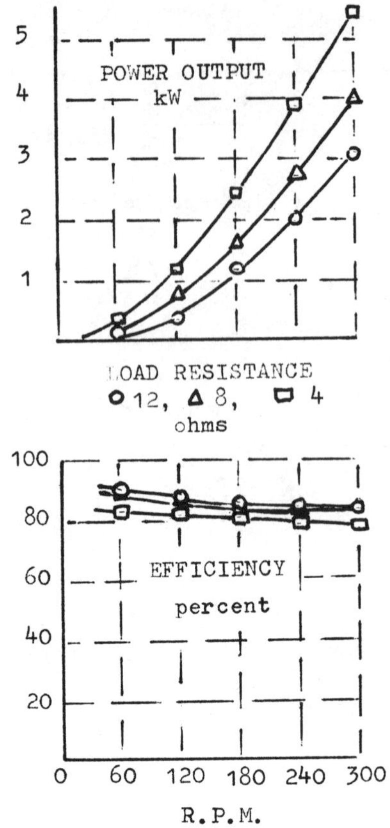

Fig.4. Power output and efficiency of alternator

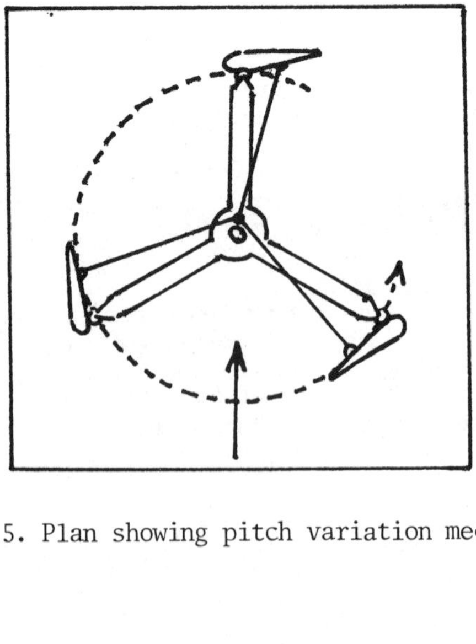

Fig.5. Plan showing pitch variation mechanism

Wind Resource Evaluation in Ireland

E. Kelledy

Department of Statistics, Trinity College, University of Dublin, Dublin 2, Ireland

ABSTRACT

In this scoping study, approaches to the evaluation of the Irish wind resource were investigated. Data requirements of potential users were considered and gaps in the existing data identified. In particular, the statistical aspects of data collection, analysis and presentation were examined.

KEYWORDS

Wind resource evaluation; meteorological data; statistical modelling; interpolation/ extrapolation; short term measurement program.

INTRODUCTION

Growing interest in wind power applications has prompted the Irish Government's Department of Energy to commission a three year program of wind resource evaluation. The approach to this task is interdisciplinary, involving such groups as the Department of Energy itself, the Meteorological Service, the Electricity Supply Board and a team of statistical consultants. The culmination of the program will be the production of guides to the national wind resource.

The Statistics and Operations Research Laboratory, at the Department of Statistics, Trinity College, were asked to carry out a three month scoping study, aimed at the definition of the role of the statistical consultants in the overall program of wind resource evaluation.

THE SCOPING STUDY

The terms of reference of this scoping study included the review of national and international techniques of wind data collection and resource evaluation, the identification of potential users of guides to the Irish wind resource, together with their data requirements, and the evaluation of a number of approaches to the enhancement of the existing data base. Detailed recommendations for work to be carried out over the first year of the program and outline proposals for the subsequent two years were made. These proposals related mainly to 'paper studies'; it was recommended, for instance, that responsibility for the selection and installation of equipment for monitoring and recording wind data should rest with the Department of Energy and the Meteorological Service.

DATA USERS AND THEIR REQUIREMENTS

The scoping study attempted a preliminary identification of potential markets for wind power. Users were arbitrarily classified as small (requiring wind turbines with a rated power of 15-20 kW), medium (100-200 kW) and large (1-5 MW), corresponding loosely to individuals, communities and utilities. Within each category four possible applications of wind power were identified:

(i) direct water pumping;
(ii) direct water heating;
(iii) D.C. electricity generation;
(iv) A.C. electricity generation.

At the level of the individual farm, for example, water might be pumped for human or animal consumption, or a wind turbine and pump might be used in drainage or irrigation. Domestic water might be heated directly, using a Joule bucket, or indirectly, via a D.C. generator. The energy demand profiles and the characteristics of the machinery required would be application-dependent, with implications for data requirements.

However, particularly at the lower end of the market, the potential users would not necessarily have a direct interest in wind resource data. For example, it is unlikely that the individual farmer would be interested in detailed estimates of wind energy availability; rather the users of data of this nature would include agricultural advisors who would have to make recommendations on the sizing of wind turbines, and manufacturers involved in the construction and installation of turbines and ancillary equipment.

THE EXISTING DATA

The major sources of wind data for Ireland are the Meteorological Service and the U.K. Meteorological Office. Observations of wind speed and direction have been made, hourly, over a long period at the synoptic stations maintained by these agencies. Summaries of these observations are published on a regular basis, and the raw data are available in computerized form. Other sources of data include the ERA survey carried out in the early 1950's, which provides estimates of long term annual average wind speed at a number of Irish hill tops.

These existing data were assessed in the light of the requirements of the various potential users. The major inadequacy appeared to be in relation to geographical coverage since meteorological and other measuring stations are not generally sited with regard to the optimality of the wind resource. Furthermore, these data generally refer to a standard reporting height of 10 m, whereas large wind turbines might be mounted at a hub-height of 50-80 m. Other inadequacies, from the point of view of resource evaluation, relate to sampling and reporting intervals, accuracy of measurement and reporting, and forms of data analysis and presentation.

APPROACHES TO ENHANCING THE EXISTING DATA BASE

Three approaches to overcoming these inadequacies and to extending the existing data base were identified in the scoping study: -

(i) Further analysis of existing data, with a specific orientation towards wind resource evaluation.
(ii) Interpolation and extrapolation to other locations and heights.
(iii) Collection of data at new sites.

These approaches are not mutually exclusive. Any measurement program must necessarily draw heavily on the analysis of existing data, in the development of techniques for relating short term observations to long term wind regime. Analysis of the existing data might also form an input to the development of techniques of interpolation between sites and these techniques might later be validated using observations made at new sites.

The development of techniques of vertical and horizontal extrapolation and interpolation involves a knowledge of boundary layer climatology and physical wind flow modelling. It is suggested that the Meteorological Service are best qualified to tackle this problem. For this reason, it will not be discussed any further here; the other approaches are described in more detail below.

Further Analysis of Existing Data

Wind speed and direction data are routinely analysed in accordance with standard meteorological practice, rather than in forms useful to those interested in wind energy. However, much useful information can be gleaned from further analyses of these data. For example, the calculation of measures of wind speed variability would be useful in choosing the best of a number of sites; if two sites had comparable mean wind velocities, the location with the less variable wind regime (fewer extreme speeds) would obviously be preferable from an energy production point of view. The analysis of the variability of wind speeds through time (e.g. conditional distributions, serial correlations, run lengths) would be useful in studies of predictability and the determination of the need for auxiliary energy.

Analysis of the existing data would also be of value in theoretical studies, as background to the development of techniques of interpolation/extrapolation and in the exploration of methodologies for relating short run measurements to long term wind regime. The concepts of statistical modelling and parameterization would be very important here. For instance, the Rayleigh distribution is commonly used to model wind speed frequency distributions. Using this model, the entire frequency distribution can be approximated, given only the mean wind speed at the site in question. If the Rayleigh distribution were found to fit at all sites for which data is currently available, then it would be reasonable to presume that it would also apply to new sites. The implications for data collection are obvious. The exploration of the statistical relationship of wind velocities at two or more sites would form a useful input to a program of short term measurement. For example, if wind speeds at pairs or groups of nearby sites were shown to be highly correlated, then long term data from existing stations could be used to convert short term observations at new sites to equivalent long term estimates. The exploitation of this statistical structure could significantly reduce the length of the measurement period required to estimate long term average wind speeds with a given degree of accuracy.

Collection of New Data

This approach would obviously be the most costly. Moreover, since wind displays considerable annual and seasonal variation it would be necessary either to make observations over a very long period, or to develop techniques for relating shorter term measurements to long term average wind regime. For many users, such techniques would be, in themselves, a valuable output from the three year program; any large scale investment in wind turbines, by a community or utility would, probably, be preceded by detailed wind monitoring, over a period of one to two years, at candidate sites. Output from a short run measurement program of this nature would be very difficult to interpret with confidence, in the absence of the type of techniques proposed above. The scoping study recommended, therefore, that considerable effort be devoted to the development of such techniques.

It is envisaged that mobile measuring stations, incorporating automatic data loggers, would be used in the monitoring of wind speeds and directions at new sites. Site identification would be made by the Meteorological Service, in consultation with the other experts. It is recommended that they should prepare detailed isovent maps, indicating their best estimates, based on current data, of mean wind speeds. These isovents should be accompanied by some statement of possible error bounds or levels of confidence. One possible technique would be to vary the width of the isovent line to indicate the degree of certainty. Using these maps, the mobile anemometers should be sited in those areas where the error band is wide, but where the wind regime is likely to be exploitable by at least one of the groups identified above. An objective of the three year program would be to refine these maps on the basis of the statistical analysis of the existing data and those collected at these new sites.

PROPOSALS FOR THE THREE YEAR PROGRAM

A program of action for the three year program was recommended in the scoping study. The proposals put forward bring together the recommendations made above; the overall objective is the production of guides to the Irish wind resource: -

Year One - Ascertain data requirements for end users and for interim studies. Anticipate form and nature of final reports. Assemble and analyse existing data. Develop a program for short term data collection.

Year Two - Site and utilize mobile anemometers. Apply methodologies developed in year one to the initial analysis of new data. Refine methodologies.

Year Three - Major analysis of new data. Preparation of final reports.

The first year is seen to be the most important for the statistical consultants. The work carried out will be of a developmental nature and decisions made in this year will have implications for the activities of the subsequent two years. Year two will see the implementation of the measurement program proper and will provide an opportunity to assess the proposals made in the first year. Refinements to the program are to be expected at this stage. In year three, data analysis will be mainly of a routine nature and the major task will be the preparation of the guides to the wind resource.

ACKNOWLEDGEMENTS

The scoping study which forms the background to this paper was funded by a research grant from the Department of Energy whose permission to publish is acknowledged. Thanks are due to Mr. John Haslett, of the Department of Statistics, for his work on the scoping study. I would also like to express my gratitude to the Meteorological Service and the Electricity Supply Board, for their assistance.

Small Vertical Axis Wind Turbine—Design and Experience

W. S. Bannister, S. Gair, J. D. Fyvie and M. N. Millar

The Wind Energy Group, Napier College, Edinburgh, UK

ABSTRACT

The Wind Energy Group at Napier College has now completed a year of very active participation in wind energy research. A theory has been developed for predicting the characteristics of vertical axis-wind turbines; the turbine test facility has been further developed; the instrumentation has been built; the power generation and control system has been designed; a bank of anemometer towers has been constructed. In addition to this practical development, collaboration with other research workers has markedly increased and the group will be enlarged to take account of the increased activity.

KEYWORDS

vertical-axis; straight-bladed; low Reynolds number; permanent magnet generator; wind-power plants.

INTRODUCTION

In September 1980, a paper was presented at Inverness on the wind energy research work being undertaken at Napier College. Until that date, only two members of staff had been involved in the project. However, on receiving funding from the SRC, a Wind Energy Group was formed and this group has now concentrated its efforts on increasing its expertise on wind energy and on developing the existing test facility at the College. As a result of attending conferences, such as at Inverness, the group has not only established contact with other researchers but has carried out collaborative work for other establishments and also visited research workers at various locations. A great deal of experience has been obtained and collaborative effort is being continued. Progress made over the past year includes detailed design and analysis of the wind turbine, electrical power generation and control, development of instrumentation and finally collaborative research effort with other establishments in the UK. This report summarises the salient features of the work completed, outlines difficulties encountered, methods of solution and details the programme of research at this establishment for the foreseeable future.

RESEARCH ACTIVITIES

1. <u>The Napier College Wind Energy Group</u>

1. The Napier College Wind Energy Group

During 1980, activities on wind energy at Napier College increased rapidly after the award of the £30,000 SRC Grant. In September of that year a Wind Energy Group was formed and this group has been meeting regularly as the work progresses and record is continuing to be made of activities, to coordinate a total effort, both from the engineering and economic aspects of wind energy research. At present, the group consists of four mechanical engineers, two electrical engineers, one economist and one accountant. Another mechanical engineer will shortly be joining the group to continue the computer work associated with the prediction of turbine characistics and analysis of experimental results. Details of the Wind Energy Group are shown in figure 1.

As the work progresses, results will be reported and members will attend conferences and maintain contact with other wind energy research workers. As part of this plan, five members of the group recently completed a very successful week long tour of research establishments in England and Wales. A report of the tour will shortly be published in BWEA Windirections.

2. The wind turbine (see Refs 1 - 4)

The small straight-bladed vertical axis wind turbine has been constructed in order to investigate the performance of various configurations of wind turbine. The maximum dimensions of the turbine to be tested will be 5.0 m diameter, with blades of 0.25 m chord and 2.0 m span. Any of these three dimensions may be altered to values below those quoted. The number of blades on the turbine will be 2, 3 or 4 Thus a range of test configurations may be tested on the same basic facility.

3. Theoretical prediction of turbine performance

A theory developed from the work of Strickland (5) and Sharpe (6 and 7) has been further advanced by Bannister (8). Using known aerodynamic characteristics of the aerofoil sections a computer programme is being developed to predict the performance of a range of turbine configurations. Experimental results from the actual turbine will subsequently be compared with the theorectical predictions.

4. Detail Design

(a) Blades. The original configuration used two rough wooden blades of approximately NACA 0012 section. Foam filled fibre glass NACA 0015 blades are now available with a chord of 0.25 m. These fibre-glass blades are heavier than we would like and also tend to be inconsistent, ie. blades are not of identical weight. Negotiations are underway with the supplier to improve the manufacture in the next set. A special wrap-round clamp has been designed in order to fit the blades to the ends of the cross arms without holes being drilled in the fibre-glass. An adjustable fitting has also been designed for the ends of the cross-arms which will allow the blades to be pre-set over a range of incidence of $\pm 11^{o}$. Extruded aluminium blades, supplied by Geoff Watson of the Northumbrian Energy Workshop, will be used at a later stage. These will be of 6 in chord NACA 0015 section and 9 in chord NACO 0018 section.

(b) The brake. The wind turbine is not run unattended hence automatic starting and stopping devices have not been incorporated at this stage. However, as a safety precaution a simple mechanical caliper, with a 10 in diameter disc, has been fitted. This is cable operated and can be used to prevent the rotational speed becoming excessive.

(c) The wind tunned model. In addition to the test site facility, a wind tunnel model is being built. This will be similar to the actual wind turbine in that the main dimensions may be altered. Initially the model will have 6 in chord extruded aluminium blades. Blades of different chord size will be used as these become available. The cross-arm and fittings will allow the blade incidence to be pre-set and the central hub arrangement is such that the number of blades can be varied from 2 to 6. The intention is to test this model in the laboratory in front of a small open jet wind tunnel and then in the open air on the test site.

(d) Anemometry. Three anemometer masts, approximately 8.5 m high have been built. These will be used with a Cassella W1402 combined direction, and velocity indicator and also two small lightweight Brookes and Gatehouse D8102 anemometers. With the two small lightweight anemometers it will be possible to change the height of the anemometer mountings as required. A further use for the bottom 3.8 m length of mast is that the small-scale model turbine can be fitted to the top for field trials. A fourth mast is being purchased. This is a 12 m telescopic mast which will be used for additional wind velocity measurements.

(e) Slip Ring Unit. This unit has been purchased and is being fitted to the bottom end of the main turbine shaft. This is a 12 way unit which will allow an earth connection plus eleven signals to be transferred from the rotating turbine. These signals will include the outputs from several strain gauges located both on the cross-arms and on the blades. This will allow measured stress levels to be compared with theoretical predications. The variation of stress level with rotational position of the blades will thus be obtained over the operational range of the wind turbine.

5. Instrumentation

A measuring unit has been developed to collect data during the wind turbine tests. The purpose of this unit is to store information being transmitted from 16 different transducers, each transducer producing an analogue output. The unit is basically a large storage area which is filled in a predetermined sequence by a reading unit.

The reading unit consists of a 16 channel multiplexor with integrated analogue to digital converter and amplifiers. Each input is memory mapped and this gives flexibility to the way in which the data is entered, ie for slow changing data fewer readings are necessary. The decision is made by the software and since the programmes are usually very short and simple, a fast turn around time can be achieved.

Once the storage area is filled, the data is then down loaded to a small tape recorder and may be processed at some later date using more sophisticated software. The main advantages of this system over a multichannel tape recorder is the cost, the unit is about 20% of the cost of a good multi-channel tape recorder. Also since the data is programmable before it is taped, the output will be easier to process.

The basic parameters to be measured are shown in figure 2, along with the information to be derived.

The testing in the main, will use the Sharpe (6, 7) technique of loading the turbine with its own inertia and allowing it to accelerate through a permissible speed range Dynamic operation of the turbine/generator into a stepped resistance load will also be investigated.

Frictional torque and parasitic torque due to cross-arms will be determined by testing at zero wind velocity and with the blades removed.

After analysis, results will be compared with theorectical computer predictions using the extended multiple stream tube theory outlined in Reference 8.

6. Electrical Power Generation and Control

Considerations for electric power generation and control are being looked at from the point of view of satisfying two separate operative strategies for a small stand alone wind turbine;

(a) operation at fixed tip speed,

(b) operation at fixed tip speed ratio

The choice of generator and electrical control required has been based on the need to convert, as efficiently as possible, using standard technology, the power output of the wind turbine and simultaneously control the load in such a manner as to always present the wind turbine with a power proportional to (rotational speed)^3 loading characteristic. The final load demand being stipulated as nothing more sophisticated than a space heat requirement in which the actual voltage and frequency output of the generator is of no consequence. To achieve this aim, a permanent magnet type generator has been selected. In the absence of field control for the generator two alternative control strategies are currently being investigated.

(i) load control
(ii) electrical power flow control

(i) - load control - with this type of control switched resistance load units are used to approximate to the power proportional to (rotational speed)^3 characteristic of the wind turbine. The greater the number of switched resistance load units the closer will be the approximation to the wind turbine characteristic but obviously there has to be some trade off against increased complexity, dynamic torque transients and hysteresis of switching in and out loading units. In operation the maximum current loading of the generator can be set and the discrete resistance values selected so that operation of the turbine is held near to the Cp maximum nor is the turbine ever stalled by the electrical load. The control is achieved by electrical sensing of the wind speed which can be carried out at the generator terminals by measuring the output frequency of the voltage produced by the generator (this is directly proportional to the shaft speed of the turbine since frequency = shaft speed x number of pole pairs). This signal when converted to a voltage level by a frequency-to-voltage convertor is used to drive a level detector which in turn performs the load switching function. This method of control has the merits of being cheap, simple and effective for this particular type of electrical load.

(ii) Power Flow Control - a more effective technical solution to the problem of obtaining a wind turbine load match using a small desk top computer is also being investigated. The arrangement is shown in Figure 3. Here, a controlled rectifier situated in the transmission line between the generator and its load has its firing angle, (and hence its power throughtput) controlled by a simple programme run on the computer. By suitable variation of the firing angle of the thyristors in the rectifier unit, it is possible to follow a $P \propto v^3$ law provided signal measurement and speed of response can be achieved within the time taken for the wind turbine to react to wind changes. This system is primarily aimed at evaluating the merits and increased energy yield (if any) of operating the turbine at variable speed in a controlled manner against other operational strategies.

7. Collaboration. Successful experimental work has been carried out at Napier College for Freddy Evans of St Andrews University and also for Geoff Watson of the Northumbrian Energy Workshop. Work is continuing for both of these establishments and as a result of the recent tour made by the Napier College Wind Energy Group

collaborative work will be undertaken for the establishments shown circled in figure 4. In the first instance, the collaboration is taking the form of sharing and exchanging equipment thereafter cooperation in the use of test facilities will be developed.

8. Future Programme. Over the next year it is anticipated that the wind turbine will be completely fitted out with instrumentation and initial tests completed and reported. Main parts have now been completed and the work of the next few months will entail assembling and calibrating the instrumentation and completing erection work on the test site. With a bit of luck results will become available at the end of the year. Thereafter, research work will continue into automatic starting devices, automatic overspeed control, continued development of power generation and control devices and ultimately aerodynamic augmentation.

CONCLUSIONS

The past year has seen a very rapid growth in the efforts at Napier College on wind energy research. Thanks to many other fellow researchers this work has been enthusiastically encouraged. It is the intention of the authors to maintain progress notwithstanding limitations of time and resources. It is confidently anticipated that the coming year will yield a further step forward in our experience and expertise.

ACKNOWLEDGEMENTS

The authors gratefully acknowledge the many contributions made by colleagues on the Wind Energy Group at Napier College and to other workers in the field of wind energy research. Thanks are also due to the Science Research Council for their support via Grant Ref GR/B52717. The work is being carried out at Napier College, Edinburgh and is being supported by Lothian Regional Council.

REFERENCES

1. Bannister W S and S Gair (1981) The development of a straight-bladed vertical axis wind turbine. Conference on Energy for Rural and Island Communities. Inverness (September 1980). Ed. Twidell, J.W. Pergamon Press, Oxford.

2. Bannister W S (1979) Aerodynamic studies of a straight-bladed vertical axis turbine. Proc. 1st British Wind Energy Association Wind Energy Workshop (April 1979)

3. Bannister W S and S Gair (1981) Wind turbine research at Napier College Proc. 3rd BWEA Wind Energy Workshop (April 1981)

4. Bannister W S and S Gair (1981) Design aspects of small straight-bladed vertical axis wind turbines. BWEA International Colloquium on Wind Energy, Brighton (August 1981)

5. Strickland J H (1976) A performance prediction model for the Darrieus turbine. International Symposium on Wind Energy Systems, Cambridge, (1976)

6. Sharpe D J (1977) A theoretical and experimental study of the Darrieus vertical axis wind turbines. 2nd BWEA Wind Energy Workshop, Cranfield (1980)

7. Read S and Sharpe D J (1980) An extended multiple streamtube theory for vertical axis wind turbines. 2nd BWEA Wind Energy Workshop, Cranfield (1980)

8. Bannister W S (1981) A theoretical analysis of small vertical axis wind turbines. International Symposium on "Applications of Fluid Mechanics and Heat Transfer to Energy and Environmental Problems' University of Patras, Greece (July 1981)

FIG 1 NAPIER COLLEGE WIND ENERGY GROUP

Name	Department	Involvement
Mr Scott Bannister	Mechanical Engineering	Group Coordinator. Aerodynamic and Mechanical Design
Dr Alan Watson	Mechanical Engineering	Head of Department. Course Co-ordinator for BSc Energy Engineering. Resources
Mr Sinclair Gair	Electrical and Electronic Engineering	Power Generation and Control
Mr Terry Frankland	Mechanical Engineering	Stress analysis and dynamics
Mr Jim Fyvie	Electrical and Electronic Engineering	Instrumentation
Mr Bill McIntosh	Business Studies	Economics
Mr George Brown	Accounting	Economics
Mr Maurice Millar	Mechanical Engineering	Research Assistant
and from September 1981:		
Mr Neil Hay	Mechanical Engineering	Computer Analysis

FIG 2 PARAMETERS BEING MEASURED

Blade angular position Blade angular velocity Blade acceleration		Instantaneous torque and power. Overall torque power and drag co-efficients
Wind velocities Wind direction		Wind characteristics
Strain gauge outputs	-	Blade cross-arm and structure loading
Generator power output	-	Overall turbine performance

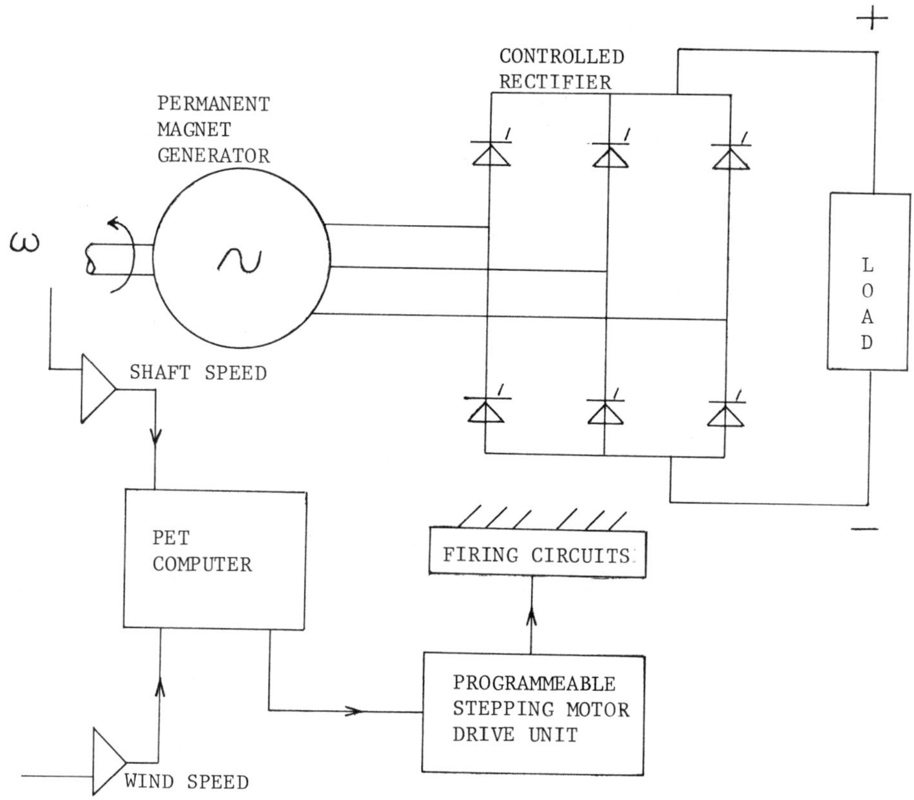

FIG 3 - POWER TRANSMISSION CONTROL

1. Edinburgh - Napier College Wind Energy Group
2. Hexham - Northumbrian Energy Workshop
3. Consett - Trimble Windmills
4. Boroughbridge - Sir Henry Lawson-Tancred, Sons & Co Ltd
5. Coventry - Warwick University
6. London - Imperial College of Science and Technology
7. Kingston-on-Thames - Kingston Polytechnic
8. Leatherhead - Central Electricity Research Laboratories
9. Arlesford - P.I. Specialist Engineers Ltd
10. Reading - Reading University.
11. Didcot - Rutherford Laboratory
12. Exeter - Exeter University
13. Exeter - Windtech.
14. Machynlleth - Centr for Alternative Technology.
15. Swansea - University College
16. Newcastle - NEI Clarke Chapman
17. (St Andrews - F C Evans
    (Papa Stour
18. Glasgow - James Howden & Co
19. Edinburgh - Wind and Water Power Ltd
20. Pitlochry - N.S.H.E.B.
21. South Ronaldsay - NSHEB 'Windmatic
22. Burgar Hill, Orkney - NSHEB
23. Glasgow - University of Strathclyde

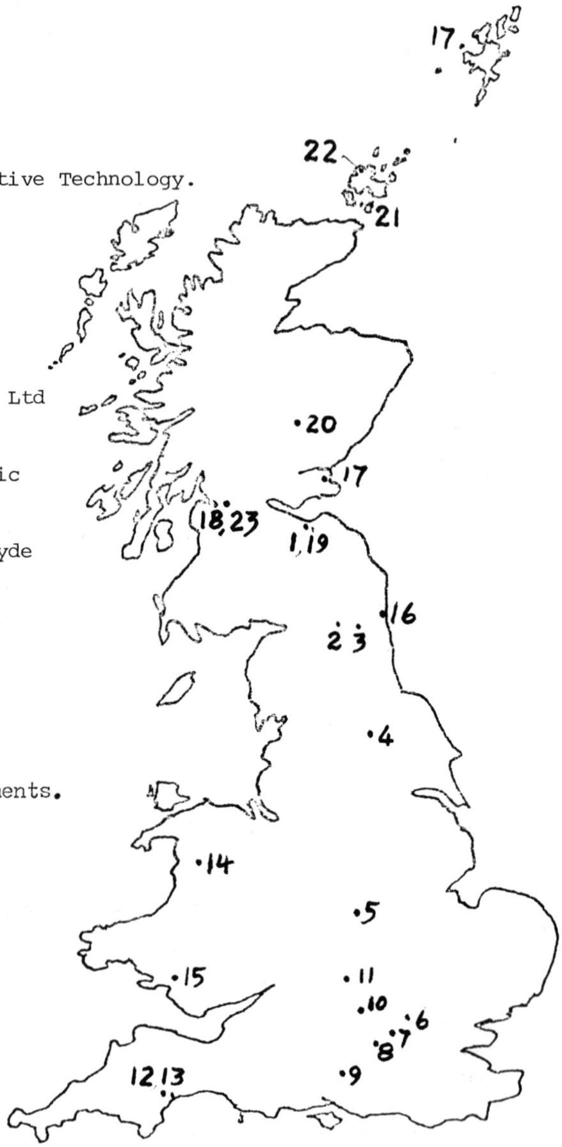

FIG. 4. Wind Energy Establishments.

Commercial Experience with 10 kW Rated Electricity Generating Windmills in Scotland

J. M. Galt and I. T. Findlay

Wind and Water Power Ltd., 1 Huntly Street, Edinburgh, Scotland, UK

ABSTRACT

Six 10 kW rated, horizontal axis, electricity generating windmills have been installed. A typical installation is described here. The upwind windmill has three twelve feet long airfoil section blades with a centrifugally operated pitch control mechanism. A hinged tail and offset rotor allow the mill to turn automatically out of the wind in high winds and a hydraulic damper on the tail hinge prevents excessive yawing in gusty conditions. The mill is mounted on a forty foot, self-supporting tower. The self-excited AC generator gives its rated output at a windspeed of 25 m.p.h. and, on a site with an annual average windspeed of 15 m.p.h.,will generate over 30,000 kWh per year. The variable voltage power is used for heating and to simplify use, a control unit is fitted to match electronically the load to the power available in the wind. Wind and Water Power has designed a Scottish built, 10 kW rated 'Osprey' windmill specifically for Scottish conditions and it is now completing a successful test period. The first will be installed at a customer's site before the end of 1981.

KEY WORDS

Wind power: windmills: electricity generating windmills: alternative energy.

INTRODUCTION

As a successful wind power installation depends entirely on correct siting, Wind and Water Power will not sell an electricity generating windmill without first visiting and assessing the site. In the past two years, over 100 assessments have been carried out for farmers, estate owners, regional councils, hospital boards and a member of our Royal Family. The main demand is for machines of not less than 10 kW to supplement existing heating, although recently there has been an increasing interest in the use of induction generators to operate in parallel with the mains supply and replace electricity used for all appliances.

Over twenty of these surveys have resulted in orders and six 10 kW rated windmills of American manufacture have now been installed (Fig. 1). This presentation describes one of these installations near Glasgow and the progression from site assessment to commissioning.

Fig. 1. 10 kW electricity generating windmill.

ASSESSMENT AND PLANNING

In early 1980, a wind power assessment was requested at a hill farm near Glasgow. The open moorland site, at approximately 1000 feet above sea-level, had few obstructions and there was a suitable site for a windmill 200 yards from the farm buildings. Meteorological Office data for the area, corrected for altitude and terrain type, indicated an average annual windspeed of 14.7 m.p.h. For confirmation, a cup-counter anemometer was installed and readings recorded for three months for comparison with Meteorological Office figures from Glasgow Airport for the same period. After appropriate corrections, these readings indicated a higher average annual windspeed of 16.0 m.p.h.

A report was sent with this information, concluding that a 10 kW rated electricity generating windmill on this site would produce between 29,500 and 34,500 kWh per year, representing 70 to 90 percent of the domestic heating load. The client decided to proceed with the installation and a planning application was made on his behalf by Wind and Water Power. Approval was granted after three months.

The windmill was dispatched from America in late October: industrial disputes delayed delivery until mid-December by which time ground conditions made it impossible to take a crane on to the site. Conditions finally improved in April, 1981.

MACHINE DESCRIPTION AND INSTALLATION

Foundations and Tower

A fixed template is used to position twelve three feet long anchor bolts in the substantial reinforced concrete foundation. The heavy duty, forty foot tall, self-supporting tower, manufactured by British Steel near Glasgow, is delivered and installed in two sections. It has three tubular steel legs and bolted cross-members: each side of the base measures 4.7 feet. A one inch thick mild steel plate, bored to match the nacelle base, is bolted to the top of the tower and the whole structure is hot dip galvanised.

Nacelle and Tail

The nacelle, of heavy welded steel construction, sits on a stub tower containing the large yaw bearings and slip rings which are fitted to prevent the cable twisting when the windmill yaws. A plate, bored to accept the rotor hub, is fitted to the low speed shaft of an offset, parallel axis, two stage gearbox with a step-up ratio of 23.8. A flexible coupling joins the high speed shaft of the gearbox to the shaft of the self-excited AC generator designed to run at 1800 rev/min. The generator has no automatic voltage regulation.

By offsetting the rotor mounting from the yaw axis and hinging the tail, wind pressure on the rotor at high windspeeds will cause the windmill to turn automatically out of the wind. A hydraulic tail stabiliser, fitted to the tail hinge, prevents excessive yawing in gusty conditions. A manual furling cable is fitted to the tail so that the mill may be shut down for maintenance or during prolonged periods of exceptionally high winds.

The nacelle and tail, bolted together at ground level, are lifted on to the tower and bolted to the plate with four high tensile steel bolts with self-locking nuts.

Rotor Assembly

The three twelve feet long blades are manufactured by clamping cast ribs of airfoil section to the main blade shaft to give a sixteen degree angle of twist and then riveting an aluminium sheet over this structure. The blade spars are held in the welded steel hub by reinforced tubular collars attached to the hub sides. Within the hub, these collars are cut at an angle of forty-five degrees. Matching collars, also cut diagonally and each with a 'dog leg' attachment, are bolted to the end of each blade spar. A spring loaded plate is placed on the 'dog legs' and the hub cover bolted in place.

In operation, due to the diagonal cuts in the collars, centrifugal forces causing the blades to move outwards will also cause the blade shafts to rotate within the collars and the 'dog legs' to rise. The spring loaded plate resists this movement and so regulates the rate of change of blade pitch angle.

The assembled rotor is lifted and the hub bolted to the plate on the nacelle face by three high tensile steel bolts with self-locking nuts.

With suitable wind conditions, the erection of the tower and windmill takes about four hours.

COMMISSIONING

The tension in the hub spring is adjusted to hold the blades at the correct tip angle when the rotor is running at its rated speed of 75 rev/min. Any increase in speed will alter the blade pitch and slow the rotor. If the rotor speed exceeds 100 rev/min for any reason, for example if the generator is run without a load, within seconds the blades will take up a stall position and bring the rotor rapidly to rest.

The tension in the tail spring is adjusted so that wind pressure on the offset rotor at windspeeds above 25 m.p.h. will turn the rotor out of the wind. This, in conjunction with a change in blade pitch angle, will control power production to 10 kW in winds above 25 m.p.h.

This windmill is required to supply power for heating and, as it is convenient to leave a fixed load permanently in circuit, a control unit was fitted to match electronically the load to the power available in the wind.

Finally, the output of the generator was checked against manufacturer's specification over a range of windspeeds.

ELECTRICAL CONNECTION

All wiring and earthing were carried out to the specifications of the Institute of Electrical Engineers.

The cable from the windmill was connected to 10 kW of heating elements in a small water tank connected in parallel with the boiler of the existing central heating system. Thermostatically operated valves are used to control the circulation of water from this tank, initially to the domestic hot water system and then to the space heating circuit, overriding time clocks if necessary. A secondary, air cooled radiant heating load was also installed and is switched on automatically if the central heating system is shut down for any reason.

OPERATION

Heavy snow within days of the installation delayed the wiring for a further month. However, the windmill has operated successfully since the end of May. The only 'shut-down', for 24 hours, was due to a faulty valve in the central heating system. The windmill has operated in winds up to 55 m.p.h. and shortly after installation it withstood winds of over 80 m.p.h. with the rotor turned out of the wind by the manual furling cable.

The windmill meets manufacturer's specification of output: the electronic control unit gives a slight improvement to this at low windspeeds. In light winds, operating noises are just audible at 100 yards and in high winds, operating noise is completly covered by the noise of the wind itself.

The customer has expressed great satisfaction with the machine and is now considering buying another for his adjacent farm.

FUTURE DEVELOPMENTS

The interest in wind power has grown so rapidly in the last year that Wind and Water Power has developed a Scottish built 10 kW rated 'Osprey' windmill, now completing its successful test period. It is fitted with three 12.5 feet long, fixed pitch fibre glass blades with centrifugally operated tip brakes for automatic shut-down at rotor speeds over 100 rev/min. It can be supplied with either a self-excited AC generator or an induction generator. It has been designed to withstand winds of over 100 m.p.h. and the corrosive effects of sea air - both common in Scotland. It has also been designed for ease of installation and maintenance in remote areas where cranes are not always available.

The first production 'Osprey' windmill will be installed at a customer's site before the end of 1981 and it is certain that demand for this competitively priced machine will be considerable, establishing a strong future for wind power developments in Britain.

Electrical Power Generation and Control Options for Wind Turbines

S. Gair

Department of Electrical and Electronic Engineering, Napier College, Edinburgh, UK

ABSTRACT

In this paper the range of available choice of electrical generators and control strategies are outlined in relation to wind turbines which are not grid connected but which may be required to supply mains quality electricity or simply space heating or battery charging facilities.

KEYWORDS

Windturbines, electrical power generation, variable energy source.

INTRODUCTION

The possible contribution of wind energy in minimising the rate of depletion of nonrenewable resources coupled with the prospects of economic energy production costs for rural and remote communities is at present of great interest.

In attempting to capitalise on this apparantly free source of energy, one of the many problem areas that confronts the prospective windturbine designer/user is the many options that are available for electrical power take-off and control.

This paper sets out to look at the likely choice of electrical generator to perform a particular task within a given load power range and outlines electrical control strategies which may be employed to maximise the energy yield from the windturbine.

AVAILABLE EVIDENCE

It is well known that the total power available in the wind is proportional to the cube of wind speed and the fraction of this which a windturbine can utilise depends upon the velocities of wind and rotor.

The power P developed by a windturbine of blade radius r running in a wind stream of velocity V_W is generally given as:

$$P = \frac{1}{2} \rho \pi r^2 C_p V_w^3 \qquad (1)$$

where Cp is known as the power coefficient and ρ is the density of air. For a given windturbine Cp turns out to be a function only of the ratio between the shaft speed and the wind speed. Usually this speed ratio is dealt with as the 'tip speed ratio' λ, given by

$$\lambda = \frac{\omega r}{V_w} \quad (2)$$

The Cp against λ curve (a family of curves when the blade pitch is variable) is one of the primary characteristics of a wind rotor whether of vertical axis or horizontal axis type and together with equation (1) and knowledge of the load characteristic enables predictions to be made of the steady-state behaviour of the wind turbine system. Figure 1 is a Cp/λ curve for a typical fixed-pitch horizontal-axis turbine, and it can be seen that a single maximum in Cp occurs when the tip speed ratio takes a particular value λ' i.e., at a given wind speed there is an optimum rotor speed at which maximum power is generated, this maximum power increasing roughly as the cube of wind speed.

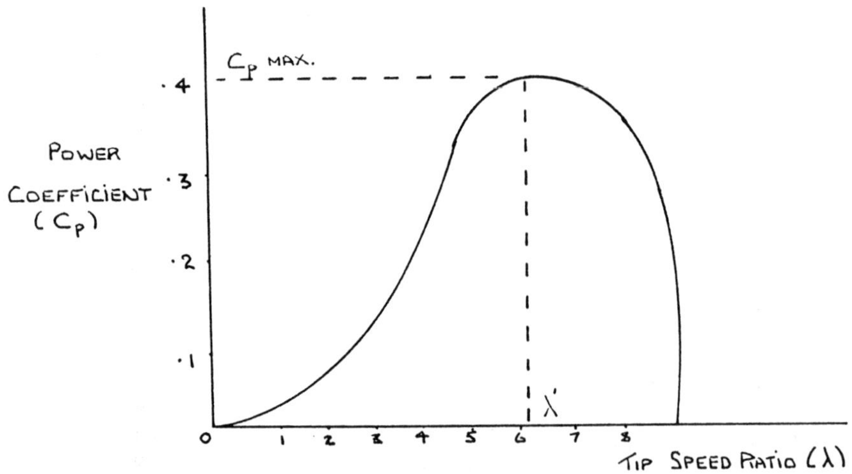

Fig. 1. Typical power coefficient characteristic of a wind rotor.

OPERATIONAL STRATEGIES

In relation to this, there are two modes of wind-driven electrical power generation. The first, constant-speed drive, has been used for large generators connected directly to the grid where constant frequency operation is essential. The second mode, variable-speed drive, is typical of most small wind generators used in autonomous applications, generally producing variable frequency and variable voltage output.

In principle it is possible to obtain a greater energy output from the variable-speed system since such a system should be capable of operating always close to the optimum rotor speed. Also there should be no need for variable-pitch control since rotor speed is determined by the speed characteristics of the wind rotor coupled to the generator with its load; electrical control of generator and load is the only requirement.

Variable-speed drive precludes the conventional alternator or induction generator modes for direct grid operation, and several modifications have been proposed, either for generating constant frequency voltage from synchronous generators or for increasing the speed range of asynchronous induction generators.

AVAILABLE CHOICE

Basically, the contenders for choice of electrical generator can be listed as (i) permament magnet (ii) d.c., series or shunt (iii) induction, (cage or wound rotor) and (iv) synchronous. Control of windpower schemes in its widest sense embraces the range of (i) "protection" of the windturbine installation in its entirity including safety of operating personnel (ii) "monitoring" speeds, stresses, power and loads (iii) "control" of start up, running, shutdown, survival and efficient energy conversion into the particular connected load.

The choice of an electrical generator and control method to be employed (if any) can be decided by consideration of the following three factors, (i) the basis of operation i.e., either constant tip speed or constant tip speed ratio, (ii) the wind power rating of the turbine and (iii) the type of load demand e.g. battery charging, sparce heating, isolated mains electricity or grid connection.

Wind power ratings can be divided into three convenient groupings, small-up to 1kW, medium - up to around 50kW and large - 200kW up to megawatt frame size.

Electrical generator types applicable to each of these ratings are:-

small - permanent magnet, d.c. generators.

medium - permanent magnet, d.c. generator, induction generator, synchronous generator.

large - induction generator, synchronous generator.

The electrical control strategy employed for any particular scheme can be designed to effect control of the generator, the power transmission link or the load, Figure 2, either independently or as in the case of some large schemes a combination of each.

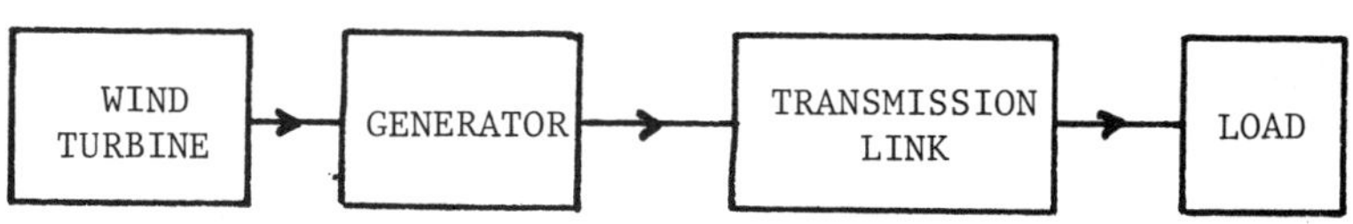

Fig. 2.

GENERATOR CONTROL

Control schemes which act on the generator alone are largely decided upon by the type of generator employed.

Permanent magnet - there is no readily available means of controlling this type of machine directly. In order to vary the torque established by a permanent magnet alternator, it is necessary to change the armature current. Thus in normal operation the output voltage and frequency must be allowed to vary with wind speed (REF.1)

D.C. generator - usually of the shunt or parallel field winding type in which a small variation of the field current achieved by means of a variable resistor connected in the field circuit will vary the terminal voltage and hence the power output. It should be noted that with a judicious choice of generator and load resistance value a fairly good natural match can be obtained between the power/speed characteristics of the generator and the wind turbine thus obviating the need for any complicated control schemes and allowing variable speed operation of the turbine (REF. 2).

Induction generator - when used to supply power to an isolated load the machine may be made to self-excite by means of a bank of capacitors connected to the terminals (REF. 3). Variation of this capacitance value varies the terminal voltage and output frequency of this system.

When a wound rotor induction generator is used for a grid connection application, power control is achieved by varying the slip energy of the rotor circuit. This may be achieved through a variety of methods such as: rotor resistance control, cascading or variable-speed slip-power-recovery schemes.

Cage rotor induction generators can be made to operate over a wider speed range by pole changing or pole-amplitude modulating the main windings to achieve one, two or three separate speed ratings.

Synchronous machine power control is readily achieved through variation of the d.c. field excitation current, the terminal voltage being a function of field current excitation, the frequency of the output voltage will be variable with wind speed for an isolated machine and fixed for a grid connected machine.

TRANSMISSION CONTROL

The most effective method of varying the power flow in the transmission link between the generator and its connected load is by means of a silicon controlled rectifier (S.C.R.) device. This unit employs power electronic devices (thyristors) whose conduction periods (i.e. on/off time) can be controlled by applying delayed trigger pulses to the gate of each individual thyristor. Some common arrangements are shown in Fig's. 3a,b,c.

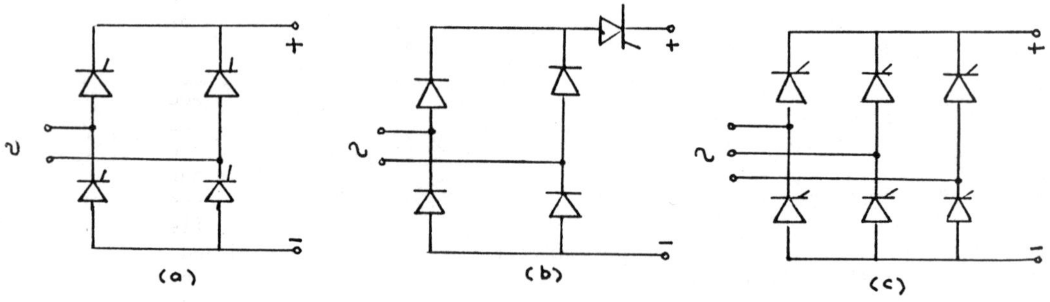

Fig.3(a) 1-phase controlled bridge
(b) 1-phase uncontrolled bridge with d.c. chopper
(c) 3-phase fully controlled bridge

Such schemes enable smooth control of the power flow in the system to be carried out but suffer from the disadvantage of being expensive since they must be rated to carry the normal full load power flow plus short time overload capability.

LOAD CONTROL

Schemes employing switched load resistors enable a stepwise load approximation to be made to the power $\propto$ (rotational speed)^3 characteristic of the wind turbine. Operation within the maximum current loading of the generator can be set and the discrete resistance values selected so that for a particular wind speed, operation of the turbine is held fairly close to the Cp maximum value. This method of control is ideally suited to small stand alone wind turbines and has the merits of being cheap, simple and effective where the load demand is for resistance heating.

TYPICAL SCHEMES

(i) Small Systems $<$ 1kW - Permanent magnet generators are eminantly suitable for small systems due to their high efficiency and low maintenance requirements. For battery charging applications their variable voltage, variable frequency output must be rectified and regulated.

Car alternators and d.c. dynamos with their regulator units are quite applicable for battery charging applications although their brush maintenance, lower efficiency and need for gearing make them less attractive than permanent magnet types. Such battery charging schemes can achieve a reasonably close fit to the wind-turbine characteristic.

(ii) Medium scale - In the power range up to 50kW permanent magnet generators can be used to provide a space heating requirement and/or battery charging application. When a dual load function is being carried out, control of the system for optimum efficiency becomes difficult to achieve. For a single load (e.g. space heat application) optimum efficiency (i.e. operation at Cp MAX) can be attempted either through (a) load switching using a speed sensing circuit operated from the output of the generator, or (b) power flow control in the transmission link by sensing wind speed and shaft speed of the turbine. Both schemes are aimed at producing a maximum energy yield from the turbine for any given wind speed. New designs of permanent magnet generators for slow speed operation, thus eliminating the need for gearboxes, together with the machines inherent high efficiency make such schemes extremely attractive for variable speed operation of stand alone windturbines up to 30kW.

The same load requirements under the same conditions of variable speed operation of the windturbine at high Cp (fixed tip speed ratio) can be met by using a d.c. shunt connected generator. Here the need for using either transmission control or load control can be eliminated simply by using the generator field current as a regulator of output power. Thus, a controller with inputs of wind speed and turbine shaft speed and having an output controlling the generator field current can be employed to operate the system in a variable speed mode (fixed λ). The power range of d.c. generators can extend well into the 100kW range but the machines high cost, brush maintenance requirement and need for a separate power electronic inverter if an a.c. output is required limit the application to space heating up to about 50kW. Stand alone schemes employing robust induction generators suffer from the drawbacks of requiring switched capacitors and a controlled rectifier to allow variable speed operation of the turbine. If an a.c. mains output is required then an inverter and transformer must also be included in the power line thus making the entire system prohibatively expensive and technically very complicated.

A slightly better alternative for an isolated system requiring mains electricity is to employ a synchronous generator and operate the turbine at constant speed. Such a system would have to employ an expensive pitch control mechanism on the turbine and would reduce the energy capture available from the wind by virtue of its fixed speed of operation.

A compromise for an isolated load requiring mains electricity would be to employ an induction generator shaft driven from the windturbine and electrically connected to a synchronous generator. Such a scheme is quite feasible (Ref. 4) and would be cost effective considering the number of diesel generator units currently installed in remote communities. A typical system layout is shown in Fig.4.

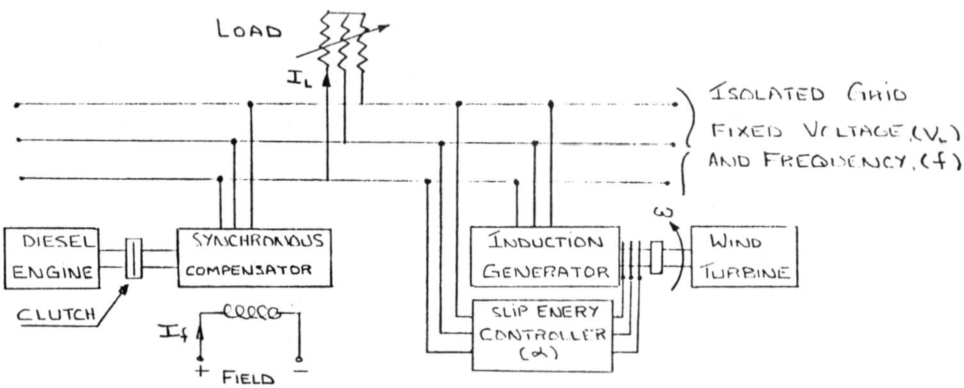

Fig. 4. Induction Generator Synchronous Compensator System

Start-up of the system would be as follows:-

(i) start diesel engine.
(ii) increase field current of synchronous generator and thereby start induction machine as a drive motor (useful if wind turbine is a fixed pitch vertical axis machine and not automatically self starting).
(iii) wind turbine operates as an aerodynamic machine and raises the shaft speed of the induction machine so that it enters the generating mode.
(iv) de-cluth and shut down diesel engine.
(v) in operation the synchronous compensator supplies reference magnetisation current to the induction generator which in turn, in addition to supplying the load power, delivers real power to the synchronous machine, to replenish windage and friction losses and maintain its rotation at synchronous speed.

Control of the system voltage and frequency is brought about by variation of the synchronous compensator field current (I_f) and some form of rotor power control in the slip-ring induction generator. The system is thereby capable of sustaining regulated voltage, regulated frequency from a varying wind source, in addition to a varying electrical power demand and makes effective use of (a) a lightly loaded existing synchronous machine, and (b) a robust induction generator which can operate over a wide speed range depending upon the control method employed.

REFERENCES

1. Bolton, H. R., and V. C. Nicodemou (1979). Proc. 1st British Wind Energy Association Wind Energy Workshop. April 1979.
2. Bolton, H. R., and V. C. Nicodemou. Proc. IEE, Vol. 126, No.9, Sept. 1979.
3. Watson, D. B., J. Arrilaga and T. Densem. Proc. IEE, Vol. 126, No.12, Dec. 1979.
4. David, R. A., and B. T. Ooi. Proc. IEE, Vol. 126, No.9, Sept. 1979.

A Utility Wind Energy Programme

M. L. Hally

Project Department, Electricity Supply Board (ESB), Dublin, Ireland

ABSTRACT:

The reasons for the recent revival of interest in Wind Power are reviewed and wind power is put into context with other energy sources available to the utility planner. Reference is made to Ireland's well known potential for wind power generation due to its location as an island bordering the Atlantic. Details are given of the previous steps taken by the Electricity Supply Board to identify the country's wind energy resource and of its experience with its first wind turbine project in 1956.

Details of the four medium sized machines to be erected in the Summer of 1981 are presented. The objectives of the test and demonstration programme are stated and explained. This programme is an important development of wind power in Ireland and will enable the Electricity Supply Board to gain first hand experience of the operation of Wind Turbine Generators, this is most important as their successful application is dependant on the chosen environmental and network conditions. Once this first hand information is available the special technological requirements may be clearly identified. This programme is part of a larger development programme by which it is intended to correlate experiences on a national and world wide basis to improve operational feed back for future Wind Turbine Generator designs. Such improved designs should reduce the potential investment risk to the prospective utility customers and bring Wind Turbine Generators into line with the other normal commercial investments. The present test and demonstration programme is an initial step in the Electricity Supply Board's policy of "readiness" for new and developing technologies which will permit it to plan for future Wind Power Projects on an informed basis and contribute to world wide development of wind as an energy resource.

Keywords: Windpower, utility programme, network connected WTG'S

INTRODUCTION:

The Electricity Supply Board, Ireland, is embarking on a Wind Power Test and Demonstration Programme as part of its overall policy for the future generation of

PROGRAMME:

In 1980 it was decided to proceed with a new Wind Energy "Test and Demonstration Programme" as the Electricity Supply Board wishes to test within the prevailing environmental conditions on its own utility power network and sites, the reliability, efficiency and economics of Wind Turbine Generators (WTGs) in the range of those commerically available. This programme will permit the collection of first hand background information to be gained at minimal cost. On embarking on this programme, the ESB recognised that it cannot be sure whether the collection of wind energy can be successful or acceptable economically, technically and environmentally. However, it recognises the fact that the potential of wind energy must be evaluated in a dispassionate scientific manner. All supplementary energy resources, no matter how small their contribution, will have to be examined because as technologies develop diversification appears to be the best policy.

Wind energy collection systems have now emerged as one of the more promising supplementary energy sources with a potential to supply energy in the medium and long term. Electrical power networks have now been recognised as the current most effective method to distribute large amounts of the collected energy. In order to make a significant contribution, wind turbines will have to be applied on a massive scale in large numbers and be integrated with the existing utility network. The programme selected can be considered to be the first step of a process of the evaluation of wind power's significance to the ESB in future Energy Planning.

Tenders were invited world wide and four medium sized Wind Turbine Generators have been ordered for installation on four individual sites around Ireland. Medium sized machines between 10 - 120 kW output were selected as being representative of those commercially available machines, the wind power industry being an infant industry, at present. The machines are from Ireland, France, Canada and Denmark/England - Irish manufacturers will supply components as appropriate. These will form the nucleus of the current Electricity Supply Board Wind Power Demonstration and Test Programme.

Alternative Energy Limited (Ireland)	-	10 KW
DAF INDAL (Canada)	-	50 KW
Windmatic /NEI Clarke Chapman (Denmark/England)	-	55 KW
Aerowatt (France)	-	120 KW

The objectives of the programme are:

- Demonstrate the present status of designs and the energy potential of WTG's under Irish Conditions.
- Collection of Wind Data and WTG Power Outputs. Data collection should provide more detailed information about wind characteristics, as the "average" type of information collected at 10 meters is not detailed enough either for WTG designers or for studies of integration of the power collected to the utility network.
- Experience of testing procedures.
- Experience of Electrical, Mechanical and Aeronautical designs of WTGs.
- Experience of the effects of Wind Power Generation on one's own utility,network designs are different in different countries.

electricity which is based on a strategy of "readiness" for the use of new main and supplementary energy sources. The main reasons for the recent revival of interest in wind energy is directly related to the cost and availability of the other conventional energy sources available to the utility planner. Wind energy is a form of Solar energy in which the atmosphere performs like a large heat engine and air travels from hot high pressure areas to cold low pressure areas in the form of wind. Wind energy is a very diffuse and variable source even in the best areas and by its nature will always be expensive to harness. It is also a 'non firm' source due to its availability. Various attempts have been made to quantify the power in the atmosphere and in 1948 a well known wind engineer named Putnam calculated it at 10^{14} Mega Watts. Even only if a portion of this could be collected reliably and economically it would be most attractive to the utility planner in exercising his energy source diversification policy. At the present stage of development is is not considered that wind energy can make a significant contribution to the national demand. Successful experience in design, costs and reliability will improve the energy contribtuion from wind power. There are, however, severe limits on the power system capacity tolerance for such an energy source. Initial detailed experiences of utility connected WTG's reported and published by the operators will contribute to the refinement of the necessary integration studies by adding realistic wind power inputs, WTG output power and other required system parameters. Once successful models have been developed, initial experience with the WTG hardware is successful and costs brought to acceptable levels utilities can then proceed with programmes to utilise wind power as a realistic supplementary energy source option.

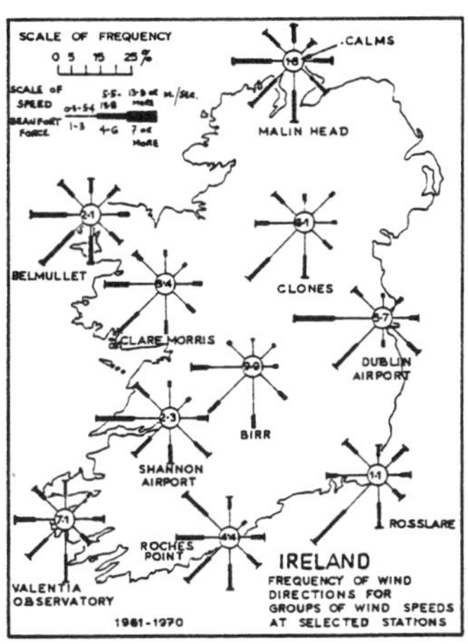

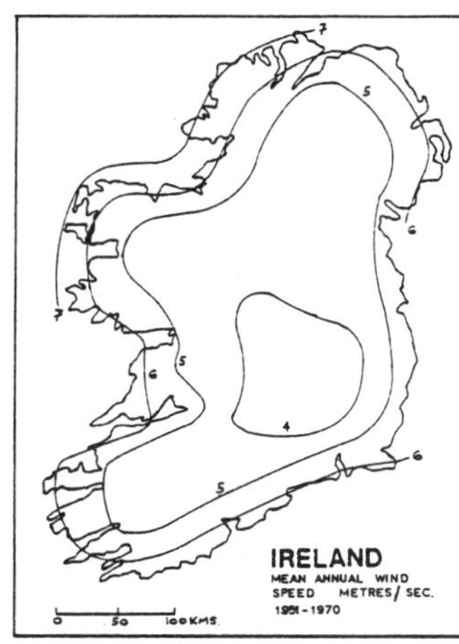

A brief study of World Wind Energy resources studies indicate the potential of Ireland for Wind Power Generation due to its location as an island bordering the Atlantic. The present programme is a further step in the Electricity Supply Board's investigation into wind power. Initial steps were taken in the early 1950s and resulted in the publication of a paper entitled "Prospects for Wind Power Development in Ireland". A test machine rated at 10 kW was installed in 1956 near Dublin and four years operational experience gained before the total destruction due to gale force winds in 1960. The development programme did not continue due mainly to lack of investment because of low oil prices at that time. Since then a continuous desk study has been in operation.

- Experience in developing regulations and tariffs for network connected WTGs.
- Creating of a data bank of wind information with other agencies in Ireland engaged in Wind Energy development.
- Exchange of information with other utilities and international bodies.

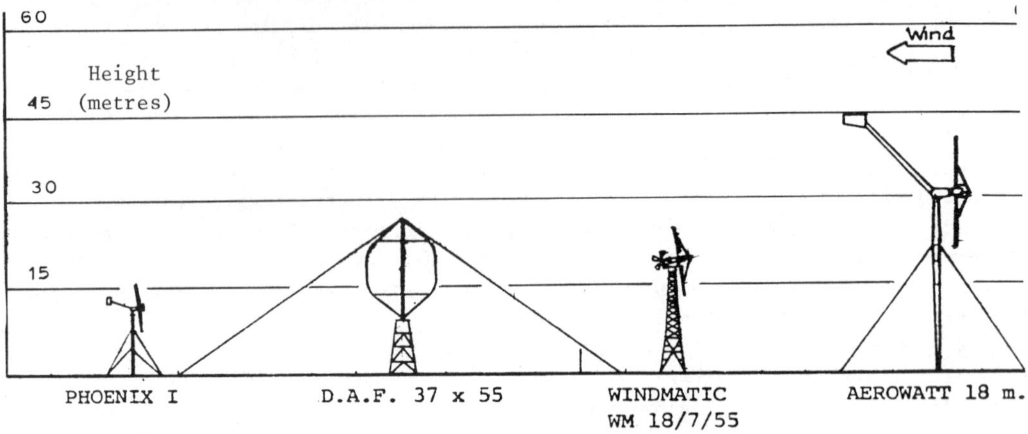

The programme includes machines of different designs and constructed of different materials because the ESB recognises that at present there is no concensus of opinion on either subject within the Wind Power Industry. The different designs should maximise the experience gained in this intial test and demonstration programme. Different sites were selected as the programme objective is not primarly to test machine designs to the limit but is to evaluate the wind energy resource, energy conversion and to study the network integration of the collected energy.

The programme could be termed as "the application of a developing technology" to a country which is well placed for the generation of power from the wind. This is most important as the potential output of a Wind Turbine generation is dependent on the environmental conditions pertaining at the chosen site. The wind regime in Ireland is described as being more constant and has higher mean speeds than most other European areas.

A main part of the programme will be to collect important data on the operation of the machines and on the environment. Data obtained will be recorded by a micro computer data logging system and stored for processing and analysis. Data to be recorded includes wind speed, direction and the corresponding mechanically induced forces and electrical output, samples of blade and tower loadings, operation of the control system, power system and other related parameters. Environmental and social acceptance will also be monitored.

Basic Machine details are as follows:

Parameter	Phoenix 1 A.E.L.	D.A.F. 37 X 55	Windmatic 1 WM 1817155	Aerowatt 18 m
Type	Horizontal Axis	Vertical Axis	Horizontal Axis	Horizontal Axis
No. of blades	2 Upwind		3 Upwind	2 Upwind
Rotor Dia.	7.6m/24.9 ft.	11.4 m	14 m (46 ft).	18 m (59 ft.)
Rotor Axial Height	-	17.6 m	-	-
Tower Height	11.2 m (36.7 ft)	17.5 m (57.4 ft.)*	18 m (59 ft.)	30 m (98 ft.)
RPM	120	20	77	82
Rated Output	10 KW at 10.7 m/s	50 KW at 18 m/s	55 KW at 12 m/s	120 KV at 16 m/s
Generator Type	D.C./ Invertor	Induction	Induction	Induction

* Height to centre of rotor

The Electricity Supply Board's Programme described is concerned only with utility connected machines. Other Agencies in Ireland will investigate the use of wind power for horticulture, land drainage, water heating and water pumping. A total of ten machines will be erected and monitored under present programmes at an overall cost in excess of £1 m. Information collected will be used to provide more exact data on the Wind Energy resource and the machine experiences will provide the necessary information required for future feasibility studies.

FUTURE PROSPECTS

Great care must be excerised when considered the purchase of a large wind turbine generator. World experience so far is that ten machines have been built in the size range between 100 KW - 1 MW and seven have been built in the range of 1 MW upwards. Machines in both these ranges are prototype and experimental. It is estimated that it will not be before the mid 80's before sufficient operational information becomes available to satisfy utility investors. Full information of the safety, reliability, availability, load factors, output characteristics and maintenance requirements of WTG and a detailed assessment of local wind energy resources are basic prerequisite required to meet the normal utility investment critical for further investment. Factors including the cost and practicality of construction of low density energy collection devices (WTG's) in remote locations will have to be fully assessed from information collected from initial test programmes. In the meantime considerable improvement in wind resource data is required to convince power system planners to assign wind power a capacity credit. Wind Power can only be considered a fuel saver till such data is available. Preliminary studies of the Wind Energy resource application will have to be refined in the light of operational experience gained and combined with realistic generating system response criteria. Operation experience includes such items as reliability, performance, operation and maintenance costs, operational constraints, component life limitation and network integration data. By the time these studies have been concluded there should be general agreement on the choice of blade material, small or large machine size and vertical or horizontal machine. Production runs should also be ready to commence thus resulting in firm prices in a real market situation.

CONCLUSION:

If the initial programme is successful, the measurements and service experience gained should produce background material for the next step of advanced investigations concerning the technology, economics and environmental factors for a larger wind turbine unit. This is the next logical step in a further long term development programme. Wind Power is possibly the best developed of the supplementary sources at present. Therefore, engineering designs have to be tested costed and the service reliability of the quipment and its integrity with existing distribtuion systems assessed. There is an abundance of designs based on the devasting combination of technical brillance and complete disregard for cost and reliablity.

The ESB Wind Power Programme described is one step in the selection process in narrowing the options available to the Power System Planner and Wind Turbine Generator Designs can be evaluated sensibly for possible further commercial development and use.

References:

"Prospects of Wind Power Development in Ireland" H. Munroe, Proceedings of the Institute of Civil Engineers of Ireland pp 129 - 267 (1953).

"The Assessment of Actual Wind Power Availability in Ireland" j. Haslett and E. Kennedy, Energy Research Vol. 3, p 333 - 348 (1979)

"The Climate of Ireland" by P.K. Ronan, Irish Meteorological Service.

Wind/Diesel Integration for Electricity Supply to Isolated Communities

N. H. Lipman* +, P. D. Dunn +, P. J. Musgrove +, B. E. Sexon + and G. Slack +

*Rutherford and Appleton Laboratories, Chilton, Didcot, Oxfordshire OX11 0QX, UK
+ Department of Engineering, Reading University, Whiteknights, Reading RG6 2AY, UK

ABSTRACT

Wind energy now appears to be economically viable, especially for some isolated communities where the cost of electricity generated using conventional fuels is high. This paper discusses the problems encountered when trying to integrate a highly variable power supply, such as wind, into a small electricity supply system, and describes various possible integration strategies.

KEYWORDS

Alternative energy; wind power; electricity generation; integrated power systems; isolated communities.

1. INTRODUCTION

Wind now appears to be one of the most promising renewable energy sources; particularly for the conditions of the British Isles. At last year's Inverness Conference I described wind integration at National Grid level. I discussed the modelling studies of the Reading University/Rutherford and Appleton Laboratories' team and showed that one can hope to absorb considerable amounts of energy (ie up to 20% or 30%) from a wind turbine system without too much difficulty or too much energy wastage.

This year I want to discuss similar studies on the integration of wind power for small isolated electricity systems. Here we are talking about power levels from about 5 kW to 100 kW, with the back-up power being provided by diesel generators. The Reading/RAL study is SERC supported and has been going on for just under two years. In this short time I believe that we have developed some very valuable insights into the problem, as I shall explain shortly.

The Reading/RAL team have recently obtained further funding from SERC to develop an actual hardware demonstration. This hardware work will be carried out with real diesels, storage batteries, etc, but with the windmill represented by a "windmill simulator" which will be a DC motor under computer control! The advantages of such a scheme are that a wide range of operating strategies can be studied with real hardware, but under controlled laboratory conditions. It is proposed to take the fully developed system, at a later stage, to the Rutherford and Appleton Laboratories' site and link it to an existing windmill.

A new collaborative team is being formed in the UK, involving RAL, Imperial College, University of Strathclyde, and two industrial firms, which hopes to build a 50 kW wind/diesel integrated system, which will include a wind turbine of 20 m diameter. This collaboration will make use of the experience already gained by the Reading/RAL team. The pilot experiment will be carried out at the RAL site. It is hoped that the study will then progress to the installation of the system at an isolated UK site (possibly a Scottish Island, if an interested community should come forward).

In the USA some experimental work is going on in the area of wind/diesel integration. Work that we are aware of, includes:

(1) The Clayton, New Mexico, Mod OA machine (rated power = 200 kW) integrated into a fairly small local diesel/gas electric grid.

(2) The Block Island Mod OA machine (rated power = 150 kW) which feeds into a very small diesel electric grid.

(3) The wind turbine generator integrated system at Cuttyhunk Island.

We are also aware of some early work going on in Holland and in Italy. There may well be a number of small projects in other parts of the world, which have not yet received any publicity.

The point that we now wish to make is that the problem of integrating wind into a small local system may sound quite trivial but we can assure you that it is not! To the best of my knowledge no one has yet demonstrated an economic and fully satisfactory integrated system. We believe that it can be done and this is the object of our continuing studies. We have investigated, in our computer model, a number of different systems and a range of operating strategies. It is our hope that one of these will be fully successful when demonstrated with real hardware in a pilot project. We hope that we will be successful in the UK in developing a system to meet the potentially very large world markets! The ERA, in a study commissioned by DoI, assessed a market in excess of 650 000 units for wind/diesel integrated systems.

2. DEFINITION OF THE PROBLEM

We have told you that this is a very difficult problem to solve; let us now try to explain the reasons why.

The wind, like many of the other renewables, is highly fluctuating in nature, and indeed this is why it was abandoned in favour of the steam engine and the internal combustion engine a century ago. These latter two sources you can turn on and off when you want, whereas the wind comes as and when it chooses. In Fig. 1 we show typical wind fluctuations in minute, hour, and day time-scales.

(a) We see that there are quite large fluctuations in wind speed (± 5% to ± 10%) in the minute time-scales for a so-called "steady wind". Wind power varies with velocity cubed and hence power fluctuations of ± 15% to ± 30% (or more), can be expected on a minute by minute basis. In the case of the National Grid study this did not worry us so much because we were thinking in terms of clusters of hundreds or thousands of windmills, and we would get smoothing of these minute to minute fluctuations of individual machines. An isolated system on the scale that we are discussing will contain one or perhaps two to three wind turbines, and hence this minute by minute power variation is a serious worry. One minute the windmill may be more than covering the load, two minutes later it may be in deficit by 20% - what can we do about this!?

(b) In the hourly time-scale there can be large increases or falls in wind power. In many ways, this is less worrying, as one can bring in diesel back-up for substantial periods of time.

(c) In a similar way, several days of little wind power is not a problem operationally as this can be dealt with by running the diesel for the period in question.

Our work shows that one can still hope to save a substantial amount of fuel during the year; perhaps as much as 50%! It might be argued that one can deal with all wind power fluctuations by the simple means of keeping the diesel electric set running all of the time. Hence, providing that it can be made to respond fast enough, it will be available at any moment to make up sudden wind power shortfalls! At Block Island in the USA (which is a summer holiday resort) they found that the wind power could reach penetrations as high as 100% during the winter months. They found it necessary to limit the wind power to less than 50% of load at any time, because of problems that the diesels had in following the wind power fluctuations. However, this was a problem of the particular diesels and their governor controllers, and need not set a limit to a properly engineered system.

The finding from our model is that you should not run the diesel all of the time, otherwise you defeat the object of the exercise which is to save fuel. A typical curve for fuel consumption of a diesel engine (50 kW) is shown in Fig. 2. We see that optimum efficiency occurs at about 80% load. At very small loads (< 20%) efficiency becomes very poor. Diesel makers advise us that you should not run below 30% to 50% of full load, as otherwise you get rapid carbonning; in addition, partially burnt fuel is deposited in the exhaust, and can cause exhaust/stack fires. We are not quite sure how precisely this advice must be followed, but it does suggest that strategies involving lengthy diesel runs below 30% to 50% load should be avoided.

The difficulty, then, is that the wind power can let you down at any moment (on minute time-scales) even when it appears to be running at well in excess of the load. If you have a diesel continuously running to meet sudden shortfalls, then it will spend much of its time running at practically zero load, which is undesirable economically and unacceptable technically.

2.1 A Numerical Example

This problem is illustrated a little more forcibly by the following numerical example.

Let maximum electricity load for the island be	100 kW
Average load figure (typical situation)	50 kW
The island "government" chooses a diesel with some expansion potential: a maximum output of	150 kW
Hence average diesel load (if no wind power) = 33.3%	

This is as small an average load as you would wish to see. In fact some isolated island communities choose to switch off their diesels overnight to keep their average load as high as possible.

Now let wind power supply half the island's electricity need (on average)
Hence average diesel load is reduced to = 16.6%

This is evidently creating an unacceptable (average) load situation for the diesel.

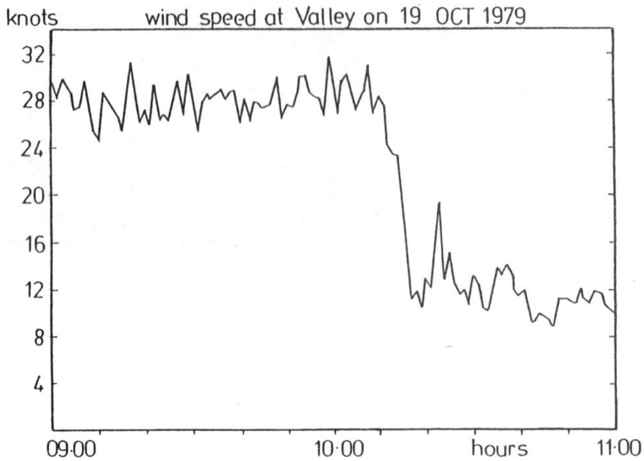

Fig. 1a. Minute-by-minute variations in windspeed.

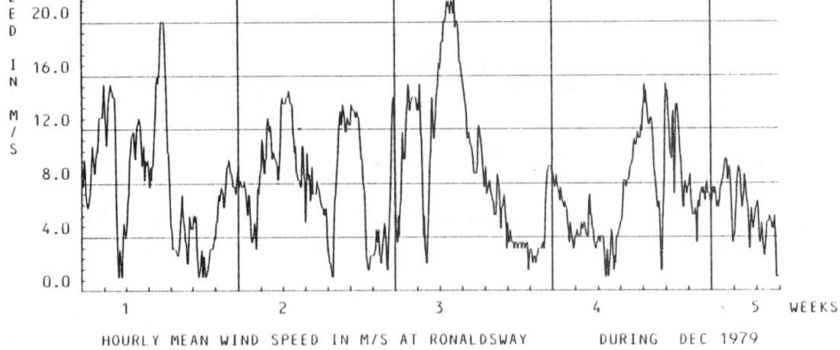

Fig. 1b. Hour-by-hour variations in windspeed.

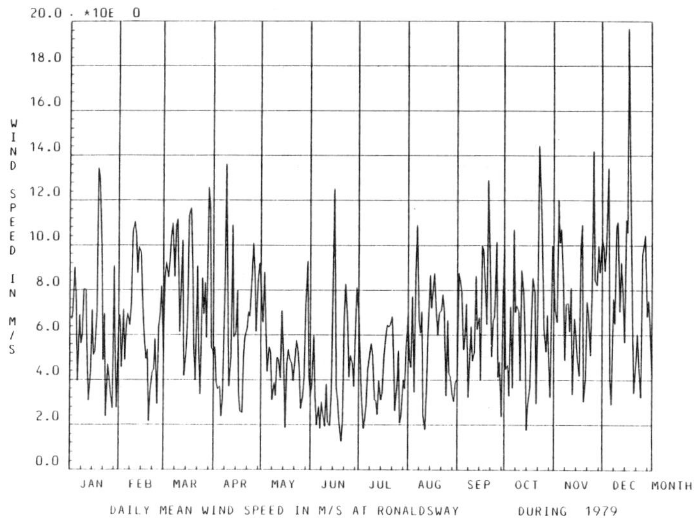

Fig. 1c. Day-by-day variations in windspeed.

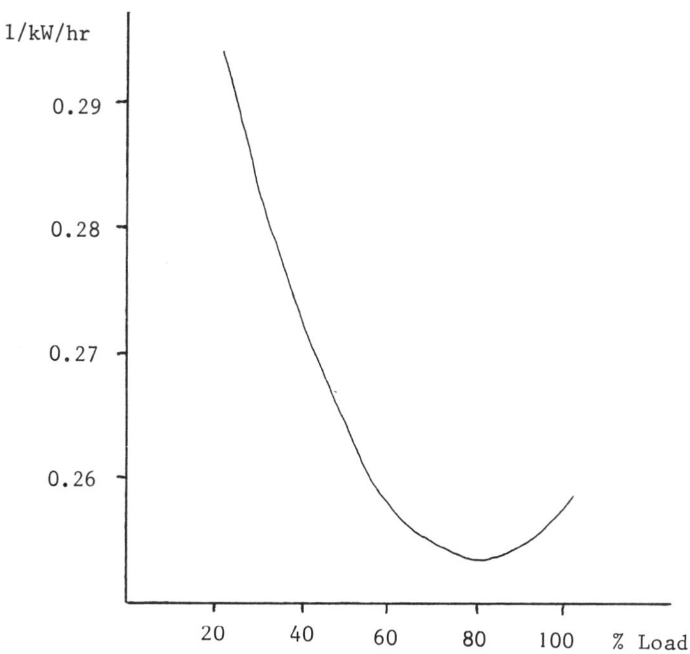

Fig. 2. Fuel consumption for a typical four cylinder diesel engine.

3. SYSTEMS STUDIES USING A COMPUTER MODEL

Impressed by this simple logic, we have examined a number of possible solutions which will improve the operating conditions for both the windmill and the diesel set. Unfortunately, all of these schemes increase the complexity of the system. But several do look very promising and we propose to pursue them further.

The computer model uses minute by minute wind data and examines the working of an assumed system into which we put certain boundary conditions and a defined system operating strategy. We follow the system (on the computer) through many weeks for up to a year and ask a number of questions:

(1) How much fuel was used?
(2) Hence, what fuel saving did the wind turbine provide?
(3) How many starts and stops did the diesel make?
(4) What was the typical running time of the diesel?
(5) What was its average load?
(6) How often did the windmill start and stop?
(7) How much of the energy was handled via a storage medium (if included)?
(8) What was the average efficiency of the storage medium?
(9) How much of the storage medium lifetime was used up in a year's operations?
(10) Finally, what was the overall cost of the electricity for the system?
(11) What would be the electricity cost for a pure diesel system operating over the same period and meeting the same loads?

The schemes that we have studied/are studying, fall into four main categories.

3.1 The "Either/Or" Wind Plus Diesel System

A simple wind plus diesel system in which the diesel is turned off as soon as the windmill is ready to meet the load. We call this the "either/or" system, the load being met wholly from the diesel or wholly from the wind turbine, but never from both simultaneously. The advantage of this scheme is its simplicity; electricity costs look acceptable. The major snag is that the diesel has to start/stop some 200 times per week (ie more than once per hour on average). An average diesel running time (for one computed example) was 28 minutes, and many very short diesel runs (of less than ten minutes) occurred.

In Fig. 3a fuel use per week is shown as a function of windmill rated power and windmill cross sectional area. We see that fuel savings in excess of 50% are achieved. Increasing windmill size increases the fuel saving, but as a diminishing return.

In Fig. 3b we show diesel starts/week. This is an insensitive function of windmill size and a fairly rapid function of rated wind speed (V_R/V_M). The sudden drop off in diesel starts per week for small V_R/V_M arises because the wind turbine seldom meets the load.

Figure 4 is of importance; it shows the distribution of diesel run lengths. We see that a great many runs occur with duration less than 40 minutes, with a sharp rise for times of less than 10 minutes. This result indicated to us that a storage medium of quite short duration might help the operating strategy greatly.

Finally, we must ask whether a system can produce electricity economically! The three lines in Fig. 5 show electricity cost as a function of windmill size. For three different average wind situations, namely, $\bar{V}$ = 4.22, 5.43 and 6.95 m/s, we see that there appears to be an optimal windmill size, although the minima are fairly shallow. We may well wish to work with a windmill area slightly above the

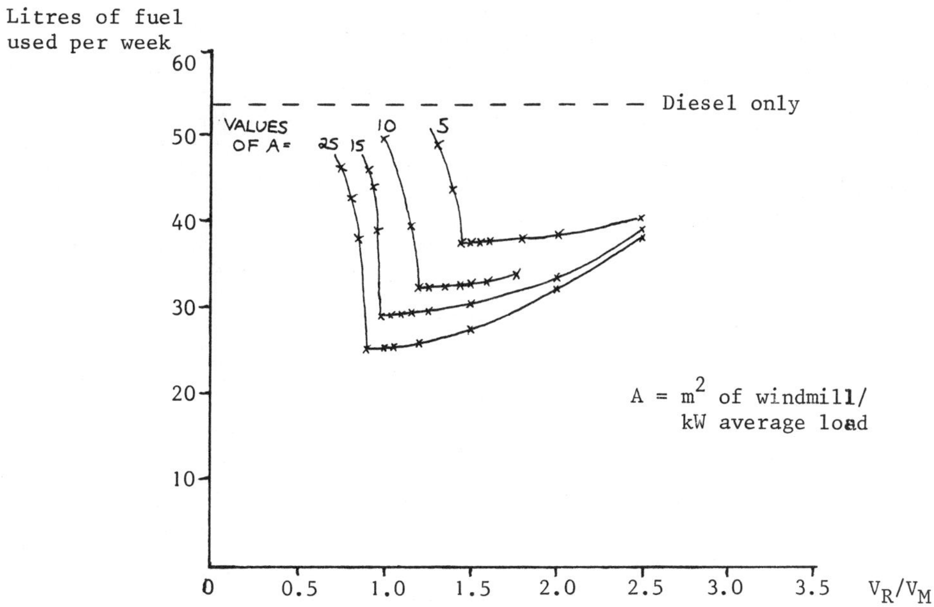

Fig. 3a. Fuel consumption per week.

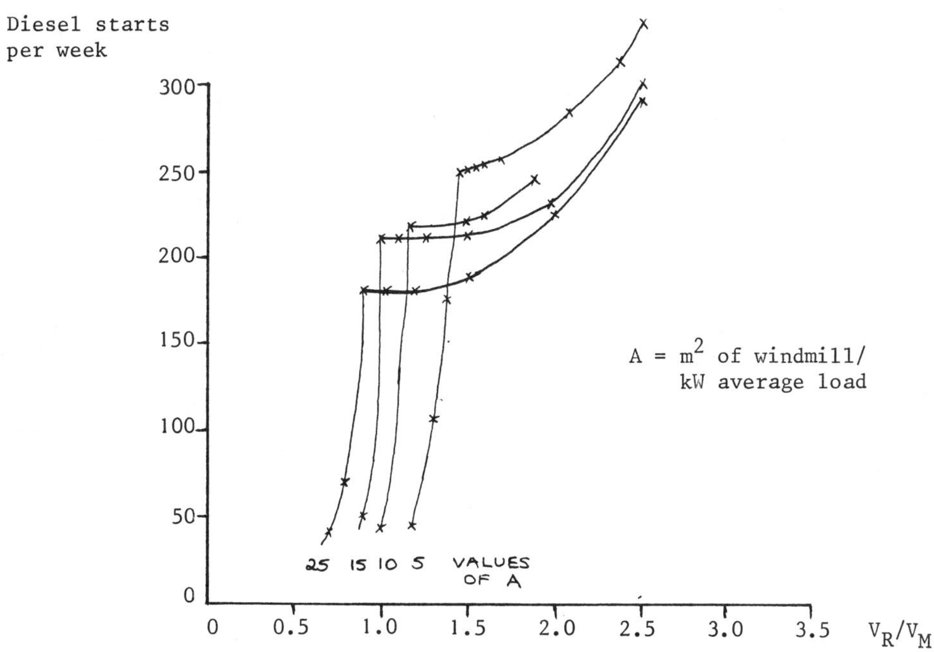

Fig. 3b. Number of diesel starts per week.

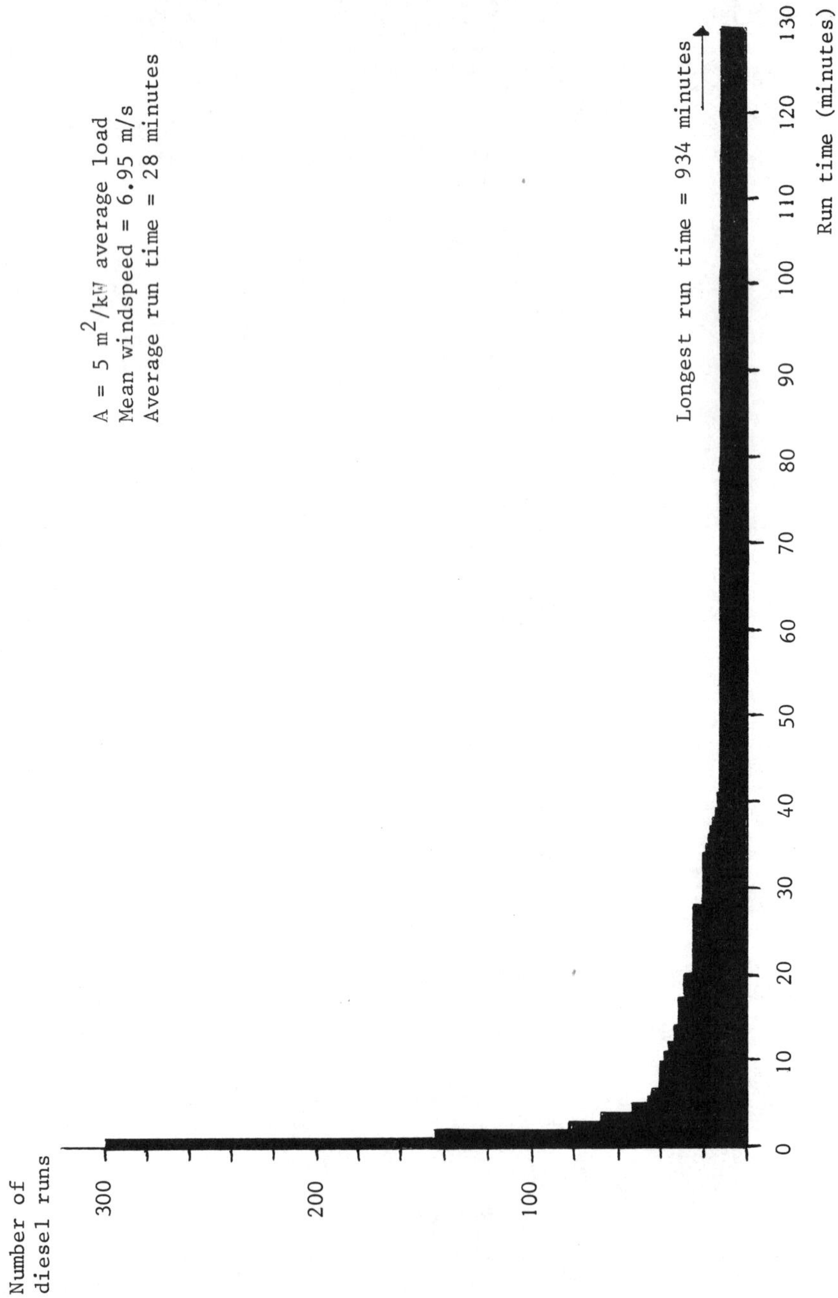

Fig. 4. Diesel runs duration curve.

electricity cost optimal in order to effect a best reasonable fuel saving. For example, a choice of 10 - 15 (m^2/kilowatt average load) looks a sensible choice, promising fuel savings as high as 50%.

Most importantly, we see that the electricity costs for all three cases are comparable with the pure diesel case (possibly slightly cheaper). Hence, if we believe these results then we see that a wind/diesel system could be economically attractive based on present day prices. Rapid fuel cost inflation could make the economics look exceedingly favourable in a few years' time.

Now we must give a few words of caution:

(1) These computations are very sensitive to the assumptions made! These include the cost of the diesel, maintenance costs, money costs and lifetime, also similar cost assumptions for the wind turbine and, of course, the fuel costs. Please remember that in the energy field it is easy to criticise a set of assumptions that have been adopted; less easy to make a reasonable choice yourself.

(2) Although this simplest system appears to be economically viable, we do not consider it acceptable technically. It implies frequent starts and stops of the diesel engine and exceedingly short runs. We doubt whether such operation would be found acceptable.

In the models that follow we have attempted to overcome these difficulties; but at the cost of increased complexity. Nonetheless, some of the schemes look very promising.

We shall deal with the further schemes much more briefly, having outlined the principles.

3.2 The "Either/Or" Scheme, but with Short Term Storage as Back-Up to the Wind Turbine

We have looked at a number of back-up strategies to improve the wind/diesel system's operating regime. Figure 4 showed that there would be a preponderance of very short runs of the diesel (less than 20 minutes). We have investigated the use of batteries which offer storage times of between one hour to ten hours. These may offer a reasonable solution in spite of operational problems with regard to maintenance and lifetime. We look briefly at batteries in the next section. More recently we have become attracted by the idea of using a compressed air store and hydraulic motor/compressor. Hydraulic systems are currently in use for a wide range of applications and the technology of hydraulic motors/compressors is well developed. These devices have a very fast response and efficiencies are quite high, comparing more than favourably with battery storage. In Fig. 6 comparison is made of wind/diesel systems with battery and hydraulic storage. A hydraulic system would have an overall efficiency (energy fed in/energy returned) of about 52%. The battery system will usually be about 40% efficient overall. The hydraulic system appears exceedingly costly compared to batteries and we are forced to think in terms of minutes of storage. An estimated cost is £500 for 6.1 kilowatt minutes. This compares with £100 per kilowatt hour for heavy duty lead acid batteries (to this cost must be added the cost of rectifiers, invertors and other power handling equipment). But the comparison is not so simple as this, since the batteries give you several hours of storage due to limits on charge and discharge rates, whether you require hours or not. But we believe that our modelling has shown that a few minutes of storage is all that is needed to produce a sensible and acceptable operating strategy for the system! In a number of worked examples, we find that the inclusion of just five minutes of storage reduces the number of diesel starts

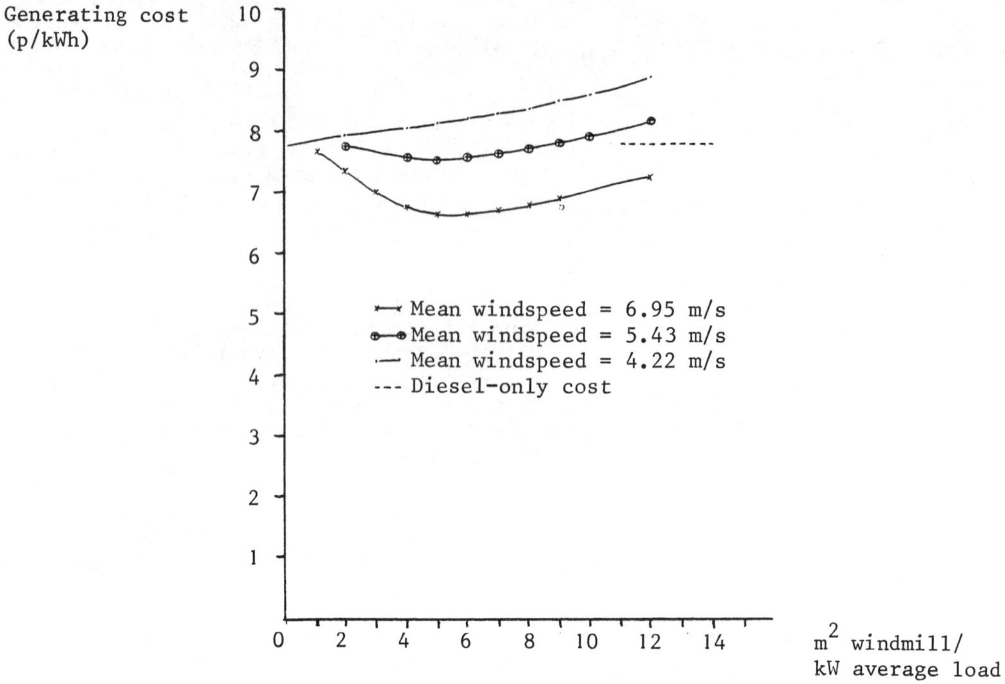

Fig. 5. Generating cost for various windspeeds.

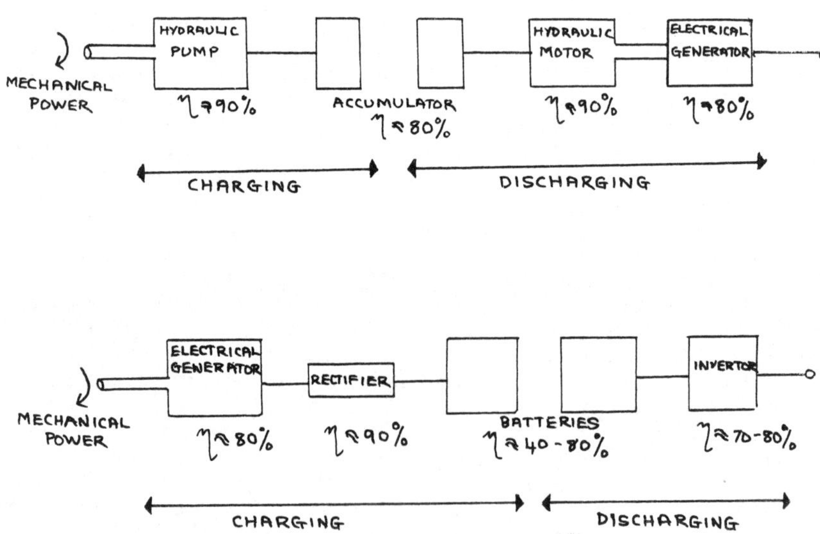

Fig. 6. Comparison of hydraulic and battery storage systems.

per week from some 250 to about 50. The average diesel run now comes out at some two hours, which we consider perfectly acceptable. This very modest storage slightly increases the amount of wind power that can be accepted by the system (by about 10%) which, in part, pays for the cost of the storage.

Some results from our modelling of a system with hydraulic storage are shown in Figs. 7 and 8.

In Fig. 7 we show starts/week and fuel use/week for systems with and without storage. In this example the windmill size is 5 m^2/kW of average load. The storage is 6.1 kW minutes/kW of average load. The mean wind speed is 6.95 m/s. We see that the storage reduces the diesel starts from some 200 - 250/week to about 50/week. Fuel use is reduced by some 10%. A wide range of windmill sizes have been considered. Increasing windmill size beyond 5 m^2/kW average load further reduces fuel use, but the benefit is decreasing. For a windmill size of 12 m^2/kW of average load, the fuel requirement for the community will have been halved compared to a system without wind power!

In Fig. 8 we examine the economics for systems with and without storage. Graphs are shown for two different mean wind speeds, namely, 6.95 m/s and 5.43 m/s. We see that the cost of electricity is very slightly higher with our hydraulic storage system compared to no storage. But the difference is only 0.4p in 6.6p (at the optimum of 6 (m^2/kW) windmill size, and for 6.95 m/s average wind). Hence, storage has a negligible effect on electricity cost, but it does give us a potentially viable operating strategy. We find this result very encouraging. Please note once more that the cost curves are fairly shallow. Thus, although a windmill size of 6 m^2/(kW average load) appears to give an electricity cost minimum, it may be preferred to have a considerably larger windmill size in order to minimise the isolated community's oil imports. We suggest that 10 m^2/(kW average load) may be of optimal overall benefit, giving approximately 50% fuel saving.

3.3 Wind/Diesel With Battery Storage

This is the first system we considered. It still looks quite attractive, although we had some misgivings with regard to battery lifetime, battery maintenance, invertor costs and overall efficiency. Yet it seems quite possible that a viable and economically competitive system can be built around a system involving battery storage! Batteries impose serious boundary conditions on charge and discharge rates. Lead acid batteries have a much reduced capacity when charged and discharged in under ten hours. Ultimately, they will have a reduced lifetime and will sustain damage. It is difficult to quantify precisely these factors. Nickel-cadmium batteries can accept one hour charge/discharge rates, but they are much more costly. Batteries require a conversion to DC during storage and back to AC when the stored energy is needed. The required invertor is costly and also has a poor partial load efficiency.

On the positive side, one hour to ten hours of battery storage permits a very good operating strategy for the diesel generator and the wind turbine. As an example, consider one hour of nickel-cadmium storage.

(a) This permits, from a fully charged state, a run of at least one hour (usually much longer) in which the windmill and storage will meet the load.

(b) Once the battery is discharged the diesel is brought back on. It can run in unison with the wind turbine jointly meeting the load. The diesel provides excess power so that the battery is recharged in one hour or longer. Hence, the diesel is guaranteed a long run at over 50% loading.

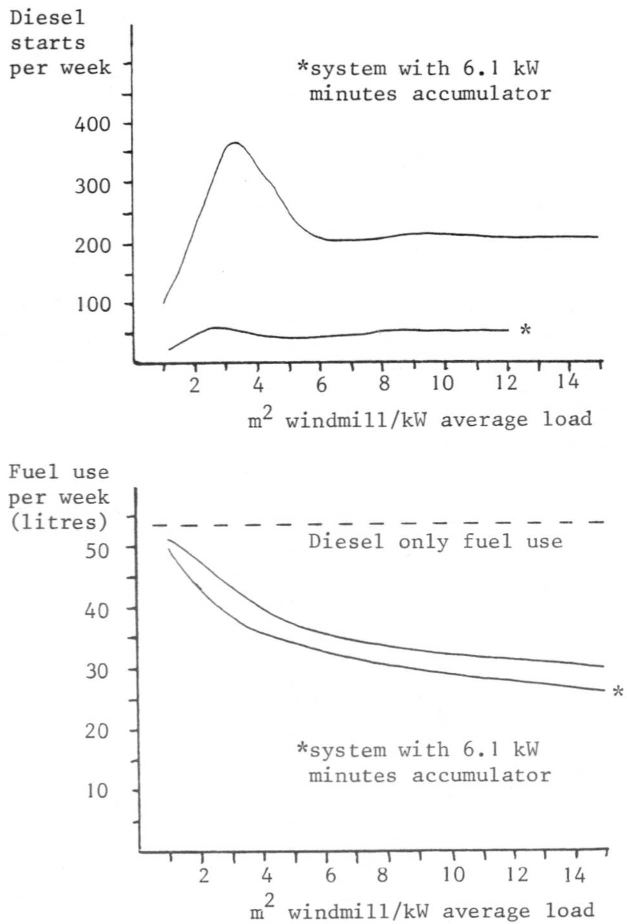

Fig. 7. Fuel use and starts per week.

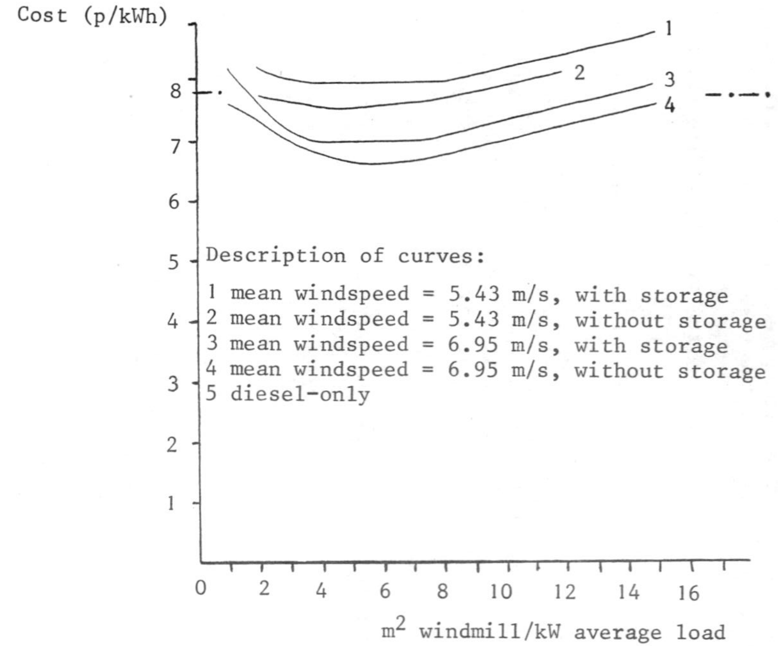

Fig. 8. Generating cost against windmill size for two wind regimes, with and without 6.1 kW minutes storage.

Thus, battery storage offers a very attractive operating strategy. The diesel is run under good operating conditions; all of the windmill's output is utilised; but, of course, only about 40% of the stored energy is recovered to be supplied to the user. By way of illustration, we show one graph from our modelling studies (Fig. 9). Here we have a ten hour lead acid battery store. Wind + diesel + storage are operated together; a minimum load of 50% is permitted for the diesel. Fuel saving is plotted against V_R/V_M for a range of windmill sizes (0.1 m^2 to 50 m^2/kW of average load). We see that fuel savings as high as 75% are possible with such a system, compared to diesel alone. We will not include the electricity cost curves here, but costs comparable with diesel alone do appear possible as in the previous case! The big uncertainty appears to us to be the battery lifetime that one can obtain when operating in the difficult conditions of wind energy. Battery monitoring and care may also prove costly.

3.4 Wind/Diesel and Other Strategies

We are investigating further strategies which do not require a storage medium. This is certainly difficult, for the reasons outlined above. It must be remembered that if there is no storage medium, then steps must be taken to prevent the supply from exceeding the instantaneous load. This must be done from one second to the next, and even shorter time-scales may be significant. Whilst the wind turbine + diesel are supplying the load jointly, the diesel can be controlled to take up the difference at any moment. This is what the Americans are doing at Block Island and at Clayton, New Mexico. Modern rapid response diesel controls will be required to meet this need. However, if/when wind power exceeds the load (and the diesel is stopped) some means of dumping excess wind power is required. Methods we and our colleagues at Imperial College are considering (eg for the proposed pilot 50 kW project) include rapid electrical switching of surplus power into a dummy load. This dummy load could be made to meet a useful purpose. One example would be to generate hydrogen for use in the "island's" transport system. Another example being considered by the Irish Electricity Authority (Eire) is to dump surplus electricity into immersion heaters in each dwelling of the isolated community. Load control appears to be a promising option for dealing with power surpluses. It also offers prospects for keeping the diesel's load (when the diesel runs) up to a sufficient value (> 30%) for reasonable operation. This is a different strategy to 3.1 ("either/or") as we now permit wind and diesel to run together in conjunction with load control. We have done some modelling of such a system, but this is still at an early stage and we will not show any of our graphs. Power dumping as a means of dealing with power excesses and of keeping the diesel well loaded, appears to be a promising approach. Yet there will be many occasions when the wind turbine is more than meeting the load, and it seems unfortunate that one would probably need to keep a large diesel running at 30% to 50% capacity into a (useful) dummy load in order to have very rapid response back-up to sudden drops in wind power. We believe that this "non-storage" approach can be made much more attractive by having a second diesel of relatively small size (eg ¼ of the size of the main machine). This second machine would run all of the time and would maintain the power balance and the rapid response back-up. The main machine would only be brought on when the windmill + small diesel + power dump was starting to get overloaded. It remains to run such a system through our model to look at the rate of starts and stops for the various components and to investigate the various possible operating stratagems. We are hopeful that such a system will be technically and economically viable, but we await to see the results of our model before we become more confident.

4. CONCLUSION

(1) We have studied isolated electricity systems involving large wind power inputs and have found the problems of integration very difficult.

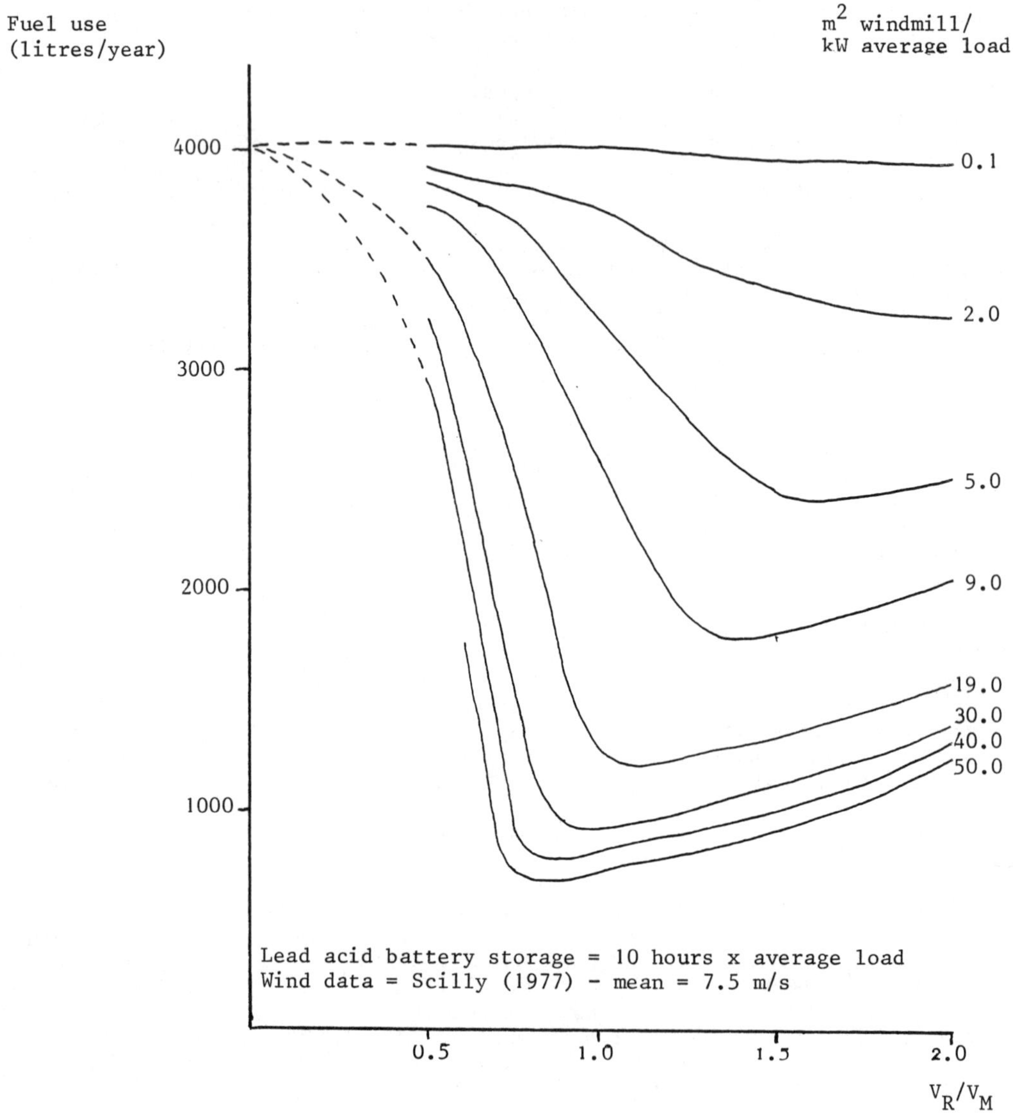

Fig. 9. Fuel consumption against V_R/V_M values for different windmill sizes.

(2) A system involving 1 windmill + diesel appears unacceptable as the diesel would be starting and stopping too frequently, and its runs would be very short!

(3) Inclusion of a short-term storage medium such as a compressed air/hydraulic device appears to help greatly, cutting down diesel starts to some 50 per week. This approach appears to be economically attractive in spite of the high cost of the storage system, in conventional terms, of £/kWh. This is because only very short storage times ($\sim$ 6 minutes) are absolutely necessary. Fuel savings as high as 50% may be achieved.

(4) An alternative approach involving storage would use lead-acid or nickel-cadmium batteries. Here storage times of one hour to ten hours would be possible permitting very good running conditions for the diesel engine. The big uncertainty is the lifetime of the batteries and the operational/maintenance costs. Yet such a system appears economically viable and may be particularly applicable to smaller systems (eg less than 20 kW).

(5) Operation without any storage medium appears problematical because of the fluctuating nature of the wind and the need to meet sudden drops in wind power. However, an approach involving a small, plus a large, diesel engine, plus power dumping into a useful auxiliary load looks promising. Such a system needs to be investigated numerically through our model. The "useful auxiliary load" could be a hydrogen generator, producing fuel for use in the island transport system.

Let me finish by apologising to Dr John Twiddell, the Conference Organiser, for not dealing with isolated community energy needs in the broader sense (ie electricity + heating + transport, etc). However, this has not been the direction of our study up to now. I hope to get together with John to expand the studies in this direction in the near future. I hope that we shall find at this Conference, representatives of communities and Scottish authorities prepared to get involved in these studies.

ACKNOWLEDGEMENT

This paper is based on the work of the Reading University/Rutherford and Appleton Laboratories small wind systems' team and carries the names of all those involved. In addition, I would like to thank my colleague Mr J Halliday at RAL who has helped me greatly in the proof reading of the paper and in discussion of its contents. I am, once again, indebted to my secretary, Miss Joan Mackie, who has worked long and hard to get the paper out in time for the usual tight deadline.

BACKGROUND PAPERS ON THE SUBJECT

Bossanyi, E.A. (1981). The frequency of wind turbine shut-downs. Proc. BWEA Conference held at Cranfield, April 1981.

Bossanyi, E.A. (1981). Wind turbines in a turbulent wind: energy output and the frequency of shut-downs. Wind Engineering, Vol. 5, No. 1.

Calnan, P.G. and E.A. Moughton (1980). Export market prospects for small and medium wind turbine generators. ERA Report No. ERA 80-15.

Halliday, J. (1981). Notes on two seminars given by Dr T. Reddoch (Oak Ridge National Laboratory, USA) on the subject of wind power integration into electricity utilities. Unpublished - ERSU Internal Report.

Lawson-Tancred, Sir H. (1979). 1 MW aerogenerator with hydraulic drive. Unpublished Report.

Lawson-Tancred, Sir H. (1980). Use of flywheels in wind energy systems. Unpublished Report.

Lawson-Tancred, Sir H. (1980). Aerogenerators operating in an autonomous mode. Unpublished Report.

Lawson-Tancred, Sir H. (1980). Aerogenerators operating in a semi-autonomous mode. Unpublished Report.

Lipman, N., P. Musgrove, P. Dunn, E. Bossanyi, G. Whittle, J. Halliday (1981). Wind energy systems integration studies by the Reading University + Rutherford and Appleton Laboratories' Group. Proc. BWEA International Colloquium on wind energy held at Brighton, August 1981.

Sexon, B., G. Slack, P. Dunn, N. Lipman, P. Musgrove (1980). Wind turbine/diesel electricity generation for small remote communities. Unpublished Report prepared for the United Nations.

Sexon, B. (1980). A model for a wind turbine/diesel system. Unpublished - Reading University Internal Report.

Sexon, B. (1981). Notes on the computer analysis of a 5 kW system. Unpublished - Reading University Internal Report.

Slack, G., B. Sexon, R. Collins, P. Dunn, N. Lipman, P. Musgrove (1980). Wind energy systems with battery storage and diesel back-up for isolated communities. Proc. BWEA Workshop held at Cranfield, April 1980.

Slack, G. (1980). Further optimising of wind/diesel/battery system model. Unpublished - Reading University Internal Report.

Slack, G. (1981). Modelling of wind/diesel system. Unpublished - Reading University Internal Report.

Slack, G. (1981). Modelling of wind/diesel system with hydraulic accumulator. Unpublished - Reading University Internal Report.

Walker, J.G. (1960). The automatic operation of a medium-sized wind-driven generator running in isolation. ERA Report No. C/T 122.

Wind Energy System Design for Remote Applications

G. R. Watson

Northumbrian Energy Workshop Ltd., Tanners Yard, Gilesgate, Hexham, Northumberland, UK

ABSTRACT

The main methods of obtaining wind data are compared and the results used to provide wind turbine performance comparisons. The effect of cut-in and furling speeds are shown with their implications on turbine design. Several systems are compared and their control and storage requirements discussed.

KEYWORDS

Wind; microprocessor data logger; Weibull parameters; cut-in; rated windspeed; furling; storage; control; monitoring.

Resource Assessment

For optimum system design it is essential that a resource assessment be carried out, unless the system meets a low-cost low-demand requirement when the user can usually be left to estimate the probable output using Beaufort or similar estimates.

To obtain more accurate information requires the use of some anemometry and the simplest unit is the Met. office standard "wind-run" anemometer. On these units the cups drive a counter (similar to the mileometer on a car) and thus a weekly reading provides an average windspeed for that week. By using available Met. office data from a local weather station (if one exists) then it is possible to correlate a relatively short term site monitoring exercise with longer term Met. data. Accuracy is usually poor (see Fig. 1) and interpretation requires a large amount of intuition. A longer term monitoring exercise improves accuracy but without extremely frequent readings there will never be sufficient information about the wind frequency spectrum to ensure effective resource use.

The next stage in complexity is the use of some form of chart record. The Met. office "Dines" anemograph is still commonly used, monitoring windspeed by means of a pitot pressure tube and recording this and direction onto a chart drum to be changed daily or weekly. The Dines is a stand-alone system (chart drive is clockwork) but electrically driven systems with different anemometry are now commonly used where a mains supply is available. We generally use such systems where instantaneous recordings are necessary but for long term monitoring work the main problem is data interpretation. Chart analysis is a lengthy operation and again involves a large amount of experience in obtaining useful information due to the data "spread" caused by rapid windspeed fluctuations.

To reduce this analysis problem various "scanning" systems are available which can be preset to a scan interval, monitor the windspeed at that instant, and then record this information onto magnetic tape. Data analysis is straight forward provided a fairly powerful computer is available with appropriate software. Power consumption and tape length can be a problem for remote applications but the main inaccuracy will be the fact that the readings are instantaneous and separated by a scanning interval, usually set by tape capacity.

The latest type of datalogger benefits from the inevitable microprocessor to provide an "on-line" data analysis system. On such a unit the windspeed and direction are monitored continuously and the microcomputer can be preset to average information over a time-span (useful for matching rotor inertia) and then provides a full frequency spectrum, usually displayed for manually recording but can also be recorded on magnetic tape (Fig. 2). Such a system is low-power stand-alone and, being on-line, reduces data analysis either to zero or to the level where a small PET-type microcomputer can provide all requisite information.

Resource Analysis

Having obtained a record of the windspeed over a suitable period and converted this into a useable form the next objective is to interpret this into a projected energy output from a wind turbine on the selected site.

An integrating or wind-run anemometer can tell nothing about wind variation, a steady wind or very high velocities linked with calm periods could both produce the same average windspeed. Assuming a more sophisticated data acquisition system has been used then it is possible to obtain a "shape parameter" for the frequency distribution and this can then be used to provide the required information. Generally the frequency distribution is of the form:

$$\text{TIME} = \exp(-U^k)$$

where U = mean wind velocity
k = Weibull exponent

For the particular case where $k = 2$ then this is the familiar Rayleigh bell-shaped curve but a consideration of Figure 3 shows the variations possible and each can make a substantial difference to turbine performance.

Wind Turbine Selection

Figure 4 uses windspeed frequency distribution information to compare two wind systems operating at various mean windspeeds. Turbine outputs are based on manufacturer's data and the energy output is at the generator terminals. Manufacturer's data becomes notoriously vague at high windspeeds and the programme is only intended to provide indicative information so rounding errors are large.

The table illustrates the importance of an accurate estimate of the Weibull exponant but also indicates how it is essential to use this information in turbine selection. The higher cut-in windspeed (the windspeed at which the turbine starts producing electrical energy) of the Dunlite will incur significant penalties at a low flat open site, although this will be less of a problem on a coastal hill-top. For sites with low average windspeeds the use of a larger rotor and the associated reduction in rated windspeed (the windspeed at which the turbine produces its nominal output) will provide great benefits, not least in the low windspeed summer months when the smallest amount of electricity is invaluable for a stand-along site. Such a design generally requires that the rotor be turned out of the prevailing wind direction at times of high wind velocity, thus a high mean windspeed site could produce a significant reduction in annual output as the turbine would be furled for too high a proportion of the time.

Any form of high windspeed control, as required for furling, inevitably introduces reliability and maintenance problems and detail design has to be carefully examined. All wind turbines will require rotor protection at very high wind velocities and the maximum windspeed is very dependant on local turbulence conditions.

The majority of systems assume the application of a brake but several have more unorthodox rotor orientation mechanisms and one or two use cheaper blades and allow a blade to fail. All have their advantages and disadvantages and selection becomes very site dependant.

Energy Provision in Calm Periods

All wind systems will suffer from windless periods and the best method of accounting for these depends, to a large extent, on the end use of the energy. Systems providing heat only must assume adequate thermal storage but this is more easily arranged than the storage of electricity.

Electricity can generally only be satisfactorily stored in DC form and numerous types of batteries are available to do this. All have their particular advantages but those most commonly used are Nickel-Cadmium and Lead-Acid. The Ni-Cad cell has the advantage of being able to survive periods of total discharge which would destroy a Pb-acid cell. It has, however, a lower terminal voltage and the increased cost usually weighs in favour of Pb-acid. The original lead plate was developed by Plante to an optimum form and Plante cells give the longest life and best capacity. Tubular plate cells are a second-best, but much cheaper, alternative and these are the most widely used.

The main system problem is providing the battery store with charge in the form it requires (Fig. 5). Battery sizing must allow for maximum allowable charge rate but all systems have some form of control to prevent over-voltage charging. The simplest uses a voltage-sensing switch to load the turbine with heating coils, the more sophisticated have some method of thyristor clipping on the turbine output waveform. The final design is inevitably a compromise as matching ideal battery requirements will forfeit a large amount of useful energy.

To reduce battery capacity on sizeable continuous demands it is often convenient and economic to incorporate a start-on-demand diesel generator into the system. Where practicable this generator feeds the load direct, with battery charging tap, and the change-over is commonly manually controlled.

For systems requiring continuity of supply, or being totally automatic, a second controller senses battery condition and available input, using the generator for battery charging as required. The controller must contain a certain amount of logic if the diesel is to operate reliably for a long unattended period.

The DC battery output (generally 24 v or 110 v) has usually to be passed through an invertor to provide 240 v 50 Hz electricity. Inverters are expensive (depending on output waveform requirements) and can be inefficient on part-loads. Modern electronic developments have improved the situation, however, with reliability and performance improving most markedly.

Should a grid network be present then this can act as a 240 v 50 Hz storage of up to virtually infinite capacity. Wind turbines can then be fitted with induction generators and, although operating at constant speed, the control aspects are much simplified. This has led to research (Ref. 1) into using synchronous generators on diesel sets to form a local grid for remote community applications. The diesel decouples when wind energy output is adequate, but the synchronous generator allows an induction machine on the wind turbine to continue producing AC power. Such a system is still experimental and the most common method of linking a variable speed wind turbine to a grid is via a synchronous inverter, yet another electronic controller which allows the load to be met from either grid or wind and puts excess wind energy into the grid.

System Monitoring

As each wind energy system has to compromise with site conditions and end use requirements it is important to continue monitoring after installation. All controllers can be adjusted, as can the rotor of the turbine and any actuating mechanisms. One of the more sophisticated data recording systems used for wind monitoring can also be used for this work and for telecom applications it is common to use radio signals to indicate various operating modes. In domestic applications the user generally carries out system "tuning" and he has the advantage of being able to allow for seasonal variations. Extracting the last few watts of energy depends very much on how valuable they are to the user!

Ref. 1: Wind Turbine Research at Napier College
W S Bannister & S Gair
3rd BWEA Wind Energy Conference, Cranfield 1981.

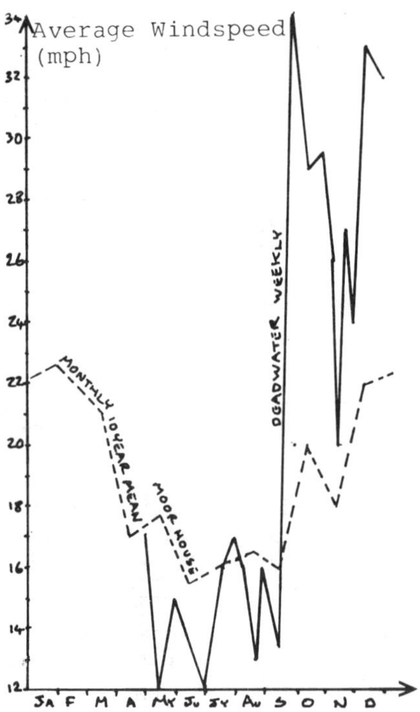

Fig 1 Deadwater / Moor House
Weekly average windspeeds

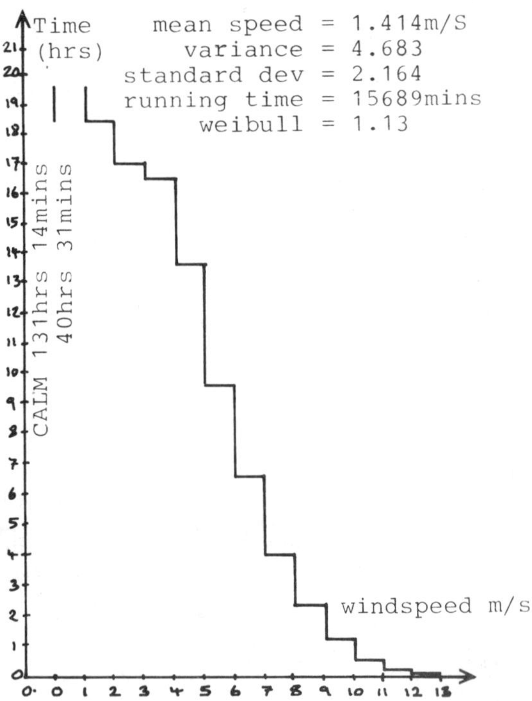

Fig 2 Windlogger Output

General flow Character	Wind Regime Type	Weibull Exponent
Irregular	Polar, wind shadow	1.0
	Mediterranean, desert	1.5
Isotropic	Temperate maritime and continental	2.0
	Roaring fourties	2.5
Constant	Trade wind	4.0

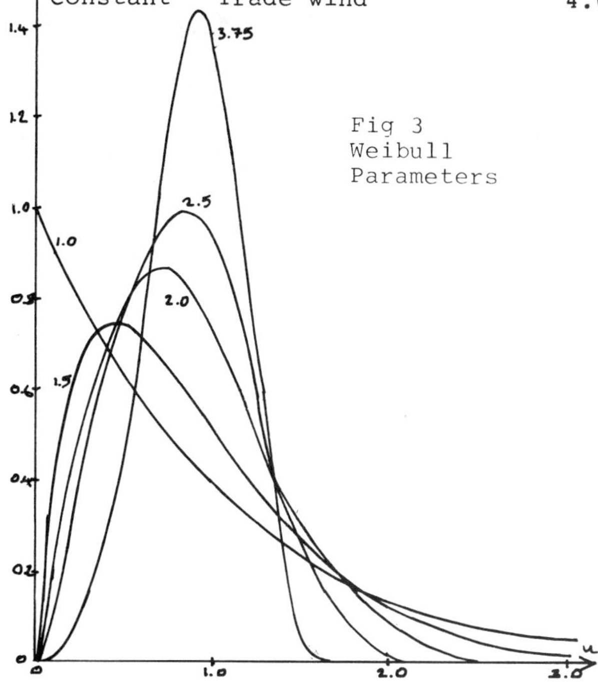

Fig 3
Weibull
Parameters

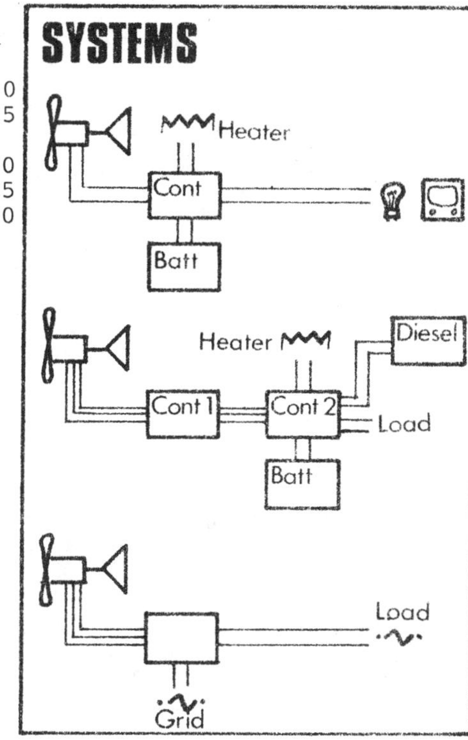

Fig 5 Typical Systems

Fig 3 Weibull Parameters

Fig 4
Comparison of
two Wind Systems

WINCO SYSTEM

Details: 200 watt Winco: 460; 10 m tower: 465; 128 Ah battery pack: 180; Controls: 90; Cabling and installation: 300. TOTAL: £1495.

AVERAGE WINDSPEED (MPH)	12 5.4 m/s		16 7.15 m/s		20 9 m/s	
WEIBULL FACTOR	1.2	2.0	1.2	2.0	1.2	2.0
HOURS LESS THAN CUT-IN	1180	445	685	254	444	162
HOURS GREATER THAN 55 MPH	0	29	1	295	29	752
% ENERGY AT WINDSPEED OF (MPH):						
Less than 10	1	0	0	0	0	0
10 - 15	7	5	3	0	1	1
15 - 20	29	19	18	7	12	8
20 - 25	43	21	30	23	32	12
25 - 30	5	17	16	15	17	12
30 - 40	1	19	11	24	19	24
40 - 55	0	7	1	21	7	32
ANNUAL ELECTRICAL OUTPUT (kWh) = E	504	1085	846	1265	1085	1244
ELECTRICITY WORTH; 91(0.0731) + (E - 91)0.0333	£20.40	39.75	31.80	45.75	39.75	45.00
PAYBACK PERIOD (YEARS)*	11	9	9.5	8	8.5	8
ELECTRICITY WORTH ALLOWING FOR INCOME TAXATION @ 30%	£26.50	51.00	41.30	59.50	51.70	58.50
PAYBACK PERIOD (YEARS)*	10	8	8.5	7	7.5	7

DUNLITE SYSTEM

Details: 2 kW turbine, 40' tower: 4600; 24 kWh battery store: 2500; 5kW inverter: 1650; Controls: 350; installation: 900. TOTAL: £10,000.

AVERAGE WINDSPEED (MPH)	12 5.4 m/s		16 7.15 m/s		20 9 m/s	
WEIBULL FACTOR	1.2	2.0	1.2	2.0	1.2	2.0
HOURS LESS THAN CUT-IN	3351	1397	2082	815	1397	531
HOURS GREATER THAN 55 MPH	0	29	1	295	29	752
% ENERGY AT WINDSPEED OF (MPH):						
Less than 10						
10 - 15	8	1	2	0	1	0
15 - 20	13	2	6	1	2	0
20 - 25	45	23	30	14	23	11
25 - 30	20	23	26	18	23	14
30 - 40	7	29	22	32	29	28
40 - 55	0	11	3	27	11	37
ANNUAL ELECTRICAL OUTPUT (kWh) = E	2290	7725	5123	10260	7725	10668
ELECTRICITY WORTH; 91(0.0731) + (E - 91)0.0333	£ 80	260	174	345	260	358
PAYBACK PERIOD (YEARS)*	13	9	10	8	9	8
ELECTRICITY WORTH ALLOWING FOR INCOME TAXATION @ 30%	£103	339	226	448	339	466
PAYBACK PERIOD (YEARS)*	12	8	9	7	8	7

* Electricity costs are inflated at 30% per annum. Actual increase in electricity costs to date from 1972 is over 500%

^
Fig 7 Dunlite in Domestic Application

<·
Fig 6 'Wind-run' Anemometer

^
Fig 8 Elektro with Tail Furled

^
Fig 9 Windlogger Microprocessor Data Logger

Microprocessor Instrumentation for Energy-related Physical Measurements

N. Lockerbie

Department of Applied Physics, University of Strathclyde, Glasgow G4 0NG, UK

ABSTRACT

The purpose of this paper is to give a practical account of how low-cost microprocessor instrumentation can aid the measurement of energy-related physical quantities. As the subject matter is of a technical nature, a specific example is given of a simple windspeed and windpower measuring and logging instrument, based on the Rockwell AIM 65 Advanced Interactive Microcomputer.

KEYWORDS

Microprocessor; Instrumentation; Windpower.

INTRODUCTION

Advantages of the Microprocessor Data-Logger

Often, whilst making even long-term observations of physical variables such as solar irradiation, windspeed, temperature etc., what is actually sought is merely some salient number; or perhaps several numbers; or at worst, a distribution of some sort: e.g. the average winter diurnal temperature-profile. Even the variability of a physical quantity takes on more meaning as a number/distribution/spectrum than as original "raw" data. In other words, for the purposes for which the measurements were made, there is often an enormous amount of redundant information in the collected data. Yet, many data-logging instruments produce as their final output "raw" data in one of the following forms: numberless pen-recorded (y-t) strip-chart, punched-tape, magnetic-tape, or numerical teletype printout. These data must then be processed either manually, or by perhaps a mainframe computer, or a combination of both.

However, both data acquisition and processing can now be conveniently carried out, relatively inexpensively,[1] by an "off the shelf" programmable microprocessor

[1]Compared with commercially-available systems that are broadly equivalent in power.

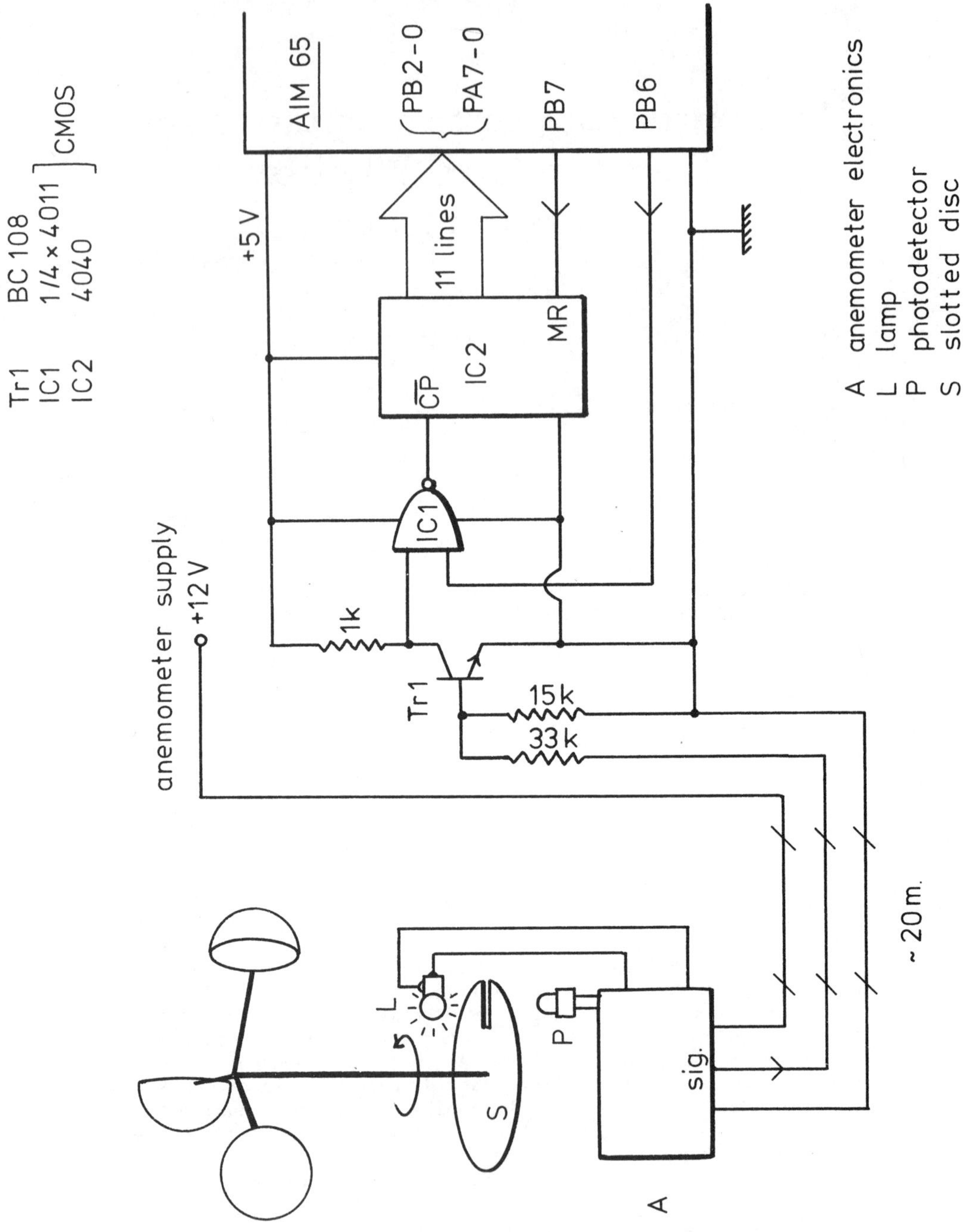

Fig. 1. The anemometer-microprocessor interface. For IC 2, $\overline{CP}$ is the clock-pulse input, and MR is the memory-reset input. The "most significant" output-bit of IC 2 is left unused.

(hereinafter called a "micro" for brevity), together with a small amount of interface circuitry. Other benefits accrue by giving the micro direct control over the measuring process itself, enabling a program to determine when to take measurements, as well as correcting for non-linear or non-ideal transducer response characteristics: for example, writing the program statement $Y = X^3$ is certainly much easier than constructing an electronic circuit with the same transfer function. But the real virtue of the micro must remain that it is sufficiently inexpensive to be used in such a dedicated fashion - often costing less than the measuring instruments/transducers with which it is interfaced.

In what follows a microprocessor windspeed and windpower logger is described, which illustrates some of these points.

MEASURING WINDSPEED AND WINDPOWER

Apparatus

For a full description of the Rockwell AIM65 Advanced Interactive Microcomputer, that was used in this work see footnote.[2] Briefly its main features were: on-board 20 column thermal printer (under software control), with a 20 character 16-segment LED display; 8k "Microsoft" floating-point BASIC, 4k read-write memory, 8k monitor with mnemonic Assembler and dis-Assembler; 6502 CPU, uncommitted user 6522 VIA for interfacing, and single-rail (+5V @ ~ 1.5A) operation - the printer running from +24V (unregulated). At the time of writing the cost of the basic unit is around £300.

The windspeed measuring instrument that was interfaced with the AIM65 was the CASELLA "Sensitive Anemometer" - (see CASELLA). This cup-anemometer, which was of the slotted-disc and light-beam type (see Fig. 1), contained conditioning electronics within its weatherproof housing which produced one short negative (~ 11 V) pulse on the signal line for every revolution of the cup-assembly: the supply for the unit was +12V at ~ 100mA. A calibration curve was supplied with the instrument in the form of "pulses per minute" vs. "windspeed in $m\ s^{-1}$" up to a windspeed of 15 $m\ s^{-1}$. In order to evaluate the logging system as a whole, the anemometer was mounted on a 5m mast on top of the University of Strathclyde Anderson Building, 3m from the South-West parapet of the building, at a height of approximately 40m above street level. At this height and location the prevailing South-Westerly winds over Glasgow were not directly obstructed, except by the building itself.

The interface circuit shown in Fig. 1, similar to that of (Gordon, 1981) comprised an npn level-shifting transistor, a CMOS NAND gate, and a CMOS 12-bit binary ripple-counter. The total cost of this circuit was ~ £1. Power to the circuit was supplied by the +5V rail of the AIM65 itself (the current drawn being a negligible 5mA), with connection to the AIM65 being made via a short (~ 20cm) 15-way ribbon-cable. Connection to the anemometer electronics was via approximately 20m of 3-core cable.

---

[2]Information can be obtained from: Electronic Devices Division, Rockwell-Collins, Heathrow House, Bath Road, Crawford, Hounslow, Middlesex. Tel: (01) 759 9911.

Method

For a run-period of a specified number of days, windspeed measurements were taken every minute by the AIM65, as described below: i.e. 1440 measurements per day (midnight-midnight). The AIM65's controlling program which was written in BASIC (see for example (Alcock, 1981)) with an Assembly Language (see for example (Leventhal, 1979)) subroutine for input, control, and timing, used these data to produce a printed summary of each day's results, with, at the end of the run, a printed summary of results for the complete run-period. These summarised quantities were:-

(a) Average windspeed in m s^{-1}.
(b) Cube-root of the mean windspeed-cubed, in m s^{-1}.
(c) Maximum windspeed in m s^{-1}, and the time at which it occurred.
(d) Power ratio: the ratio of the mean (cubed-windspeed) to the (mean-windspeed) cubed.
(e) The average available power in the wind in W m^{-2}, calculated from $\frac{1}{2}\rho\,\overline{v^3}$, ρ being the density of air at 288.16K (15°C) and 1.0133 bar, and $\overline{v^3}$ being the mean (cubed-windspeed).
(f) A windspeed distribution, given as minutes spent in each 0.5 m s^{-1} windspeed range.
(g) A normalised histogram of (f).

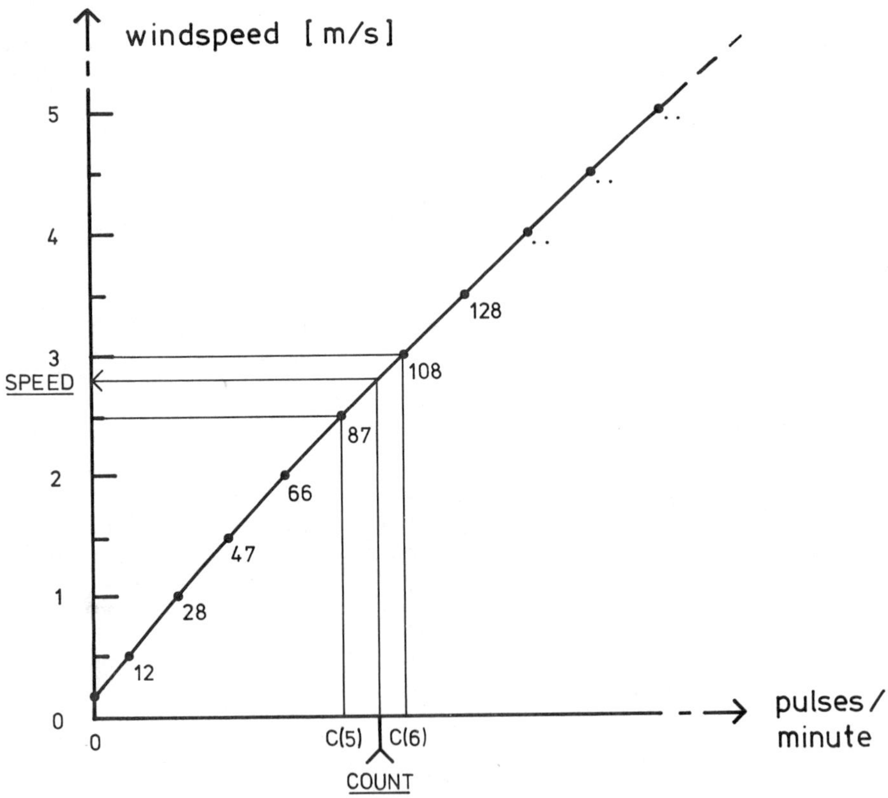

Fig. 2. Extraction of the windspeed from the anemometer calibration-curve, by linear interpolation.

In addition at any time during the day, the current windspeed was displayed on the AIM65's 20-character display as WINDSPEED: XX.XX M/S.

Windspeed measurements were made in the following way (see Fig. 1): controlling pulses on software - configured output-lines PB7 and PB6 (6μs and 1 minute, respectively) reset the counter to zero and enabled it to count pulses from the anemometer for 1 minute. At the end of the minute the accumulated count was "frozen", and read into the AIM65's memory on 11 software-configured input-lines.

The binary count was subsequently converted by BASIC to a decimal number "COUNT" in the range 0-2047, and then used to accurately determine the true windspeed - see Fig. 2. The numbers 0, 12, 28...etc. were taken from the actual anemometer calibration curve at 0.5 m s^{-1} intervals, and were stored as the elements of a 31 element array C(I); I = 0, 1, 2.... By comparing COUNT with the C(I), starting with C(1), the appropriate 0.5 m s^{-1} windspeed range could be found. In the example of Fig. 2 COUNT < C(6) and the windspeed must therefore be in the sixth (I = 6) speed range. Moreover, the actual windspeed becomes:-

$$\text{SPEED} = \left\{ I - 1 + \left(\frac{\text{COUNT} - C(I-1)}{C(I) - C(I-1)} \right) \right\} / 2 \qquad \text{m s}^{-1}$$

with I = 6. In this example the sixth "windbin" (also the element of an array) would be incremented with one unit of time for the speed distribution. Also the calculated windspeed "SPEED" would be compounded with the existing (daily and overall) running average speeds, cubed and compounded with the existing (daily and overall) running average-cubed speeds, compared with the existing (daily and overall) maximum speeds etc..

The measuring process would then be ready to repeat until 1440 measurements had been made, at which point (a) - (g) would be printed out for the day, followed possibly by (a) - (g) for the run as a whole, if it was the last day of the run.

Results and Discussion

Figure 3 shows the printout for a typical day. Clearly not all of the printed results would necessarily be "salient numbers" for a given windspeed-logging application, and the daily printout rather proliferates data - the end-of-run printout would probably suffice in many cases. Nevertheless it would be straightforward to make the necessary changes to the program. For example, to print the average available windpower above a certain windspeed instead of for all windspeeds. Figure 3 also shows that at this site a simple average-windspeed (e.g. electromechanical-counter type) logger could give a totally misleading indication of the available windpower, as evidenced by a POWER RATIO of almost 2: over 3 weeks of data-gathering, the average POWER RATIO was found to be 1.85 .

Finally, it should be noted that the general approach outlined above could equally well be applied to the measurement and logging of other physical variables, through the use of voltage-to-frequency or analogue-to-digital converters.

ACKNOWLEDGEMENTS

I would like to thank Bill Grainger for help with the anemometer, and Eric Dallman for working on the early part of the project.

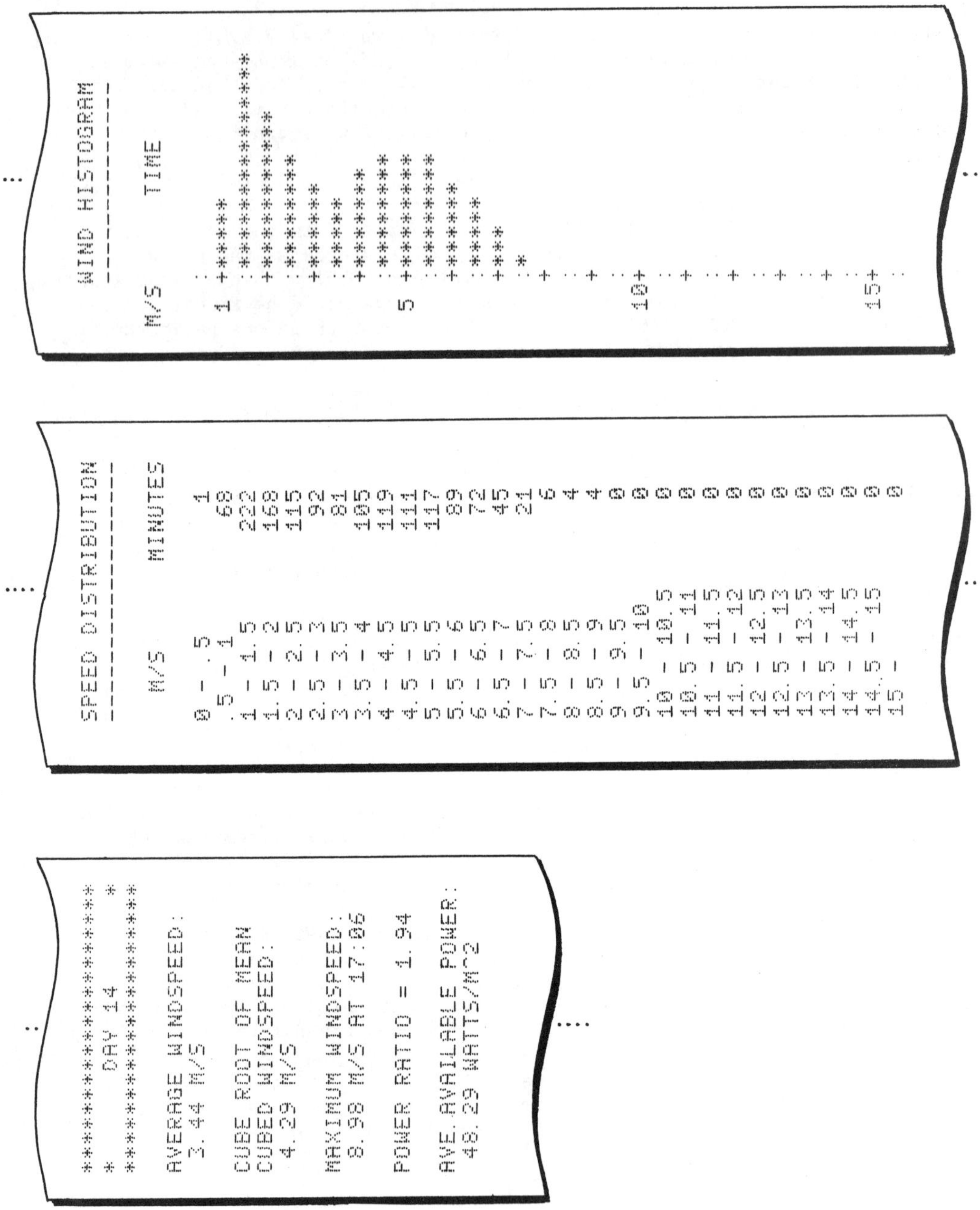

Fig. 3. AIM65 printout of one day's measurements from a 3-week run period (see text). The histogram (top) of the windspeed distribution (centre) was normalised in the program to 15 column widths at its peak.

REFERENCES

Alcock, D. (1981). Illustrating Basic. Cambridge University Press, Cambridge.

Casella, Regent House, Britannia Walk, London N1 7ND, Tel: (01) 253 8581; note that this anemometer would not be ideal for logging windspeed at natural windgenerator sites, because of its low speed limit (15 m s^{-1}). It was ideal for the experimental site chosen, however.

Gordon, J.L. (1981). Wireless World, 87, 1544, 42-44.

Leventhal, L.A. (1979). 6502 Assembly Language Programming. Osborne/McGraw-Hill, Berkeley, California.

Estimating Wind Energy Potential Using Statistical Models

A. H. Elgammal

Department of Mechanical Engineering, Paisley College of Technology, Paisley, Scotland, UK

ABSTRACT

A review of the statistical models used for describing wind duration curves which can be used in predicting wind energy availability is presented. Using data from six meteorological stations, it has been shown that an accurate assessment of wind energy can be computed by using the mean annual wind speed and a normalized empirical distribution for the wind duration profile. Furthermore such a distribution could be used to determine the optimum characteristics of aerogenerator with a view of its particular utilization.

KEYWORDS

Wind power estimates; wind statistical models; aerogenerators.

INTRODUCTION

In contrast to fossil fuelled and nuclear power installations which will continue for some time to provide the bulk of energy needs, we are concerned here with a possible alternative or supplementary source of energy, which at some time or other, may need to be exploited to form a part of the overall energy resources. However while it is possible to predict with some certainty the power produced from conventional sources of energy and its cost, the situation is different for wind power production. Neither the performance of an aerogenerator plant, indicated by its efficiency as a power unit, nor its annual output of energy can be easily predicted in advance. Nevertheless the importance of a precise assessment of wind energy potentialities in a particular location emphasises the sensitivity of the methods employed for estimating wind energy.

This paper deals with statistical models used for describing wind duration distribution which can be used as a tool in predicting energy output of an aerogenerator system. Using data from six meteorological stations, it has been shown that an accurate assessment of wind energy can be computed by using the mean wind speed and a Normalized empirical distribution for the wind duration profile. Furthermore such a distribution was used to determine the optimum rated speed of aerogenerator with a view to its particular utilization.

ESTIMATION OF ENERGY AVAILABLE FROM THE WIND

The total quantity of energy imparted by various wind velocities with known frequencies can be calculated from the equation;

$$E = \Delta t \, \Sigma (f_1 G_1 + f_2 G_2 + \ldots\ldots\ldots + f_n G_n) \tag{1}$$

where $G_1, G_2, \ldots\ldots G_n$ are the mean power imparted by air streams within different velocity intervals, and $f_1, f_2, \ldots, f_n$ are their frequencies of occurrence.

Wind Duration Curves

Wind duration curves (Fig.1) give the percentage of time during which the indicated wind velocity was exceeded. Conventionally these curves are determined from the frequencies of occurrence of various wind speeds in a certain site, which requires the use of hourly measured values of wind speed. However, in many feasibility studies such data is not often available. This has led to the development of a number of statistical models for describing wind speed duration curves using the monthly or annual mean wind speed.

Hennessey (1977) showed that the Weibull function gives a good fit to measured data. The Weibull distribution is a two-parameter function, its cumulative distribution which represents the probability of wind speed $\geq V$ is given by;

$$F(V) = \exp \{-(V/C)^k\} \tag{2}$$

where k is a shape parameter and C a scale parameter. The mean of the distribution i.e. the mean velocity V_m, is equal to:

$$V_m = C \, \Gamma(1 + {}^1/k)$$

where Γ is the gamma function.

Cliff (1977) suggested the use of the Rayleigh distribution for sites with mean wind speeds greater than 10 mph. The Rayleigh distribution is a special case of the Weibull distribution when K = 2, its frequency distribution is given by;

$$F(V) = \exp \{ -\frac{\pi}{4} (\frac{V}{V_m})^2 \} \tag{3}$$

Olsson and co-workers (1975) used for the purpose of wind energy prospecting a simple normalized shape function;

$$F(V) = 2^{-(V/V_m)^\beta} \tag{4}$$

Only one parameter, the exponent β, is needed to characterise the shape of the wind duration curve at a given average speed V_m. The exponent β varies between 1.76 to 2.2 depending on site location (e.g. inland, near coastal or costal) also it is influenced by the wind fetch terrain properties.

Widger (1976) presented a model, based on relating the average fastest wind speed to the average wind speed "V_F", using an empirical formula ($V_F = 2.38V_m + 5.314$). Then assuming that the wind speed distribution could be adequately transformed to a normal distribution by the square root parameter Z, where;

$$Z = 4.26 \frac{\{(V/V_m)^{\frac{1}{2}} - 1\}}{\{(V_F/V_m)^{\frac{1}{2}} - 1\}} \tag{5}$$

the wind speed frequency distribution is given by,

$$F(V) = 1 - \frac{1}{2\pi} \int_z^\infty \exp\{-(V/V_m)^{\frac{1}{2}}\} \, d(V/V_m)^{\frac{1}{2}} \tag{6}$$

Data from six meteorological stations in the west coast of Scotland are used for plotting the wind duration curves shown in Fig.1. The aforementioned normalized distributions are also plotted in the same figure for comparison. From the figure we may note that the measured data plotted tend to lie on a common curve once they have been rescaled by the local annual mean wind speed V_m. Furthermore both the Weibull distribution Eq.(2), and the Olsson distribution Eq.(4), are in close agreement with the measured data.

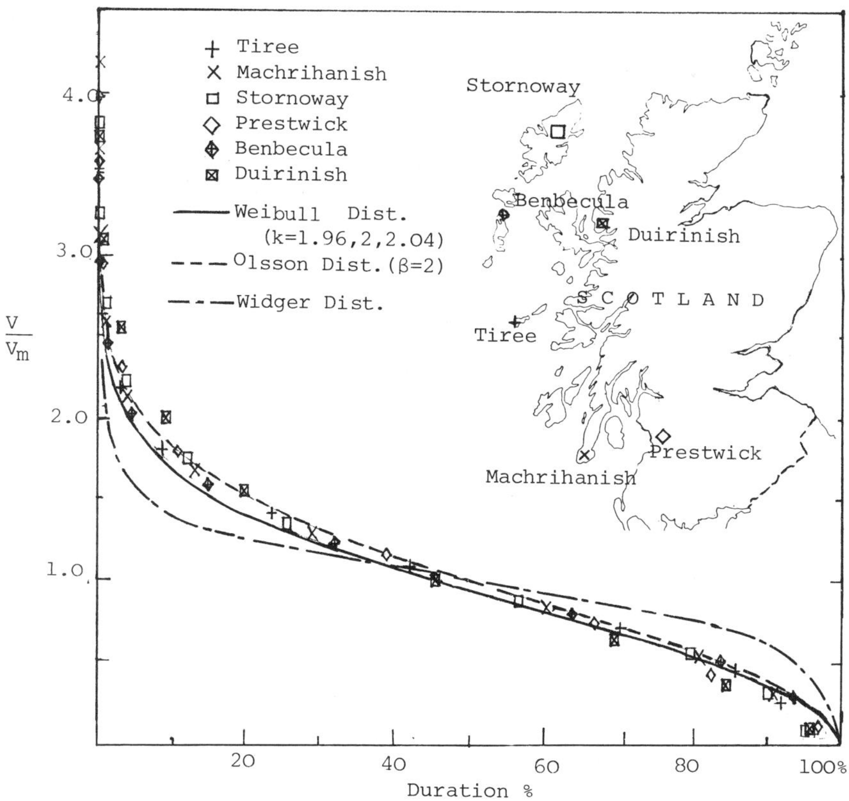

Fig. 1. Wind duration curves

Power Duration Curves

Power duration curves are conventionally obtained from the velocity duration curves. The area of the power-duration curve gives an alternative to Eq.(1) for estimating the annual energy available. Power duration curves for the six meteorological stations also for the aforementioned statistical models are shown in Fig.2. Again curves based on measured data are in good agreement with those based on statistical models. However, when computing the energy available based on the area of the power duration curves, Olsson distribution Eq.(4) showed minimum errors as indicated in Table 1.

Furthermore the Olsson distribution is characterised solely by the predetermined exponent 'β'. This potentially gives it an advantage over the Weibull since estimates of available wind energy could be achieved by installing simple totalizing anemometer which measures mean speed, or a similar anemometer.

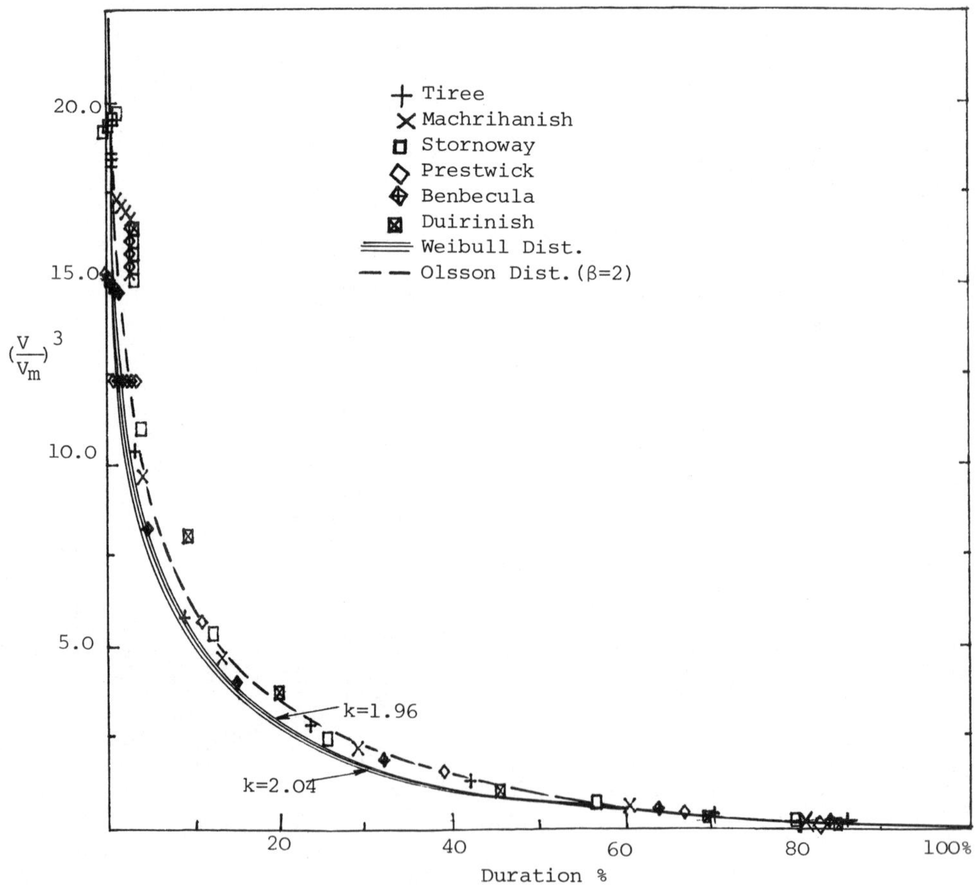

Fig. 2. Power duration curves

TABLE 1. Error in computing energy available using statistical models.

Station	Mean speed Knots	Energy available $\Sigma fG\ kWh/m^2/y$	Error % (Olsson Dist) (β=2)	Error % (Weibull Dist.) K=1.96	Error % (Weibull Dist.) K=2.04	Error % (Rayleigh Dist.)
Tiree	15.58	5959.7	+8.57	-7.85	-11.24	-9.57
Machrihanish	13.14	3699.6	+5.04	-10.82	-14.12	-12.51
Stornoway	12.58	3471.3	-1.65	-16.51	-19.59	-18.08
Prestwick	9.52	1480.4	-0.14	-15.22	-18.35	-16.82
Benbecula	13.86	4226.8	+7.91	-8.39	-11.78	-10.12
Duirinish	10.98	2501.1	-9.18	-22.89	-25.75	-24.35

ESTIMATION OF ENERGY OUTPUT FROM AEROGENERATOR.

Even when the wind energy available from a particular site has been determined, an accurate estimation of the energy which could be produced can not be achieved without consideration of characteristics of the aerogenerator likely to be used and an approximate evaluation of the losses associated with the energy conversion process.

Hence, considering the aerogenerator characteristics (e.g. starting wind speed 'V_s' rated speed 'V_r' and furling speed 'V_f')usable energy over a time T is given by;

$$E = \tfrac{1}{2}\rho\, V_m^3 \left\{ T \int_{F(V_s)}^{F(V_r)} \left(\frac{V}{V_m}\right)^3 d\,F(V) + T\left(\frac{V_r}{V_m}\right)^3 \left[F(V_r)-F(V_f)\right]\right\} \qquad (7)$$

and is of course dependent on the choice of the value V_s and V_r. Allan and Bird (1977) showed that for value of $V = \tfrac{1}{2} V_r$ ($P=1/8P_r$) the overall efficiency of the aerogenerator is about 20 percent and concluded therefore that the best relationship between starting speed and rated speed is given by $V_s = 0.5\ V_r$. Using power duration curves based on Olsson's distribution Eq. (4) the optimum rated wind speed as a ratio to mean speed has been determined for maximum usable energy. This was achieved by assuming values of V_r/V_m in the range of 1.8 to 2.8 and $V_s = 0.5V_r$ and computing the usable total energy from Eq.(7). Figure 3 shows the variation of the total annual energy computed with the variation of V_r/V_m for different values of the parameter β.

However, it must be borne in mind that optimisation of usable energy does not necessarily lead to cost optimisation because the prices of larger aerogenerator components might increase faster than the energy output.

CONCLUSION

Comparison between different methods for estimating available wind energy showed that an accurate prediction could be achieved by using the mean wind speed together with the statistical model presented by Olsson Eq. (4) for the wind duration distribution. Hence the energy potential of a certain site could be determined by installing simple totalizing anemometer for mean wind speed measurements, or a similar anemometer.

The optimum values of the aerogenerator characteristics e.g. V_r/V_m and V_s/V_m which influence the aerogenerator output could be chosen for a particular utilisation using the power duration curve based on the aforementioned statistical model.

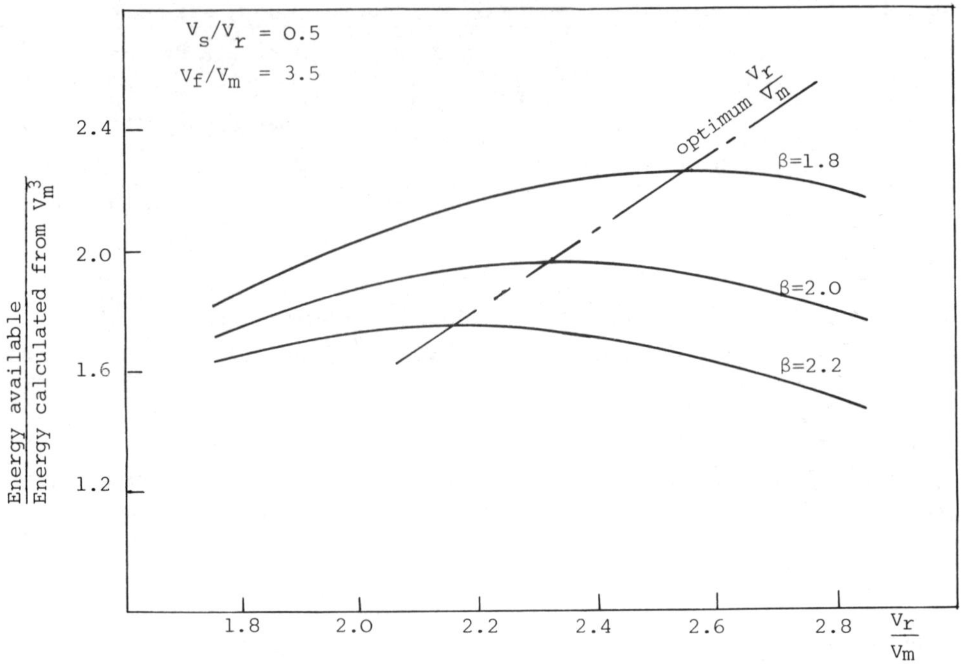

Fig. 3. Variation of available energy with rated wind speed ratio.

ACKNOWLEDGEMENT

The author wishes to express his gratitude to the Mechanical and Production Engineering Department, Paisley College of Technology for the financial support, necessary to undertake this study.

REFERENCES

Allen,J., R.A. Bird (1977). The prospects for the generation of Electricity from wind energy in the U.K. Energy paper No.21. H.M.S.O. London 1977. Published by the Department of Energy U.K.

Cliff, W.C. (1977). The effect of generalised wind characteristics on annual power estimates from wind turbine generator. PNL.2436, Battelle, Pacific Northwest Laboratories, Richland, Washington 99352.

Hennessey, J.P. (1977). Some Aspects of Wind power statistics. J. Appl. Meteorol. 16, 1977, 119-28.

Olsson, L.E., O. Holme, R. Kreig (1975). Wind characteristics and wind power generation. A three year Meteorological programme in Sweden. Proceedings of the Second Workshop on Wind energy conversion systems. Washington D.C. 1975.

Widger, W.K. (1976). Estimating wind power feasibility. Power Engineering, August 1976, 58-61.

TOPIC C

Wave Power

The Clam Wave Energy Converter to Provide Electricity for Island and Coastal Communities

A. M. Peatfield

Wave Energy Research Group, Coventry (Lanchester) Polytechnic, Priory Street, Coventry, UK

ABSTRACT

The paper describes the design and operation of the Clam wave energy converter and discusses its progress and merits in the light of the possibility of using limited numbers of devices to provide electricity for island and near-coastal communities. Devices deployed in suitable wave climates are capable of providing electricity at about 5 p/kWh.

KEYWORDS

Wave energy; renewable energy utilisation; alternative energy source; costs of electricity from wave energy.

INTRODUCTION

Sea Energy Associates Limited and the Coventry (Lanchester) Polytechnic have been involved in the United Kingdom wave energy research and development programme for over 6 years, working firstly on the Salter Duck programme and, for the last 3 years, on the development of a second generation wave energy device called the Clam.

Extensive 1/50 th. and 1/10 th. scale model tests of the Salter Duck device at both Draycote reservoir and Loch Ness, combined with full scale design studies, confirmed the inherent efficiency and viability of a floating spine based wave energy device. However the full scale design studies indicated that the cost of overcoming some of the engineering problems could be high.

Thus the Clam arose out of the subsequent re-design process which aimed at utilising the attractive aspects of a spine based device, stability and low mooring forces, whilst at the same time avoiding the known problem areas of large bearings and a complicated power conversion system. The resulting simplified design was tested at 1/50 th. scale in both narrow and wide wave tanks as well as in the natural environment at Draycote, leading to further design simplifications which should considerably reduce the device cost whilst maintaining a high overall productivity.

The Department of Energy is expected to reach a decision early in 1982 on whether to progress towards full scale testing of one of the major wave energy device designs, or to continue on a research basis with reduced funding. Apart from other political considerations, the decision is likely to be made on the costing of a nominal 2 GW station located off the outer Hebrides and delivering electricity to the national grid at Perth with a target cost of less than 5 p/kWh.

The Department of Energy approach seems to exclude consideration of a single, or small numbers of devices being used to provide electricity for island, or isolated coastal communities, thus helping to provide a new service or replacing expensive oil burning generating capacity.

This paper describes the Clam wave energy device and discusses its progress and merits in the light of the limited objectives of providing electricity for local island and near-coastal communities.

THE CLAM WAVE ENERGY CONVERTER

The operational experience at 1/10 th. scale in Loch Ness with the Salter Duck device, together with measurements of wave induced bending moments and mooring forces, confirmed the operational efficiency and structural economy of a spine based wave energy terminator; a terminator being the name given to a device which extracts energy along a front nearly perpendicular to the principal wave direction.

However the full scale implications of the huge rotating mass of the Duck floats, together with the problems of the power take of system, brought about the greatly modified design called the Clam.

The Clam is based on the reversal of the simple flap concept normally used to generate waves in laboratory tanks. It comprises a number of flaps hinged to the base of a floating spine, with air bags sandwiched between the flaps and the spine front. The waves transfer their energy to the flaps, moving them in and out, thus forcing the air in the bags through self-rectifying, Wells, turbines into an interconnecting, enclosed, duct running the length of the spine, causing interchange of air between clam elements. The randomness of sea wave patterns, combined with an angle of inclination of the spine to the wave front of about 30 degrees, allows phased operation of the clam elements, with the spine acting as a stable reference body.

Figure 1 shows an illustration of a typical Clam wave energy converter, with 10 flap-bag units on a spine of about 250 m. in length, moored at an angle of about 30 degrees to the oncoming wave fronts.

The cutaway section in Fig. 2 shows the spine with the flap, bag, self-rectifying turbine and connected generator.

The spine is the major structural component acting as a stable frame of reference in the form of a hollow floating beam capable of withstanding the wave induced bending moments and hydrostatic forces and allowing free interchange of air between clam elements.

Wave induced bending moments have been measured in model tests at Draycote reservoir and at Loch Ness. A quasi-static mathematical model has been developed and verified and can be used to predict full scale design loadings for any defined sea state.

Each of the 10 steel flaps is about 25 m. long, 15 m. high and is attached to the

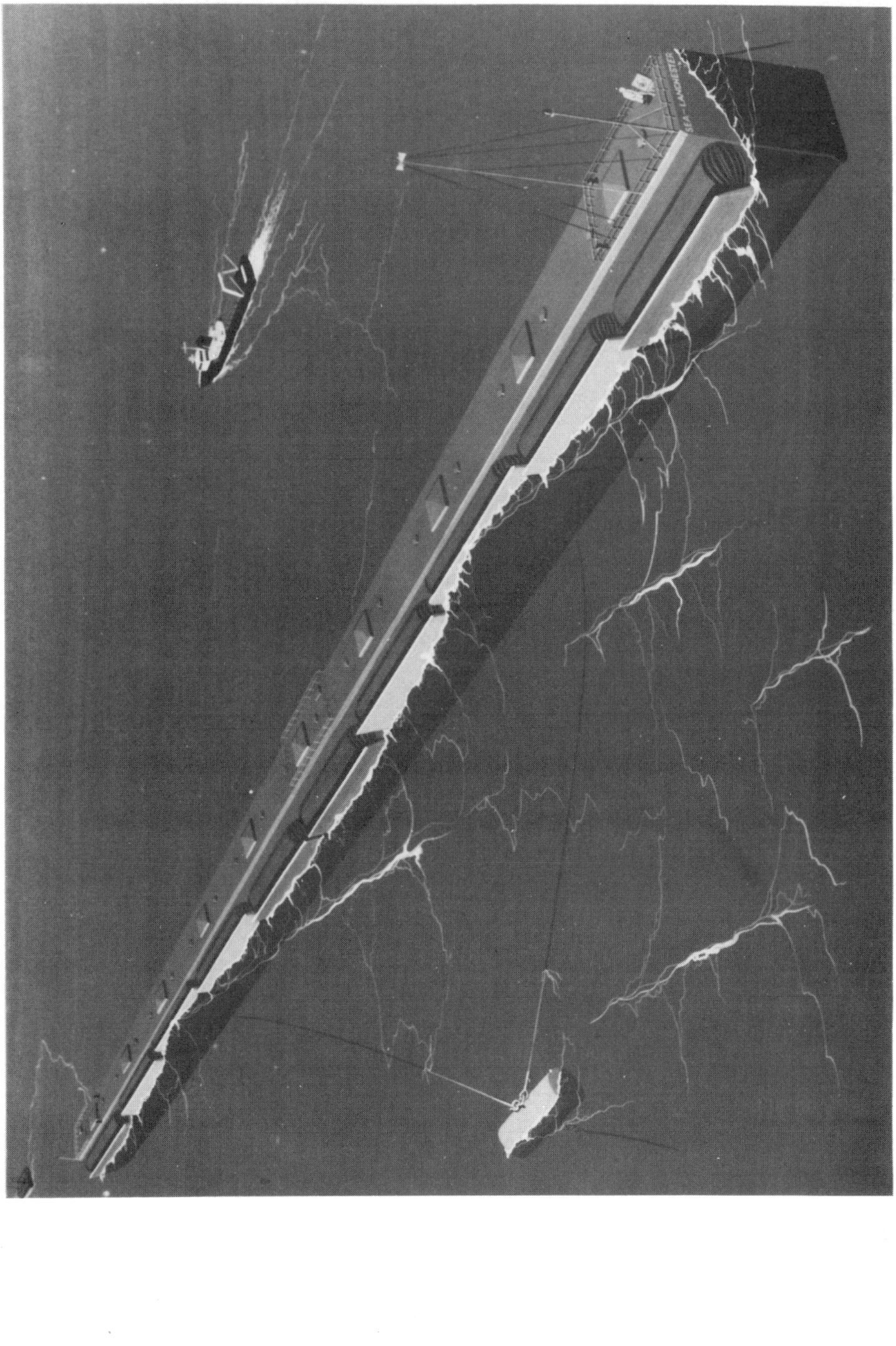

Fig. 1 Clam Device

Fig. 2 Section of Clam Device

spine at its base by up to 50 individual hinges of the type used as rollers on the tracks of heavy earthmoving equipment.

Properly tuned and damped by the design and control of the self-rectifying turbine, the flap-bag system is capable of high, broad band, wave energy absorption efficiencies in the region of 85-90%. However, the flap is an expensive component of the system and imposes considerable constraints on the satisfactory operation of the airbag thus making design and manufacture difficult and long term reliability uncertain.

The power conversion system employs a self-rectifying air turbine at each bag outlet and uses the spine as a common duct allowing interchange of air along its full length. A series of 1/10th. scale turbine performance tests have been carried out which confirm that the machine can be designed to present a linear load to the flap-bag system, be efficient over a broad power range and have low drag losses when running free. Also the inertia of the turbine-generator units ensure an important power smoothing over a number of wave cycles.

The mooring system was developed for the Loch Ness duck tests and has been thoroughly investigated in subsequent spine model tests at 1/100th. and 1/10th. scales. The front mooring for each spine consists of a single anchor point, in 60-100 m. of water, connected to a leading buoy of approximately 150 tonnes displacement and then by two rodes to the spine, in a V-yoke configuration. The V-yoke rodes are of unequal length in order to ensure that the spine takes up an angle of about 30 degrees to the oncoming wave fronts. The whole system can move round the single anchor point, automatically adjusting to any change in principal wave direction as wind directions change. Rear restraining moorings would be used to prevent excessive movements under the influence of offshore winds. Model tests indicate that the maximum mooring force likely to be experienced by a 250 m. long spine in the worst North Atlantic conditions will be less than 500 tonnes.

Unlike most wave energy devices, the Clam system has a "close-down" mode which allows inspection and maintenance to be carried out at sea with the moving parts stationary. By venting the pressurised air system, the flaps automatically close the bags and are held fixed against the spine by the pressure of external water head.

DESIGN IMPROVEMENTS

Since the presentation of the 1979 Clam design, a great deal of experimental and detailed development work has been carried out with the aim of reducing costs and increasing device productivity. Particular attention has been paid to the flap, it being a large moving part with problems of attachment and long term operation as well as representing over 20% of the total cost of a Clam device.

Recently a soft fronted bag has been designed and tested at 1/50th. scale, which not only eliminates the need for a flap but allows air to be exchanged through ducting attached to the top corners of the bags, into the spine via air-ports situated in the top, well above normal water level. Thus the requirement for only one major orifice into the spine, per bag, and the fact that this orifice can also be used as the access hole through which the turbine-generator unit can be fitted and maintained, has simplified the spine design with a further reduction in cost.

Over the spring of 1981, bending moment and mooring force measurements have been carried out on a 1/10th. scale Clam spine in Loch Ness. This spine is now being fitted out with soft fronted bags and simulated turbine dampers ready for a series of complete device tests in October-December 1981.

PRODUCTIVITY AND COSTS

Most of the research and development work on the Clam device is being aimed at assessing the productivity and costs of devices designed for a nominal 2 GW station located off the outer Hebrides and delivering electricity to the national grid at Perth.

The 1979 Clam design was costed on a productivity derived from a limited series of 1/50th. scale narrow tank tests and a very pessimistic mathematical assessment of the effects of spine movements on absorption efficiency. The resultant cost was approximately 6.5 p/kWh.

However since then improvements in design and the results of a more comprehensive series of both narrow and wide tank tests indicate an increase in overall productivity of nearly 25%.

The tremendous cost reduction due to the replacement of the flap-bag unit with a soft fronted bag, the simplified spine design and construction, together with the increased productivity makes the present estimated cost 4-5 p/kWh. at 1981 prices.

These costs are for the nominal 2 GW station off the outer Hebrides and include a large element, 20%, for the cost of transmission to Perth, as well as a reduction in productivity caused by the shielding effect that adjacent devices have on each other when the waves are from other than the ideal direction.

When considering single devices, each close to the community it is supplying, the saving in transmission costs and the increased productivity, because the device can swing to face the optimum direction without shielding effects, should offset the increased costs likely for the design and manufacture of single devices compared with the numbers involved for a nominal 2 GW station.

The presently proposed design has been tailored to suit the wave climate along the outer western and north-western coastlines of Scotland, including the Hebrides, Orkneys and Shetlands, where the mean annual power level is typically 40-50 kW/m.

A single, 250 m. long, Clam device in such a wave climate would provide a mean annual output of about 2.5 MW, with the winter average being up to nearly 4 MW.

The northern and north-eastern coastlines of the mainland have lower annual power levels but, with suitable variations of Clam design and operation, could still provide suitable sites for wave energy converters.

Some of the best wave energy sites in the world exist round the coasts of Scotland within easy reach of many isolated island and coastal communities that are at present supplied, if at all, by expensive electricity generated from oil.

ACKNOWLEDGEMENT

The Coventry (Lanchester) Polytechnic Wave Energy Research Group has been funded during this research programme by Sea Energy Associates Limited and by the Department of Energy.

Wave Energy Studies at the UK National Engineering Laboratory

G. Elliot* and G. Roxburgh**

*National Engineering Laboratory, East Kilbride, UK
**Roxburgh and Partners, Consulting Engineers, Glasgow, UK

ABSTRACT

The work programme carried out into the development of the oscillating water column from the basic concept to a major power station design is described. The validity of a small wave-power station serving a remote community as a first objective in an overall development plan for wave energy is discussed.

KEYWORDS

Wave energy; oscillating water columns; Francis turbine; floating offshore structures; fixed offshore structure; design optimization.

INTRODUCTION

In 1974 the National Engineering Laboratory (NEL) undertook a study on behalf of the Department of Energy into the economic and technical feasibility of large-scale generation of electricity from the sea waves in UK offshore waters. The study showed that wave energy had been a source of interest for many years and listed some 340 UK patents dating back as far as 1856 for devices which were claimed to be able to utilize sea wave energy.

The 1974 study (Leishman and Scobie) covered many aspects of wave energy and examined the knowledge which was available regarding energy levels around the UK coasts, as well as the impact on the environment of capturing such energy. The report identified where active research into wave energy was being undertaken and showed that, up until 1973, it had been of little interest and a low key activity. However, during the course of the study, a tremendous surge of interest was taking place caused by the action of the OPEC countries in 1973 which greatly affected the world's oil supplies.

The greatest interest was being shown in both Japan and the UK where several universities and other organizations had commenced research into the harnessing of wave energy. NEL decided to participate in the development of wave energy devices. Because of their experience in offshore activities as well as their special facilities in turbo-machinery, the oscillating water column principle was selected as being the most appropriate to NEL's expertise. Although some work was being carried out in Japan the principle was not being studied or developed in the United Kingdom. NEL submitted a proposal to the Department of Energy to examine the oscillating water column. This was accepted and resulted in a programme of work funded by the Department of Energy.

EVOLUTION OF THE CONCEPT AND DESIGN

The initial investigations involved examining the simple inverted cylinder oscillating water column concept (Meir, 1978). Tank experiments on two-dimensional models demonstrated that, irrespective of depth of immersion, the capture efficiency of such a cylinder could never exceed 50 per cent. The energy in the incident wave was either transmitted and hence lost, or reflected back from the device. This was concluded to be the case with any arrangement of multi-directional symmetry.

In order to increase efficiency it was postulated that an asymmetric device in which the trailing edge of the column was extended downward would improve energy extraction by reducing or eliminating the transmitted wave energy (Fig. 1). This proved to be the case and very high peak efficiencies were achieved.

The further development of the concept was the introduction of a bottom reflector. This had the effect of improving the coupling between the incident wave and the water column which, as well as maintaining the high peak efficiencies, had the effect of broadening the bandwidth of the response of the device (Fig. 2). This of course is a desirable feature because the incident wave energy contained in the sea is made up of a fairly wide bandwidth of individual wave frequencies. This asymmetrical column with bottom reflector was the subject of a substantial amount of experimental and theoretical work in order to optimize and understand the features of a column of this shape (Fig. 3).

FLOATING STRUCTURES

The main experimental work was carried out in a two-dimensional tank in which the basic column shape was kept static for various geometric arrangements. However, in reality, the basic column contained in a floating structure would have to exist with various degrees of freedom in a deep water situation. In order to develop a design it was necessary to examine the interaction of its motion with the efficiencies and behaviour of the water columns.

One of the fundamental problems of a floating wave energy structure is that, unlike ships or offshore structures which are hydrodynamically streamlined, wave energy structures have, by definition, to interact with and absorb the incident wave energy. It is likely therefore that very large forces associated with this absorption would be experienced and the basic design problem to be resolved was how best to react these forces.

There were three possible options available:

1 Transmit the forces to the seabed. This was likely to be practicable only with a shallow water device as it is undesirable to transmit large forces through mooring systems.

2 Balance the positive and negative wave forces between adjacent waves either along the wave or at right angles across the wave crest. This required very long structures to be created in order to span large distances and the cost effectiveness was then questionable. Extending the device along the waves was of course possible and is favoured by at least one device. However, we believe this approach does pose practical and engineering difficulties.

3 Resist the forces locally by utilizing the inertia of the structure and that of the water in its vicinity (Fig. 4). The 'added mass' would thus be

the major means of resisting the forces likely to be generated.

It was the third approach which was adopted by NEL and during the course of the hydrodynamic investigation many model shapes were tried and investigated in the wave tank.

Figure 5 shows a selection of these. The first, Fig. 5(a), is particularly interesting since this shows an early asymmetric model of the wave spanning type where the back reflector is in the horizontal plane. This has the effect of preventing wave transmission like a vertical deflector. However the best efficiency which could be achieved with this device was well below 50 per cent for a floating two-dimensional model. It is interesting to note that CEGB Marchwood Engineering Laboratories have decided to study this model in a long spine three-dimensional form whereby they claim much better performance due to the restriction of hydrodynamic response achieved by the third dimension. The second shape, Fig. 5(b), performed slightly better than the first although still only achieving peak efficiencies of less than 50 per cent. One of the major difficulties is that when a floating body is induced to move it has the tendency to generate waves which would radiate from it. The fourth model, Fig. 5(d), was the first in which the effect of wave cancellation was observed. The tendency to radiate waves had been greatly reduced and it was clearly observed that the incident wave was being completely absorbed by the device with insignificant wave radiation from the device itself.

The fifth shape, Fig. 5(e), was a further optimization whereby the overall volume was reduced, with little detrimental effect to the device performance. The option to close off the back section, shown dotted, was examined as a possibly desirable feature from a full scale constructional aspect. A two-dimensional model of this device is shown in Fig. 6.

The third model, Fig. 5(c), was the one selected as a shape which would be more readily constructed in reinforced concrete.

POWER EXTRACTION

Throughout the period when the hyrodynamic shape was being developed, it was important to develop in parallel an efficient and reliable method of extracting energy from the column. The simplest one envisaged was to use the air displaced by the water in the column to drive a machine. The characteristics of the air flow out of the column are a high velocity with a low inertia. The nature of the application calls for a high degree of reliability and robustness in the machine. In order to preserve the high conversion efficiencies of the wave to air phase it was important to ensure that the characteristics of the machine matched as closely as possible those of the column, otherwise excessive damping would have been applied to the column preventing it from operating efficiently.

A Francis turbine was selected as giving the characteristics most closely matching those of the column. Although Francis turbines are normally used in hydro-electric power stations operated by water, the machine can operate equally well using air as the fluid. The Francis turbine is a very efficient machine and its compact design makes it a robust and reliable device (Fig. 7).

Associated with the air inflow to the turbine it was necessary to conceive a ducting and rectifying arrangement whereby air could be converted from a reciprocating flow to a uni-directional flow through the turbine. A four-valve rectification system was conceived (Fig. 8) to achieve this uniform direction of flow.

A programme of work was carried out in parallel with the hydrodynamic development which examined the characteristics of a radial inflow Francis turbine at model scale together with a rectifying valve arrangement. The experimental work

(Fig. 9) has shown that this system operates satisfactorily in a pulsating airflow regime and it has been shown that mean annual efficiencies of the order of 80 per cent can be achieved from these turbine systems.

FULL-SCALE DESIGNS

At this stage the services of a consulting engineering organization experienced in civil and offshore works were required and Roxburgh and Partners were selected as members of the NEL team. The combination of the hydrodynamic work with the turbine development and with other ancillary studies examining mooring and other force regimes led to full-scale designs being produced by the team in 1978 for a 100 MW power station. Two designs were in fact carried out: one of steel construction to be built by conventional shipyard techniques and the second of reinforced concrete constructed in a manner similar to that used for gravity offshore structures.

Reference Design in Steel

The cross-sectional shape adopted for the steel design was that which gave the best overall efficiency in the tank test. The reference design (Fig. 10) could readily be constructed in a conventional shipyard. The overall dimensions of the steel design were 121.5 m long by 35 m wide by 33 m high. Each unit consisted of a bank of 6 modules each containing a column. The construction was designed to take place on a shipyard slipway although an alternative configuration for construction in a dry dock was also prepared. The specifications used were those which are current practice in tanker construction and each unit had a floating steelweight of around 12 000 tonnes.

Reference Design in Reinforced Concrete

The construction in concrete required that the most efficient hydrodynamic shape be modified to enable construction to be reasonably practicable. The shape which was used (Fig. 11) was rectangular, 35 m wide with a draft of 25 m in still water conditions, and consisted of six water columns each 12 m by 18 m cross section. The overall size of each unit was 116 m long by 35 m wide by 35 m high. It was proposed that the structure would be constructed using the dry dock technique as for large offshore production platforms where the base tray is cast in the dock, then the dock flooded, the tray floated out and moored and the structure completed by slip forming while afloat.

Offshore Construction

The 100 MW output power station designed was intended to operate off the west coast of Britain in an approximately north/south line and the 25 units needed to produce the power would occupy a sea distance of 13.5 km. The structures would be moored in an end-to-end configuration with approximately 400 m clear between each to allow the necessary compliance in the mooring systems.

The structures would be moored using a tensioned hawser system with conventional high-holding-power drag anchors and the addition of clump weights to improve the uplift resistance. The proposed mooring system was made up mainly of man-made fibre ropes using their inherent strain energy storing characteristics in preference to wire or chain, thereby providing sufficient elasticity to eliminate snatch forces.

These structures were intended to be intermittently manned and the necessary accommodation and safety equipment was provided on board each unit.

The object of carrying out the design studies was to examine the feasibility of such a design and to produce an estimate for the cost of generating electricity from wave energy.

The result of the study of the floating 1978 reference design indicated that the costs of producing electricity by these designs were high and likely to be of the order of three to five times greater than was attractive as an alternative form of energy.

FIXED STRUCTURES

Oscillating water columns fixed to the seabed were considered early in the programme but were initally discarded because the incident wave energy would be greatly attenuated in shallow water and therefore fixed columns seemed unlikely to offer a cost effective power generation system. Examination of the cost centres of the floating designs showed that, over the life of the structure, approximately 30 per cent of the capital repayment and maintenance costs related to the moorings. This led to the suggestion that a study of a bottom-standing device might show worthwhile cost savings.

A fixed structure in shallower water would have the following advantages:

Less space between devices

Lower survival wave heights

Higher energy conversion efficiency

Easier maintenance

No flexible power cable

Closer proximity to shore

Higher load factor

The main disadvantages might be:

Lower incident wave energy

Interference from seaweed

Installation difficulties

Space Between Devices

Since the devices are fixed to the seabed they can be placed in continuous lines, thereby absorbing 100 per cent of approaching wave energy in any particular location. This is in contrast to the floating devices which due to the necessity for fully compliant moorings could only present a 20 to 30 per cent absorption front to the waves.

Lower Survival Wave Heights

Depending on the characteristics of the chosen site the maximum wave which can occur is water depth limited and usually of a height less than that in deep water. In the case studied, the site depth was chosen as 18 m and the corresponding 50-year design wave was assessed as having a maximum height of 15 m.

Higher Energy Conversion Efficiency

With a fixed oscillating water column it becomes possible to absorbe the energy from both the short wind waves andthe long swell (Fig. 12). In contrast the floating oscillating water column tends to extract energy only from the shorter waves. Thus a much larger proportion of the incident wave energy is available for conversion to power.

Easier Maintenance

Any floating structure designed to extract wave energy is likely to be constantly moving. A fixed structure however, makes maintenance of itself and its onboard plant much easier, and by its fixity, stresses the plant less severely, thereby reducing wear and tear and reducing the maintenance requirements.

No Flexible Power Cables

Power transmission from floating devices using flexible cables is likely to be very expensive in initial capital cost and subject to breakdowns due to flexing or external interference. Indeed, suitable power cables have yet to be developed by industry. A fixed bottom-standing device eliminates the need for flexible power cables and the problem is reduced to one of maintaining an existing type of fixed cable such as is in common use on many underwater power links.

Closer Proximity to Shore

By siting the device in water depths around 18 m the distance offshore is reduced to about 1 to 5 km from landfall dependent on local conditions. This gives a shorter transport distance for men and materials and creates a more sheltered environment in the lee of the structure for operation of maintenance vessels. It also reduces greatly the lengths of underwater cable required to land the power.

Higher Load Factor

Moving the device inshore into a less extreme wave climate and mounting it on the bottom will result in an improved load factor. The extreme waves found in deep water (greater than 40 m depth) can no longer occur at the device since the wave height is now water depth limited to about 15 m and the general wave front is likely to be refracted by the bottom effects making it more nearly parallel to the line of devices. This is expected to result in a greater proportion of usable wave energy being available to the device.

Lower Incident Wave Energy

While a lower wave incident energy might appear as a disadvantage, the much higher conversion efficiency that can be obtained from a fixed, as opposed to a floating, structure together with the possibility that the incident energy is distributed in a regime which contains less unusable peaks, could be a positive advantage.

Interference from Seaweed

The heavy forest of kelp which is found off the west of the Hebrides was feared to be a problem in reducing incident power and obstructing the flow within the oscillating water columns. A study commissioned from the Scottish Marine Biological Association showed that the bulk of the kelp forest (identified as being Laminaria hyperborea) occurred inshore of the 18 m depth selected for the device and that neither the growing kelp nor the seasonally cast fronds or stalks were likely to pose any serious problems in respect of diminution of power output or of maintenance. The report further showed that other marine fouling species were unlikely to prove of significance from an operating or maintenance point of view.

Installation Difficulties

The emplacement of a structure of relatively modest self-weight which has a natural draft of the same order as the water depth in which it is to be founded, poses several problems. The structure must be given auxiliary buoyancy prior to emplacement, be rapidly set down on station and quickly ballasted to full stability thereafter.

The use of a purpose designed catamaran barge is the currently preferred method of installation. This also provides the working platform for rock anchoring the structure to the seabed. Its expense can only be justified in a very large development. Further development of emplacement proposals forms a major part of a continuing study.

APPLICATIONS FOR FIXED BREAKWATER WAVE ENERGY STATIONS IN REMOTE COMMUNITIES

There is no doubt that Britain holds a substantial lead in wave energy technology relative to the rest of the world.

Most of the work carried out in the UK has, however, been directed at investigating the possibility of obtaining, within a reasonable time-scale a substantial proportion of the UK energy requirements and Fig. 13 shows the NEL breakwater type wave piston designed for that purpose.

The development of wave energy may not proceed directly to this objective as, in common with most other engineering developments, a step-by-step approach is likely to prove more practical. There are many locations in other parts of the world with a less severe wave climate than that existing in north-west Britain. These locations could be served by single or multiple wave energy stations. For example, the Pacific, Atlantic and Indian Oceans, particularly in the southern hemisphere, have many coastal zones facing a long fetch. In these locations are climates which are capable of providing consistent and readily convertible wave energy without the considerable survival problems that apply in more temperate zones, and unlike UK do not have indigenous reserves of hydrocarbon fuels.

From the commercial viewpoint, the power authorities in those locations are unlikely to be readily convinced that they should build wave energy stations or buy British wave power technology if the UK did not mount a demonstration project for one of its own remote communities.

In seeking a location for a demonstration project, it would be preferable that it should not be connected to the national grid but be served by a local grid supplied by diesel generation. A compatible supply, provided by the wave energy station, could then demonstrate a direct fuel saving, whilst at the same time it could rely on the original supply as a back-up for periods of calm weather. Various UK locations have been considered, in particular island communities, that appear to meet the criteria desirable for a demonstration to be provided.

PRELIMINARY STUDY OF A SITE IN TROPICAL WATERS

The problems of constructing a single wave energy station (less than one megawatt say) are substantially different from those of a major installation of many gigawatts.

In the case of a fixed shallow water device, the construction of a small single station will rely very much on standard maritime civil engineering techniques.

This is in contrast to the mass production and sophisticated installation techniques upon which the economics of a major scheme depend.

A preliminary exercise for an island in the Southern Pacific Ocean, (Fig. 14) utilizes a sheet-piled coffer dam of elliptical plan shape.

The wave piston chamber is formed of in situ concrete within the coffer dam which also provides the housing for the turbo-generating plant.

Immediately prior to start-up, an opening is made in the front of the coffer dam so that wave power can be abstracted from the sea into the device.

A further stage in the development of the device illustrated will be a series of tank tests so that the output can be equated against costs and the economic viability confirmed.

CONCLUSION

It is considered that a wave-power station for a remote community of the order of power of 1-5 MW would have a significant place in a 'step-by-step' approach to the exploitation of the major wave energy resource, both for the UK and other parts of the world. Indeed it could provide the breakthrough needed to allow this infinite, inexhaustable source of power to be harnessed for mankind.

ACKNOWLEDGEMENTS

The authors wish to thank the Energy Technology Support Unit of the Department of Energy and the Director of the National Engineering Laboratory for permission to publish this paper and the members of the NEL Device Team past and present and the staff of Roxburgh and Partners for their help in its preparation. This paper is British Crown copyright.

REFERENCES

Leishman, J. M. and Scobie, G. (1974) The development of wave power - a techno-economic study. NEL Publication EAU M.25. National Engineering Laboratory, East Kilbride, Glasgow.

Meir, R. (1978) The development of the oscillating water column. Proceedings - Wave Energy Conference, London.

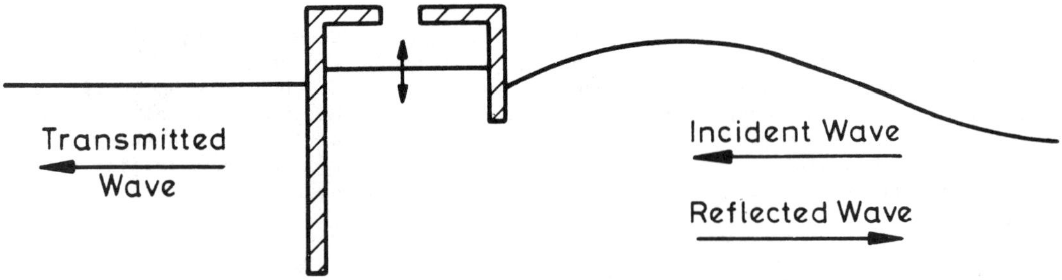

Fig. 1. Two-dimensional oscillating water column

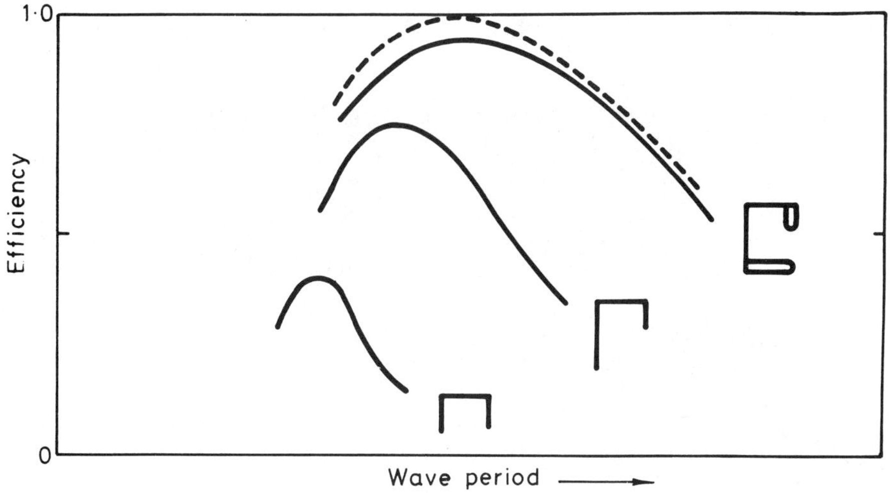

Fig. 2. Efficiency and column shape

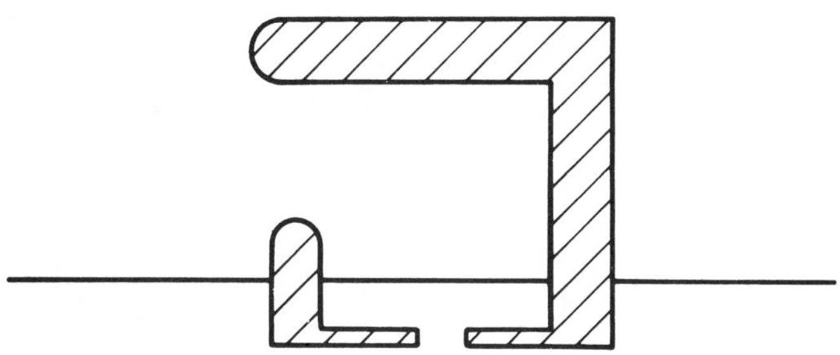

Fig. 3. Preferred column proportions

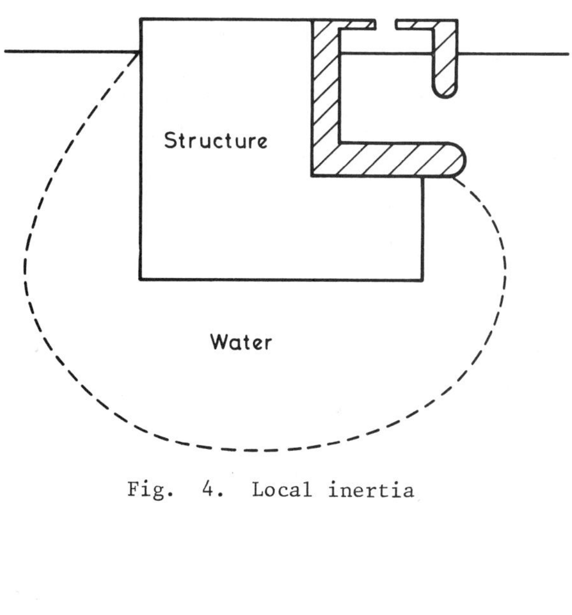

Fig. 4. Local inertia

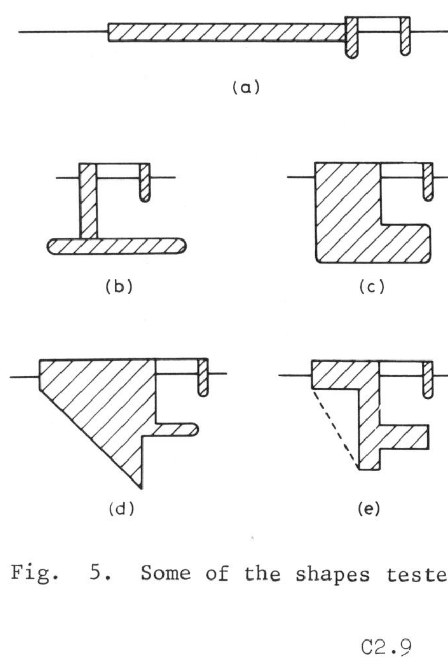

Fig. 5. Some of the shapes tested

Fig. 6. Two-dimensional model of preferred shape in test tank

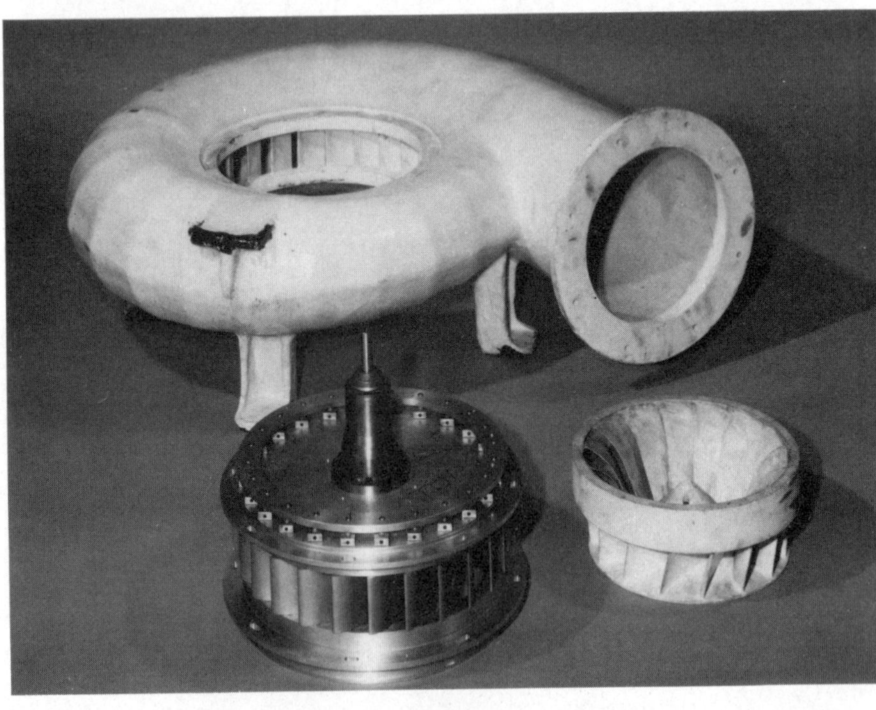

Fig. 7. Francis turbine components

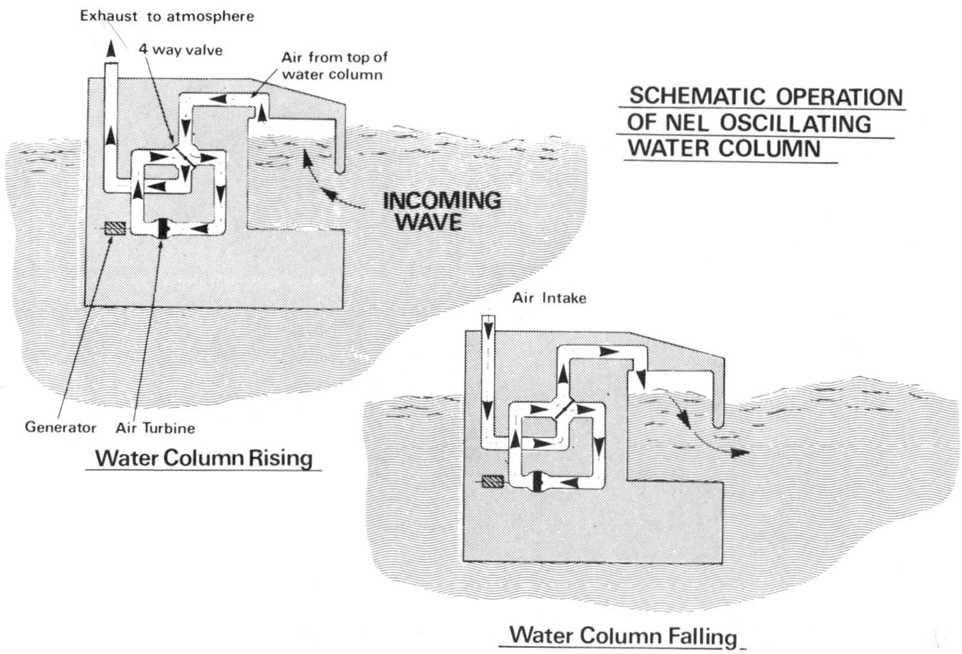

Fig. 8. Schematic operation of NEL oscillating water column

Fig. 9. Turbine test rig

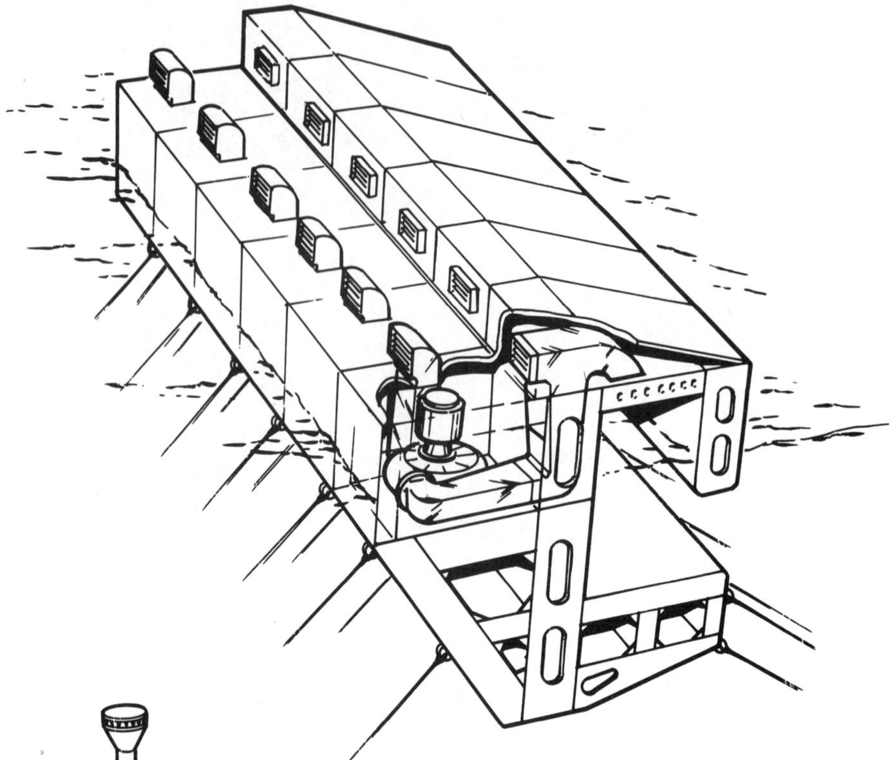

Fig. 10. Floating structure - steel design (1978)

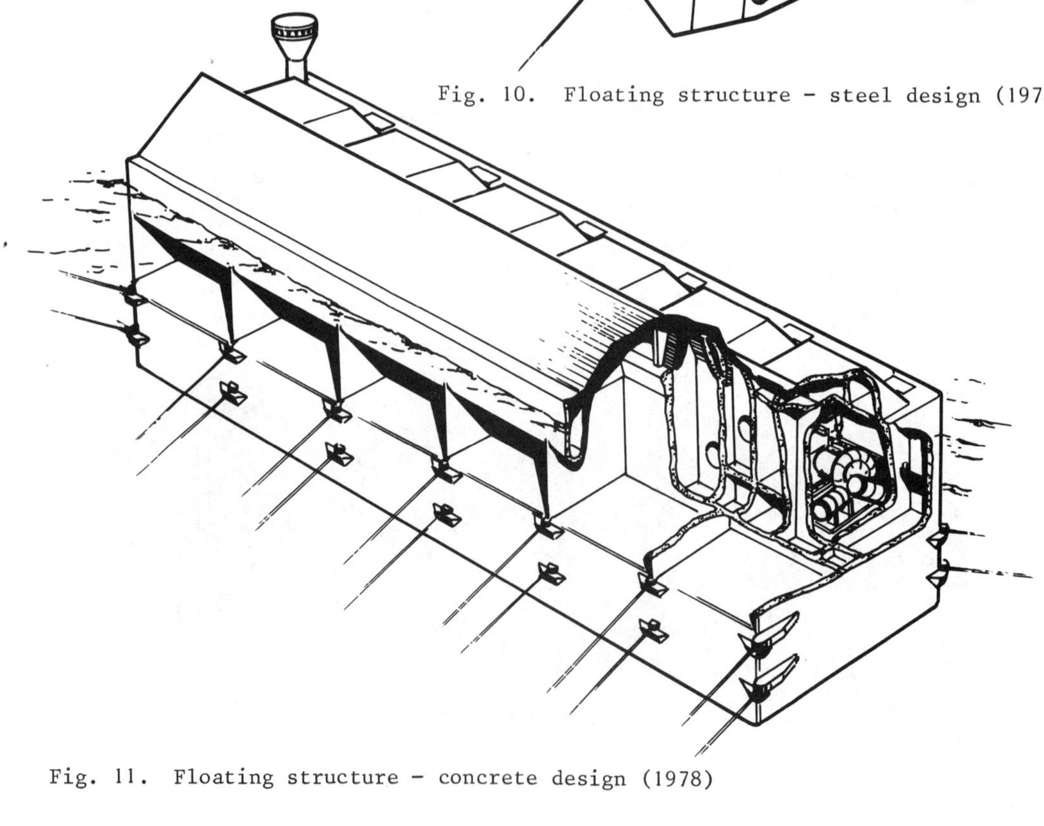

Fig. 11. Floating structure - concrete design (1978)

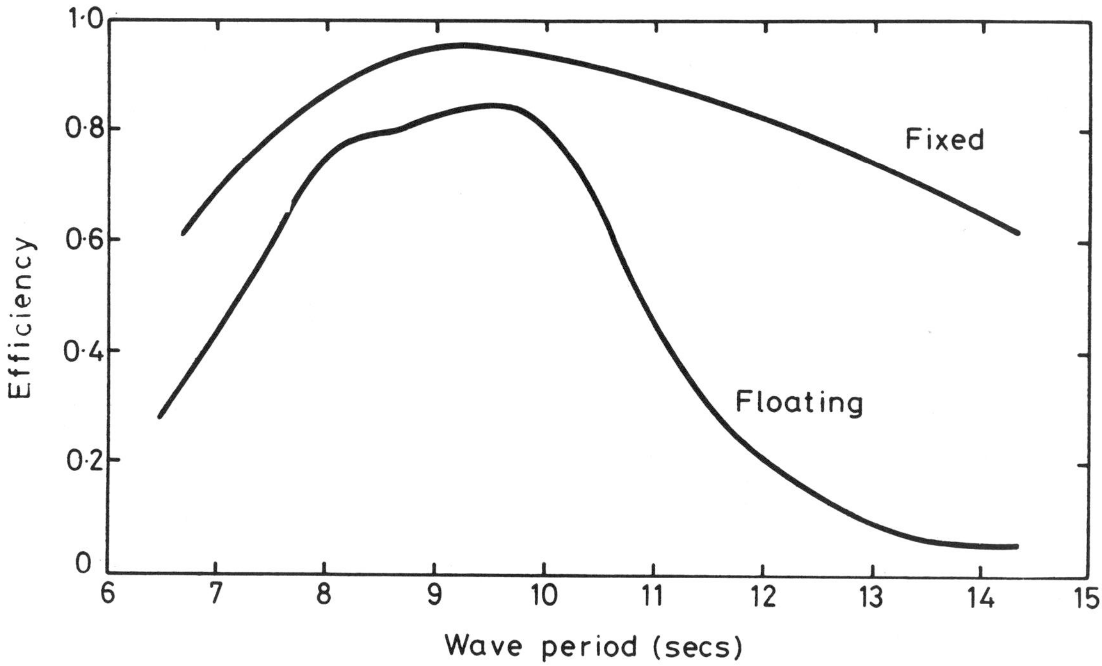

Fig. 12. Efficiency and restraint

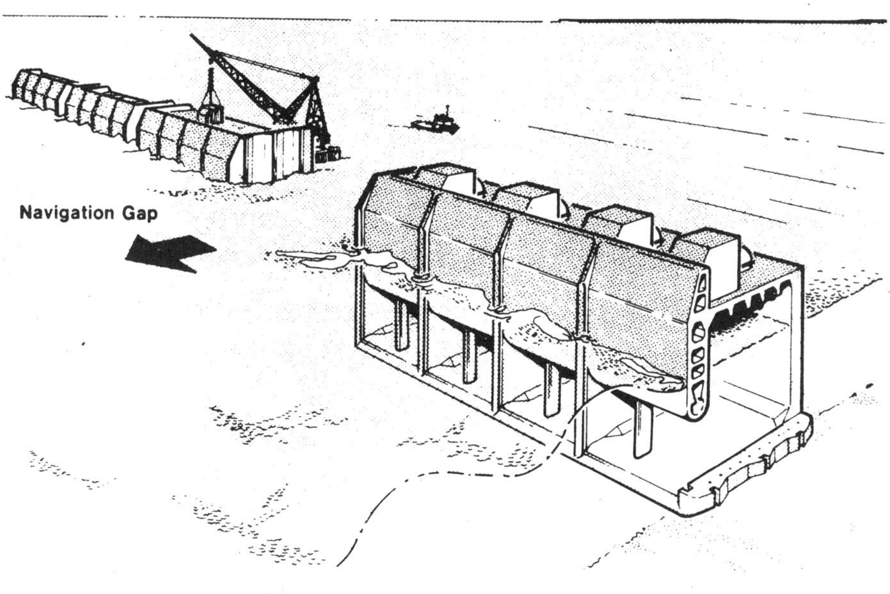

Fig. 13. The NEL breakwater type wave piston (March 1980 reference design)

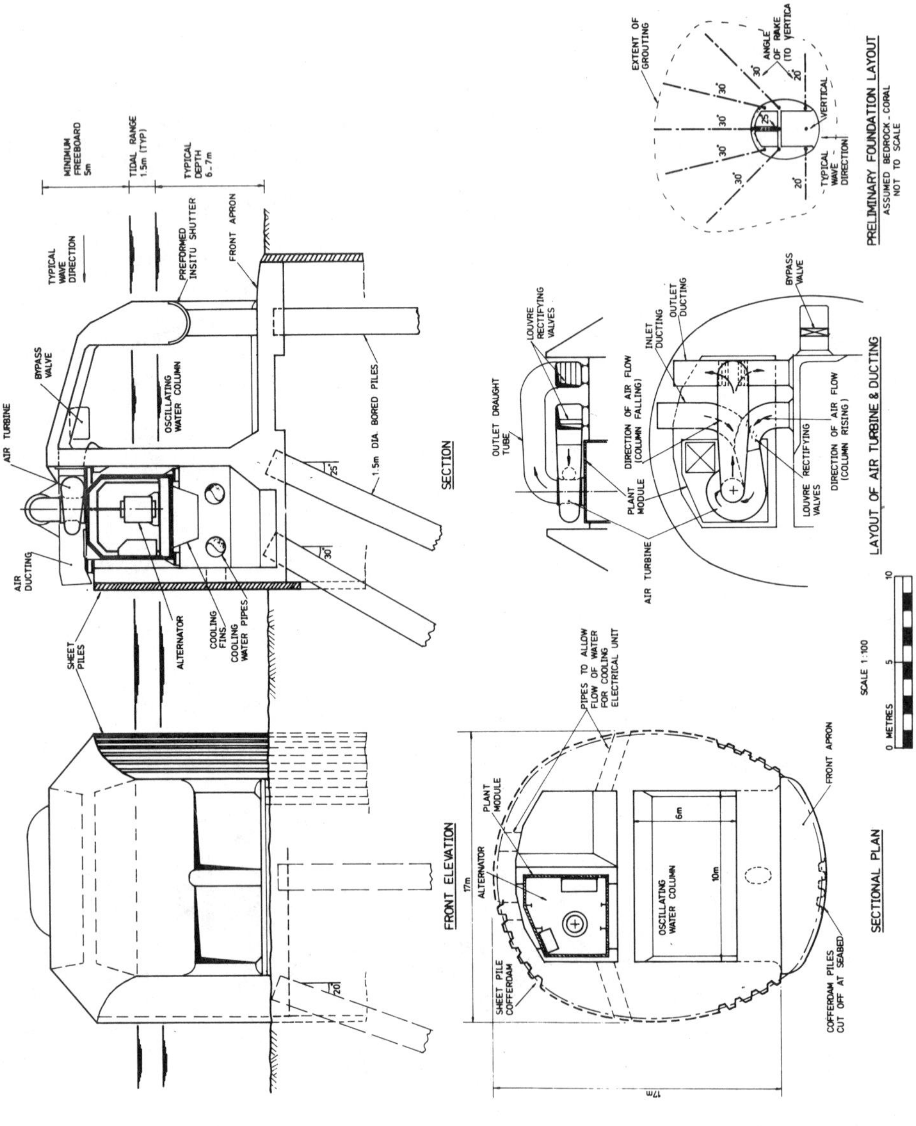

Fig. 14. Preliminary sketch of small wave power station for a Southern Pacific island.

Energy for Island Communities — Small Wave Energy Systems Based on the Air Diaphragm Principle

M. Woolley

Wavepower Limited, Carlton House, Ringwood Road, Woodlands, Southampton, UK

ABSTRACT

The general case is made for the consideration of wave energy as a possible useful and economic alternative source of power for island communities. Work in the UK to date has been concentrated on the large scale installations established at a site remote from the eventual user. It is this work that has been used to indicate the starting point for very different designs, both in scale and objective, but are required for the Island situation where the power is needed local to its point of conversion. The design presented is suitable for locations where the seabed is gently sloping and bedrock is at some depth below seabed and where modest power levels are required. Other configurations are possible to suit locations where great water depths occur close inshore and bedrock is exposed.

The principles of the floating Lancaster Flexible Bag as designed for the UK Department of Energy Wave Energy Research and Development programme are adapted using the same air diaphragm principle applied to a bottom mounted inshore small scale installation.

Costs in the region of 3-5p/kw hr appear feasible and compare very favourably with quoted costs in the region of 24p/kw hr for some Island communities at the present time.

The design is at the feasibility stage and will soon progress to a specific design study followed by a pilot construction scheme.

KEYWORDS

Energy; Island; Wave; Air diaphragm; Lancaster Flexible Bag

Where there is wind and water, there are also waves. Because of the collection effect of the water/air interface, it is also frequently true to say that where there is water there are waves, irrespective of the immediate presence of wind.

Small island communities are surrounded by sea which, depending on the degree of exposure, will contain considerable quantities of wave energy. Typical figures for the output of wave energy converters located 10-20km offshore are in the range of 10-20kw/metre of wave front intercepted. For these major installations envisaged as supplying Giga watts of power into the UK National Grid many of the design problems and their solution are quite different to those which are relevant to the supply of power to a small isolated island community. However, there are also many aspects of the former application which are directly transferable to the small scale installation and encourage its close investigation.

History is littered with inventions for the extraction of wave energy with only one thing in common - failure. It would be fair to ask what new factors have emerged in the 1980's that are likely to change the eventual outcome of the current work as compared to the dismal record of previous attempts.

This paper is not primarily written with the intention of a thoroughgoing analysis of the total wave energy scene but if what I have to say on a specific system is to be listened to, there is a need to first very briefly establish in your minds the credentials of wave energy as a sensible alternative renewable energy source. The nearest comparison is with wind energy which is more readily accepted in concept because of its historical applications well known to all. However, the relationship between the modern aerogenerator and the traditional windmill is tenuous in the extreme. Beyond using the same variable wind source, there seems little connection except the confidence engendered by prior art.

Wave energy shares with wind energy the problems of variability and unpredictability although in connection with the latter it is certainly possible to predict the presence of more power in the winter in the UK when it is wanted than in the summer when less is required. In the case of offshore installations this ratio of winter to summer power has been shown to be in the order of 4:1. Inshore the ratio may be somewhat less but is unlikely to drop below 2:1 for most locations.

Being the integral of the wind, the waves are somewhat more reliable (i.e. less variable and more consistently available) than the wind being, as they are, frequently developed from storms a long way out to sea, there is a larger catchment area relative to the island community being served. Anticipated load factors for most UK locations are in the region of 25% as compared to wind power installations which are likely to operate in the region of 15%.

Wave energy installations take up little or no land, nor do they consume a resource which might have other uses as is the case with many of the biomass solutions to the energy problem. The waves represent a hitherto untapped resource which in the past has given no benefit to the community; known only for their extreme destructive power and loss of many a life.

In certain locations there may be the additional benefit to be gained of an improved wave climate in the area of the wave energy converter because of the power extracted. It must be said immediately that this benefit will not extend to the provision of full protection as for a harbour because of the need to allow extreme storm conditions to pass over the converter. When a harbour breakwater already exists, however, the reverse may be possible; that is, the existence of usable foundation or support in a favourable location may well assist in the economic viability of a scheme if it is used to provide the main support structure for the converter.

Turning now then to the reasons for optimism in the current search for a solution to the wave energy extraction problem. They are:

1. At no time in the past has so much effort and resource been devoted to the problem. (Even so, it is a very small sum compared to that spent on R + D into other sources of power).

2. In the 1980's we are able to test our ideas very thoroughly before committing comparatively large sums of money on prototype development.

3. The choice of materials is vastly superior to that available in previous decades.

4. The low cost, weight, size, performance and reliability of modern electronic control systems which were previously not available in any form.

5. The assistance of Kevlar and flexible material technology and experience.

6. The progress already achieved with the UK Government funded R + D programme.

7. The need; which has never been greater and will increase with time.

Having spent a little time in establishing the credentials of Wave Energy and before proceeding to describe one promising solution for island communities, I should perhaps pause for a moment and make a few comments on the UK Department of Energy Wave Energy programme and those taking part in it.

The objective of the programme is to have produced by the end of 1981 a number of feasible systems for the extraction of 2GW rated power off the West Coast of the Hebrides at a projected cost of power delivered in the region of 5p per kw hr and from which one system only may proceed to prototype construction for further evaluation in 1982 onwards. The systems put forward for assessment will have been subjected to rigorous mathematical analysis and model testing at scales up to 1:50 full size. All components will have been designed in outline but some components may require further development and proving tests. There seems little doubt that the three main contenders and others will offer technically feasible systems but it remains to be seen how near they each get to the 5p per kw hr delivered target.

What is immediately obvious is the comparison with target price 5p per kw hr and that already paid for power by the Islanders of North Ronaldsay, 24p per kw hr (Ref: Barbour & Twidell, 1981) and the conclusion that the difference ought to indicate a potential market for wave energy for Islanders in similar situations.

All that I have said so far may be wellknown to the majority of the audience. I make no apology for repetition. What follows will not be so widely known and some of the concepts are being described publicly for the first time.

I should like to mention in passing that the Government funded programme carried out by Wavepower Limited and their associated Consultants, Gifford and Partners, has been supported by extensive work by major subcontractors, including such wellknown names as Sir Robert McAlpine, Wimpey Laboratories, Avon Rubber Co., A & P Appledore, GEC, BHRA and Consultants Preece Cardew & Rider.

The early (1975 onward) work concentrated mainly on the Cockerell Raft concept but for the past 18 months this has been dropped in favour of the Lancaster Flexible Bag (LFB) conceived by Professor French of Lancaster University. It is the concepts contained in the LFB which I now describe to you as applied to the smaller scale inshore installation.

The Air Diaphragm Principle

The common element and patented feature taken from the Lancaster Flexible Bag is the bag itself or, as we prefer to call it, the diaphragm. This separates the seawater from the air which is the driving medium for the turbine and hence generator which produces the electrical power. Changes in the level of the sea due to the passing wave create changes in water pressure which are converted via the flexible membrane into changes in air pressure. Individual diaphragms are manifolded into groups according to the total power requirement of the installation and flow and return ducts operating at high (0.46 bar) and low (0.44 bar) pressure respectively allow air to circulate through non return valves in the closed system between diaphragm chamber and turbine.

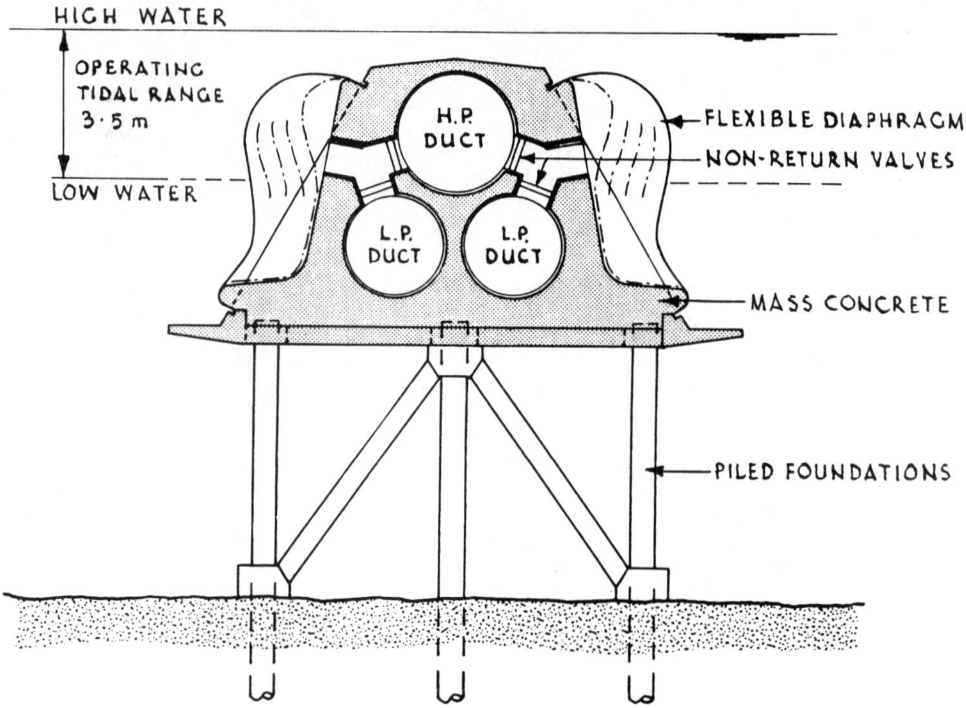

TYPICAL CROSS SECTION THROUGH AIR-DIAPHRAGM W.E.C. FIG. 1

Extensive research has established the ideal proportions, mode of operation and method of construction for the diaphragm itself. Be it offshore or inshore location, these factors remain very much unchanged except in one respect and that is in making some allowance for tidal variation. This means that the vertical proportions of the diaphragm may need to be somewhat increased so that it is efficiently operative over the full tidal ranges. Tidal variations in the region of one metre will have no noticeable effect on performance but above this figure there may be a small penalty to pay for the deeper installation required and some small loss of efficiency. For the 3.4m tidal range illustrated, an efficiency loss in the region of 10% may be expected.

An attractive feature of the flexible diaphragm module approach proposed is the degree to which it can be tailored to local requirements and yet have all the concepts of repetitious component manufacture and a maximum use of local labour in construction. Experiments have shown that individual units will respond to incident wave action whatever its direction. This enables an overall installation to make maximum use of beneficial local features without undue concern regarding the precise direction of the incoming sea.

There is the added benefit of relatively smooth power from a number of units connected in series supplying a common power unit which improves the turbine performance and thus increases the efficiency of the installation as a whole. The relatively smooth power is more easily absorbed into a small grid network

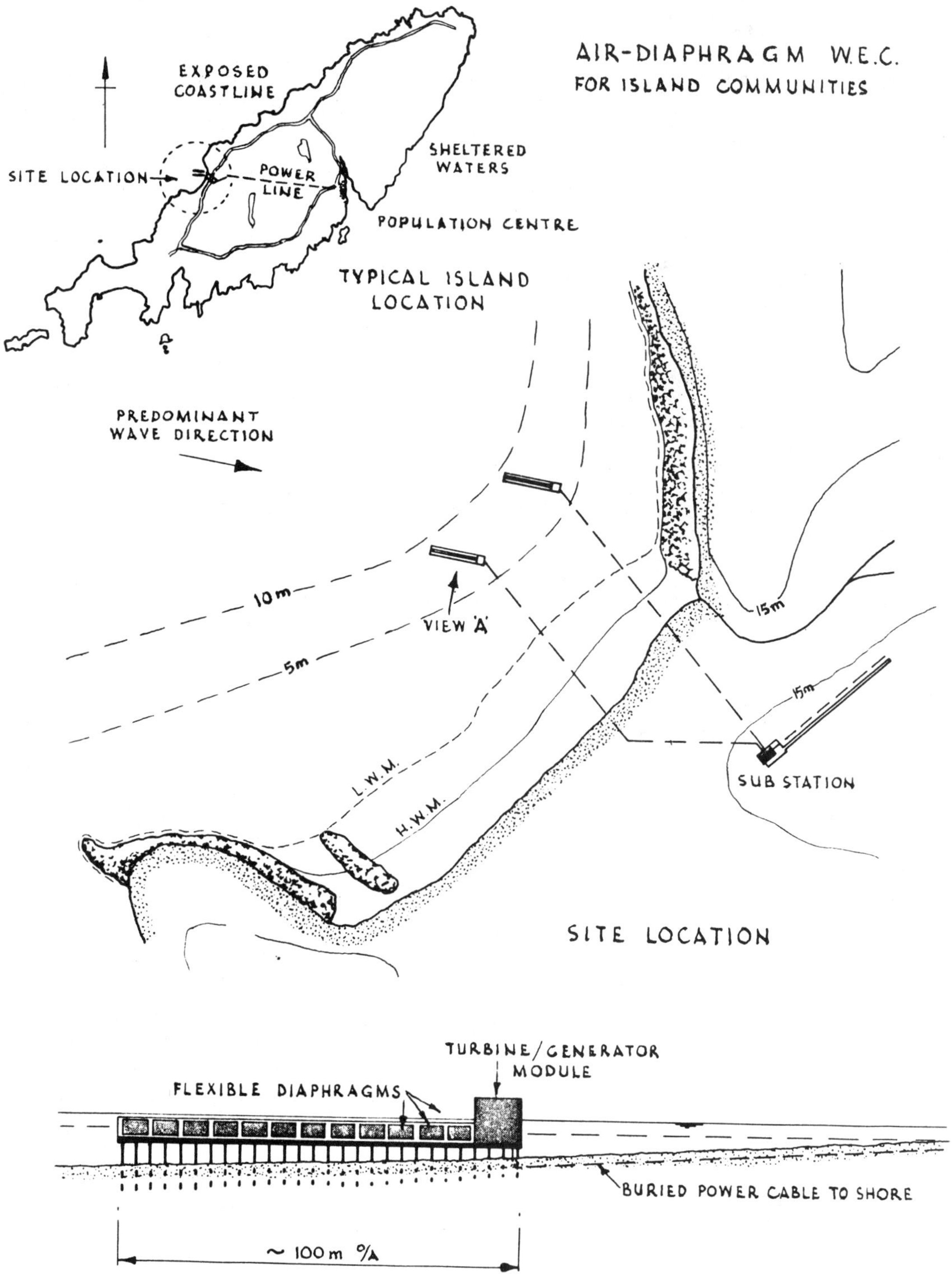
AIR-DIAPHRAGM W.E.C.
FOR ISLAND COMMUNITIES
EXPOSED COASTLINE
SITE LOCATION
POWER LINE
SHELTERED WATERS
POPULATION CENTRE
TYPICAL ISLAND LOCATION
PREDOMINANT WAVE DIRECTION
10m
5m
VIEW 'A'
15m
L.W.M.
H.W.M.
SUB STATION
SITE LOCATION
TURBINE/GENERATOR MODULE
FLEXIBLE DIAPHRAGMS
BURIED POWER CABLE TO SHORE
~ 100m O/A

VIEW 'A'

FIG. 2

than is the peaky power which would be generated from a larger number of directly coupled units.

Size and Layout of Typical Installation

A useful typical installation for a small island community is thought to lie in the region 0.2 - 0.4MW mean annual average for which a rated output of 0.75 - 1.5MW might be required. Flexible diaphragm modules are arranged each side of central spines, having a total overall length to suit the required power level. Each 100M length has its own separate ducting system to a separate turbine and generator mounted at the landward end of the structure. Foundations, dependent on local conditions, would be piled and built outwards from the shore as finger piers using plant mounted on previously driven piles as in conventional jetty construction.

A suitable location has to have reasonable landward access and yet not be unduly sheltered from the open sea. Clearly, if an existing harbour location is chosen for development, there may be a high price to pay for the likely wave attenuation as compared to the open sea. However, this may be more than recouped because of the relative ease of construction and good access. No so obviously perhaps is the likely unsuitability of headland locations because of usually non-existent landward access. This leaves the intermediate areas between the two extremes being the most likely location for a wave energy power station that does not make use of existing structure and yet takes due account of the access problems both in connection with construction and maintenance.

For useful amounts of power, the minimum depth of water required is in the region of 10 metres with diaphragms extending from water level down 5 metres. Individual diaphragm modules would be in proportion 5 metres deep by 7 metres long. Except for the turbine and generator modules, the whole installation would be inundated at high water but would break surface at low water.

Figure 2 shows a general scheme layout for a typical location. This type of layout is suited to locations where the seabed is gently sloping and bedrock is at some depth.

In typical Pacific locations where great water depths can be very close inshore there is an alternative layout running along shorewise which is more appropriate.

Components of Air Diaphragm Wave Energy Converter (WEC)

The foundations and primary ductwork are likely to be constructed in concrete in deference to the power of the sea (where self weight has its own benefits) and in deference to the long term corrosion effects of the sea water environment. Detail design of these items will be directed towards maximising local construction and minimising imported materials. In addition to saving cost, this should improve employment prospects during construction and teach local skills which might have applications for maintenance and other purposes. Relatively light and easily handleable units incorporating diaphragm, back panel and valve assembly, would be manufactured centrally and imported as a standard repeatable module and used as a former for casting concrete to form a larger element for at sea installation.

Necessarily, the turbine generator sets together with control equipment would be of specialist manufacture and imported to site as a complete unit for installation.

The Next Step

Wavepower Limited, together with their associated Consulting Engineers, Gifford and Partners, having established in fair detail the offshore concept for a 2GW station using LFB WEC's, are now at the feasibility stage for the smaller scale onshore Air Diaphragm version described.

We hope shortly to obtain funding for carrying the work forward to a specific design study followed by a pilot construction scheme.

REFERENCE

Barbour, D., and Twidell, J.W. (1981) "Energy Use on the Island of North Ronaldsay, Orkney". In Proceedings of the Conference on Energy for Rural and Island Communities, Inverness 1980 (Ed. Twidell, J.W.), 39-52 Pergamon Press, Oxford.

ARTISTS IMPRESSION OF AIR-DIAPHRAGM W.E.C.S FOR ISLAND COMMUNITIES FIG. 3

TOPIC D

Hydro and Tidal Flow

Small Scale Hydro Electric Power

E. E. Francis

Department of Energy, UK

ABSTRACT

The results of a survey of the exploitable small scale hydro electric resource potential in Wales are presented together with the main factors and examples of the economics of developing this resource. Demonstration schemes are being prepared to promote this resource and the main conditions and criteria for such schemes are outlined.

KEYWORDS

Hydroelectricity; resource size and distribution; economic appraisal; site characteristics; capital costs.

INTRODUCTION

Small scale hydro electric power is seen as the development of the potential energy available in a river or stream. Normally these developments are seen as run-of-river schemes in which the natural or regulated flow of water passing through a drop in topography is utilised. For natural water flows the output of the development will be highly dependent upon weather patterns and the local hydrology. For regulated water flows the energy output is more predictable but now dependent upon the water releases associated with the management of the river systems. Schemes can range from a few kilowatts to a few megawatts, although in the UK these are unlikely to exceed 10 MW.

This report does not consider the specific hydro schemes of the type built by the North of Scotland Hydro Electric Board or the Central Electricity Generating Board. These invariably mean the building of a dam to smooth seasonal water flows, and although they are capable of producing larger capacities and energy outputs, the capital civil costs and the impact on the environment are likely to make this type of development less attractive.

RESOURCE SIZE

<u>Resource potential of Wales</u>. The theoretical gross river potential can never be

fully exploited after account is taken of the other river uses and the developments along its banks. Recognising that the practical potential is only a small proportion of the theoretical maximum, the Department of Energy (DEN) commissioned Salford University in 1980 to establish the exploitable technical potential of Wales on a site basis. Each site considered took account of:

(i) the impact on the amenity and environment eg scenic waterfalls were not deprived of water, where new weirs were required these were always taken to be less than bank high;

(ii) the ease of access for construction eg no new roads were considered;

(iii) the topography of the site, eg gently sloping valleys requiring long leats were not considered;

(iv) the engineering difficulties eg where very large water volumes were possible only a proportion of the flow was considered.

None of the sites considered took account of their economics and the application of the energy output.

Each site was visited to establish the above factors and the head of water available. The capacity and energy output of a site were then calculated using Regional water flow duration curves and equipment efficiencies. Each site was considered either to be capable of development directly in the river channel (eg at a weir) and 1.25 times the mean river flow was used to establish capacity output, or the development would be via the use of a leat and penstock in which case the mean flow or some proportion dependent upon (i) and (iv) above was used to calculate capacity.

The above methodology implies that each site has been conservatively assessed. In total over 560 sites are capable of being technically exploited whose capacity exceeds 25 kW. These sites could give a total capacity of about 70 MW and produce about 340 GWh of electrical energy each year or save the equivalent of 85,000 tonnes of oil.

The distribution by size and the total capacity and average energy output by size is presented in Table 1. It is clear that the number of sites rises rapidly if smaller site sizes are considered, many hundreds of additional sites would result if a lower limit of 10 kW had been considered.

TABLE 1: Distribution of Sites, Capacity and Annual Energy by Site Size

	Site Size - Kw					
	25-49	50-99	100-199	200-499	500-999	1000 +
No. of sites	210	160	112	58	18	4
Total Capacity MW	7.8	11.5	14.2	17.7	13.7	7.5
Total annual energy GWh	36.0	52.5	66.0	94.0	57.0	33.0

It is equally clear from the table that although the number of smaller sites is large their contribution to total capacity and energy is not great. 210 small sites (37%) contribute only 11% of the capacity and energy outputs. The bulk of the capacity (44%) and the energy (47%) will be generated at sites in the size range 100-500 kW.

Tables 2 and 3 show the distribution of sites through the range of heads and water flow rates found in the study. These simple distributions showing more sites with low heads and low flow rates, continue even when the smaller sites are excluded. 25% of sites have heads less then 4m and 45% of sites have flow rates less than 0.4 cumecs.

TABLE 2: Distribution of Sites with Varying Head of Water

	Head of water (m)								
	<4	4-8	8-16	16-24	24-32	32-40	40-60	60-100	>100
All sites	143	67	114	35	64	27	49	48	15
All sites >50kW	92	38	51	15	40	22	35	46	13
All sites >100kW	52	20	24	4	15	12	23	27	11

TABLE 3: Distribution of Sites with Varying Water Flow Rates

	Water Flow Rate (cumecs)										
	<.4	.4-.8	.8-1.2	1.2-2	2-3	3-4	4-5	5-10	10-20	20-30	>30
All sites	239	85	34	58	45	20	13	32	9	10	7
All sites > 50 kW	113	57	19	37	28	17	13	31	9	10	7
All sites >100 kW	43	27	15	13	16	12	7	23	9	10	7

Further work. There are undoubtedly many other areas of the UK with the potential to develop small scale hydroelectric resources. The most obvious areas are those with a tradition for the use of this resource, SW England, Northern England and Scotland.

It is thought that the potential exploitable energy available in the English areas will be comparable to that of Wales. The average size of installation is likely to be smaller due to the lower topography, although this will be compensated by the larger land mass. Many of the factors effecting the development of this resource are the same as experienced in Wales; it is not therefore proposed to survey the English potential resource size at this point in time.

The conditions in Scotland are, however, somewhat different including the climate and topography, the existing extensive development of this resource by the Electricity Boards, the organisation of the water industry, the sparse population and access to potential sites. For these reasons the DEN is currently considering the possibility of surveying the Scottish exploitable potential for small scale hydro electric power. It is expected that such a survey would produce energy estimates substantially greater than the Welsh study.

ECONOMICS

Background. The economics of a small scale hydro electric installation are largely determined by the capital costs of the plant and the value attached to the usable energy output; the operating costs form only a minor component.

The capital cost of an installation is dependent upon the size of the plant, the type of plant specified and the particular site conditions. In general the larger unit sizes and the higher heads of water will give the lowest plant costs. The type of plant employed can vary from the simplest system designed to take a constant water flow with little control, up to the most elaborate system having full output and frequency control. The site conditions will set the costs of the civil works; at sites where these largely exist these might be as low as ⅓ of the plant costs, whilst at virgin sites the civil costs might be 5 times the plant costs. In addition the output has to be connected to a demand centre requiring suitable protection equipment. Lastly an allowance must be made for project engineering costs.

The value attached to the energy output is dependent upon the use to which it is put, for example it could replace a diesel generator on a farm or supplement a firm's supplies from the electricity authority or export direct to the electricity network. The value of the energy is either the cost of the competitive alternative energy supplies if it is used directly or the income received from the electricity authority if exported to them.

In general the economics of a hydro scheme will be the most attractive if:

(i) the water civil works already largely exist;

(ii) the energy is directly used by the investor;

(iii) all the energy produced can be usefully used throughout the year.

Examples of Small Scale Hydro Electric Installations. The above general observations are best illustrated by considering a number of examples; three have been detailed in Table 4. It will be noted that the capital costs range from £1,000/kW to £1,200/kW reflecting the size of plant, site conditions and head available. Example C assumes the use of an induction generator, while examples A and B are assumed to operate in a stand-along mode. All three examples have the civil works largely available. High load factors are specified in these examples; the conditions of example B imply a degree of energy wastage as the design output is based on less than mean flow conditions in order to achieve a reliable output throughout most of the year. Example A assumes a degree of load management, balancing supply with the different types of demand, eg normal domestic requirements, freezers, water and space heating. The operating costs vary from £10/kW pa to £13/kW pa reflecting the degree of in-house maintenance that can be utilised.

The results of the economic analysis of these examples are expressed either as a Benefit/Cost ratio for a Discount Rate of 8% or as a real percentage return on the investment. These have been presented for a benefit rate remaining constant in real terms over time and for a benefit rate rising in real terms as energy prices increase. The Benefit/Cost ratio represents the present value of the net

TABLE 4: Examples of Small Scale Hydro Electric Power, Site Characteristics and Economic Analysis

	Example A		Example B		Example C	
Site Conditions	Existing weir adjacent to premises		Existing weir 300m from demand centre. New penstock req.		Existing weir and leat 500m from network	
Head m	3		40		8	
Flow (mean/design) cumecs	.30/.40		.30/.24		2.20/2.20	
Capacity kW	10		80		150	
Energy pa kWh	44,000		420,000		750,000	
Load Factor %	50		60		57	
Application	Domestic supplies including storage space and water heating		Commercial application replacing oil heating		Direct export to Electricity Network	
Benefit rate	3p/kWh		Oil at 41p/th		Capacity at £23/kW Energy at 1.84p/kWh	
Capital Costs £	10,000		90,000		180,000	
Operating Costs £pa	100		700		2,000	
Annual Benefit £pa	1320	1320* rising	9,000	9,000⌿ rising	17,300	17,300⊛ rising
Benefit/cost ratio @ 8% Discount rate	1.45	1.86*	1.10	1.86⌿	1.0	1.25⊛
Real Return on Investment %	12	15*	9	15⌿	8	10⊛
- 20% reduction cf	9	12*	7	12⌿	6½	8⊛
- 20% increase in capital costs	10	13*	7	13⌿	6½	8⊛

* Electricity prices assumed to rise in real terms at 2% pa
⌿ Oil prices assumed to rise in real terms at 4% pa
⊛ Energy tariff assumed to rise in real terms at 2% pa.

annual income/savings over the life of the plant (40 years) divided by the capital costs of the investment. The real return on the investment is that interest rate that just balances the capital investment with the net present value of the lifetime benefits.

For the assumptions used the results suggest that all three examples are economic investments making real returns of 8% or more. (Real returns for many investments over the past 8 years have been negative or close to zero.) With the expectation that electricity and oil prices will rise in real terms, significantly higher returns will be achieved.

A reduction in the load factor of 20%, this could be the result of installing too high a capacity or not fully utilising the energy available, reduces the return on investment by about 2-3% points. An increase in capital costs of 20%, due to difficult site conditions, would have a similar effect. If the water civil works have to be extensively repaired or built at a virgin site then the capital costs could be significantly higher with a consequential effect on the economics.

DEMONSTRATION SCHEMES

The results of analyses of the practical potential and the economics of small scale hydro power led to the following announcement in December 1980 by Mr John Moore, Parliamentary Under Secretary of State for Energy.

> "The economics of small scale hydro electric schemes look sufficiently promising for me to authorise the identification of specific schemes to establish their economics and their suitability for demonstration of the latest technology."

Salford University have now been commissioned by the DEN to identify a number of sites to demonstrate the following aspects of small scale hydro power:-

(i) that existing technology is economically viable in commercial sized plant and that this resource can make a worthwhile contribution to energy supplies and thereby save fossil fuel;

(ii) realistic cost data and production costs and to disseminate this data to instill confidence in potential investors;

(iii) that the institutional problems can be overcome;

(iv) that small hydro resource utilisation is compatible with other river uses;

(v) the advantages of using the latest technological developments and designs;

(vi) that this energy resource can be integrated into the existing energy infrastructure for a number of different applications.

The Demonstration schemes are to be identified to take account of the following interlinked criteria:

(i) the type of plant available, both conventional and unconventional; the objective being to reduce capital costs wherever possible and to maintain or improve energy output and safety criteria;

(ii) the size of installations including variations in head, flow and different types of site conditions;

(iii) the method of utilising the energy output eg domestic, agriculture, industrial or community applications;

(iv) the use of different types of turbines;

(v) a geographical distribution such that those areas most likely to have hydro potential can demonstrate this resource.

The intention is to identify about 6-8 sites that between them meet the above objectives and criteria. Those schemes selected will be detailed up to outline design specification, budget costs and economic analysis and will be considered for grant aid from the Energy Conservation Demonstration Project Scheme. In order to achieve support from this scheme the following aspects will be considered; the national energy saving potential if all similar applications are adopted, the national cost effectiveness of the scheme to save fossil fuel, the economic viability of the project and the degree of replicability. Support approval could then lead to a cash grant of up to 25% of the capital and installation costs plus up to 100% of any necessary expenses incurred directly as a result of the demonstration eg the cost of monitoring energy savings.

The DEN is wishing to encourage the development of this resource through these demonstration schemes. Mr Moore announced:

> "although the UK cannot look to this resource to make a large contribution to energy supplies, it can however make a worthwhile contribution. Small scale hydro electric power is re-emerging as an energy source which has some attractive characteristics; it can be cost effective, it can have little impact on the environment and it is a secure localised energy source that will save the use of fossil fuel."

ACKNOWLEDGEMENTS

This paper is published with the permission of the UK Department of Energy, althouth the views expressed are the personal opinions of the author.

REFERENCES

Francis, E. E. Small Scale Hydro Electric Development in England and Wales. Paper presented to the Third International Conference on Future Energy Concepts January 1981, Institution of Electrical Engineers, London.

The Development of a Range of Small Water Turbines

P. W. Agnew

Department of Mechanical Engineering, University of Glasgow, UK

ABSTRACT

This paper describes the development of a range of small low-priced propeller turbines, suitable for developing a few kW of power on sites with up to 6m head. The available sites include many thousands of old water mill sites: in many places weirs and other hydraulic works still exist.

The main problem was that of obtaining financial support: for lack of money it was necessary to go straight from the drawing board to the site testing stage, without any laboratory testing and development work. However, the problems have been overcome and the units are now on the market.

KEYWORDS

Hydro-electric; propeller turbine; small scale; development; low cost.

INTRODUCTION

A number of events in the early 1970's caused a revival of interest in "Alternative" OR "Unconvential" sources of energy. First, a number of analyses were published, showing that our consumption of oil could not go on increasing at the current rate for much longer, because the world oil reserves are strictly limited. Then there was the Arab-Israeli War of 1973, with the subsequent rise in the price of oil: and in this country there was a succession of strikes by miners, power station workers and others which showed how dependent we are on our centralised and monopolistic energy suppliers.

There was a revival of interest in wind power. Wind power had once been importand - at one time there were many thousands of windmills in the South and Midlands of England - and there had been a brief revival of interest during the fuel shortage of the late 1940's and early 50's. However, in the days when people had to depend on wind and water power, they always preferred water power if it was available, because of its greater reliability. For instance, there were never very many windmills in Scotland, though we have plenty of wind: because in most districts there was enough water power for the purposes for which power was required. It therefore seemed to me that too much attention was being given to wind, and not enough to water power: particularly to small scale water power. Britain is not a land of high mountains and great rivers, so that the scope for large scale hydro-electric schemes is strictly limited: rather it is a land of hills and small streams. Because of our damp, mild climate, even quite a small river runs all the year and, without drying up in summer or freezing in winter. Once these little rivers drove inumerable water mills, but nowadays for the most part the power is allowed to run to waste. However, many of the old weirs and other hydraulic works are still in existence.

The first action I took was to build a very small turbine to run on the little stream running down in front of the house where I was then living. It drove a car-type alternator, to charge a battery. Unfortunately I had made one of the classic mistakes that once made can never be rectified. I had overestimated the flow in the stream, with the result that the system (designed to provide emergency lighting only) never did work properly.

At the same time I wrote a magazine article, which was published in October 1974. It extolled the virtues of water power, and invited people who thought that they had suitable sites to write to me. I thought at that time that it should only be necessary to find the available machinery, and the available sites, and to fit them together: but it soon became clear that they were a very bad fit. The power available at a water power site is proportioned to the head (the verticle fall from the surface of the head pond to that of the tail race) multiplied by the flow rate. A site with a small available power, if it has a high head, must therefore have a very small rate of flow: one with a large rate of flow must have a very low head. A high head and a small flow means a small hill stream: these often dry up in summer, sometimes for months at a time, and therefore do not provide a reliable source of power unless there is a site where a reservoir can be economically built. Unfortunately, such natural reservoir sites are rare.

A site with a low head and a relatively large flow, is in practice a drop in level on a river, caused either by a natural obstruction such as a ledge of rock or by a dam or weir. Such a site can provide a much more reliable source of power. Many of the sites used to be occupied by water mills: an overshot water wheel requires a head slightly greater than the diameter of the wheel, and the majority of wheels were 3.5 - 4.5m (12-15ft) diameter, wheels larger that that being very expensive. There are therefore a large number of old water wheel sites with up to 6m head.

When water turbines first appeared, many of the old water wheels were still in use and the turbine makers tended to concentrate on machines for heads too high for water wheels to use effectively. Also, a machine that runs fast enough on a low head to drive an electrical generator, must be a machine of high "Specific Speed" which means in practice a propeller turbine. The first propeller turbine of modern type was developed by Kaplan in what is now Czechoslovakia at the time of the first world war, and the first Bulb and Rim Generator turbines were installed in Nazi Germany. Partly for this reason, perhaps, propeller turbines have been slow to catch on in this country: the first Kaplan machines were not installed here until the 1950's, and the 2 small machines in the Awe Barrage are still the only bulb turbines in the U.K.

The demand appeared to be for machines to develop a few kilowatts on heads of from 2.5 to 6 metres or thereabouts. To cover the range of available sites, there would have to be a range of different sized machines: and each machine would have to be capable of operating on the full range of heads. It was essential to keep the cost down: and to this end it was going to be necessary to use standard quantity produced generators. In order to allow the use of standard appliances, a.c. was required at 50Hz. Now small standard alternators run at 3,000, 1,500 or

1,000 r.p.m. If a standard turbine is used on a range of heads, its optimum running speed is proportional to the square root of the head: it therefore needs to be possible to fit a range of drive ratios between the turbine and the generator. This is most conveniently done by using a belt drive.

It was, at first, intended to make a turbine somewhat similar to a Kaplan, but without guide vanes: the water was to enter tangentially into a "Vortex Chamber" and then discharge axially through the throat, in which would be the propeller. However, a small company that had shown interest in the project offered to make a prototype, provided that they could re-design and simplify the casing. The result was as sketched in Fig. 1.

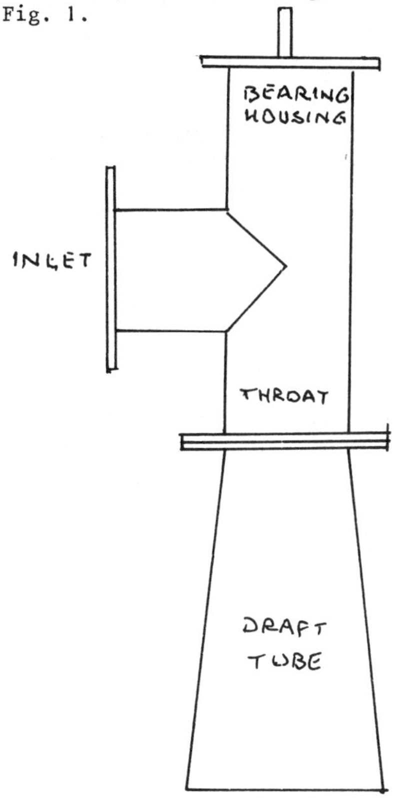

Fig. 1.

This was not a good turbine: the casing was not good hydraulically and the "throat" was not accurately circular so that it was impossible to maintain a small tip clearance. But beggars can't be choosers, and I was grateful to have anything at all for experimental purposes. It was used with a variety of propellers in a very simple laboratory test rig for a series of students projects, and at least it was possible to get some idea of what the problems are.

At this stage I sketched a design for what I considered to be a practicable turbine and tried to find some person or organisation that was prepared either to make it or to back it financially. It was assumed that the job would be done properly: that is to say, that as well as building a prototype turbine we would build an improved laboratory test rig, and that the development work would be done with the help of the test rig, before testing a turbine on site.

It proved impossible to get financial support for this proposal. It was realised that the cost of development might be reduced by cutting out the laboratory testing, and going straight to tests on site, but for this it was necessary to have a site.

I had for some time been considering buying an old house or cottage, with the idea that I might profitably use some of my spare time and money in reconditioning it, possibly for use as a holiday home. I now decided to buy an old water mill instead.

At much the same time the H.I.D.B. expressed interest in the project, and put me in touch with McKellar Engineering Ltd. The first prototype had the axis of the propeller vertical, and differed in various minor ways from the present design. The hydraulic design of the casing was much better than the original prototype, but was still not very good. The first trials were made with a fixed pitch propeller, and were reasonably successful. Later an adjustable pitch propeller was fitted: apart from some leakage of water into the mechanism, this was also reasonably successful. The adjustable pitch propeller on the 200mm turbine - the propeller dia. being 200mm and the hub dia. 90mm - is believed to be the smallest adjustable pitch turbine propeller in service. In order to change the pitch, the machine is stopped, and the large nut on top of the turbine shaft is turned as required.

These machines are normally used with electronic load governors. The propeller pitch is set so as to obtain as much power as possible from the available flow of water, and the electronic governor keeps the voltage (or frequency) constant by adjusting the amount of power fed to a "ballast load". (Agnew 1981) The turbo-generator with variable pitch propeller costs about 10% more than the unit with fixed pitch.

There were various defects in this prototype, and it was decided to build an improved version. This has an improved casing, giving an increase in power from about 3 to 5kW in the 200mm machine on 6m head. The shaft is inclined at 45° instead of being vertical. This was originally done to facilitate the use of oil-lubricated bearings, but it was found to bring other advantages, so that it has been perpetuated despite the fact that sealed ball bearings are now being used. I did not see the illustration shown in Fig. 2* until after the design had been modified in this way. Though it shows a much larger machine, there is a certain resemblance. One major difference is that our turbine has no guide vanes: there appears to be no precedent for a propeller turbine without guide vanes, but Fig. 3 shows a pump with variable pitch propeller but no guide vanes (Stepanoff, 1948). For almost every type of turbine, except the Pelton wheel, there is a corresponding pump of similar design.

Fig. 4 is an outline of the machine at this stage of development. The object on the end of the generator shaft is the overspeed trip gear. This took the form of a clutch, which de-clutched the generator if its speed exceeded a set value. The intention was that the generator would then come to rest, and that the turbine would come to no harm from overspeeding. However, there were two problems. The first was with the clutch itself: a single phase alternator (which is what we were using) does not have a constant torque reaction, but one that oscillates at twice the electrical frequency. It was found that unless the clutch was very carefully made and fitted, it rattled alarmingly and showed signs of premature wear. The fine limits necessary to prevent this made the unit expensive. The second problem was that if the turbine was allowed to overspeed, it ran into a critical speed, which caused the propeller to hammer loudly on the casing. While this was never allowed to continue long enough to cause damage, it was felt that the system was not satisfactory.

* (Pfafflin 1978)

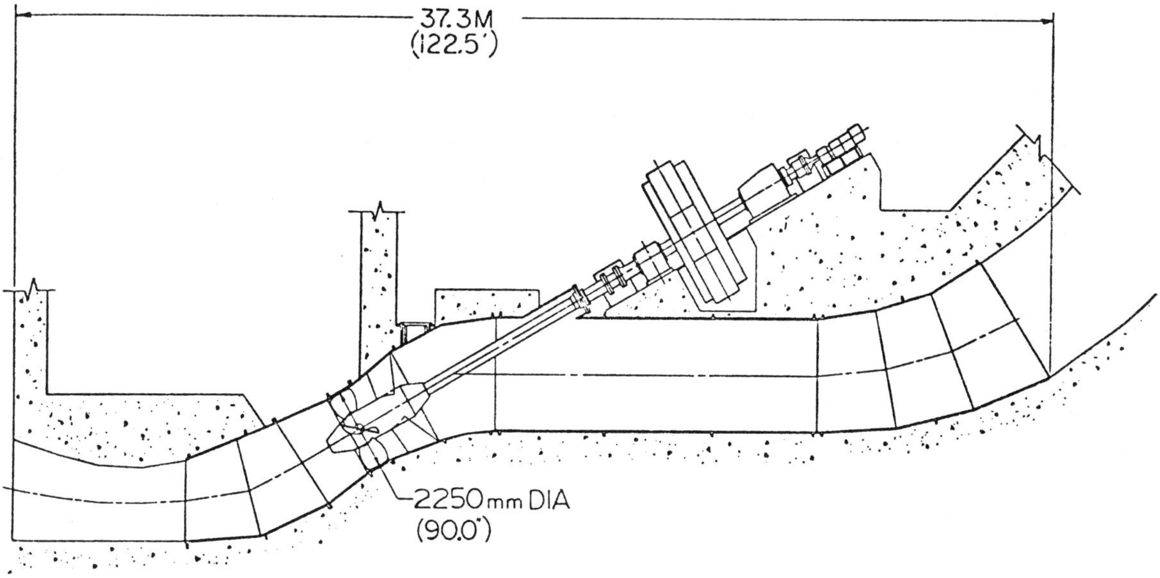

Tube turbine installation, Dolby Hydro Station, 1976/77. 2-4,200 kw, 14.m. (48') head.

Fig. 2.

The overspeed trip gear has therefore now been modified to close the butterfly valve: the chief problem in practice having been to obtain a sufficiently slow and well controlled motion of the valve. It is believed that the problems have now been overcome.

The original purpose of the development has been fulfilled, to the extent that we now have a range of turbo-generators capable of producing up to 30kW on heads up to 6m at a very competitive price. A unit is now being tested by the National Engineering Laboratory on behalf of the Overseas Development Administration, with a view to using these machines to develop power from irrigation canals in certain developing countries.

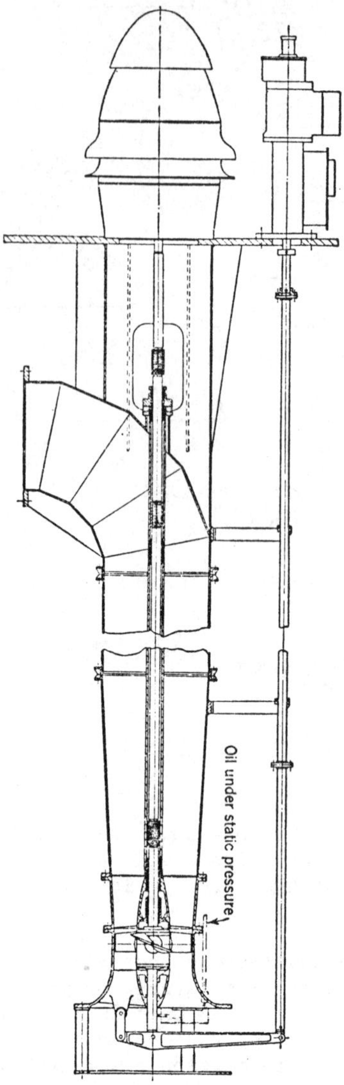

Twenty-four-inch Ingersoll-Rand adjustable impeller vane axial flow pump at Central Illinois Electric and Gas Company; 22,000 g.p.m., 24.8 ft. head, 880 r.p.m.

Fig. 3.

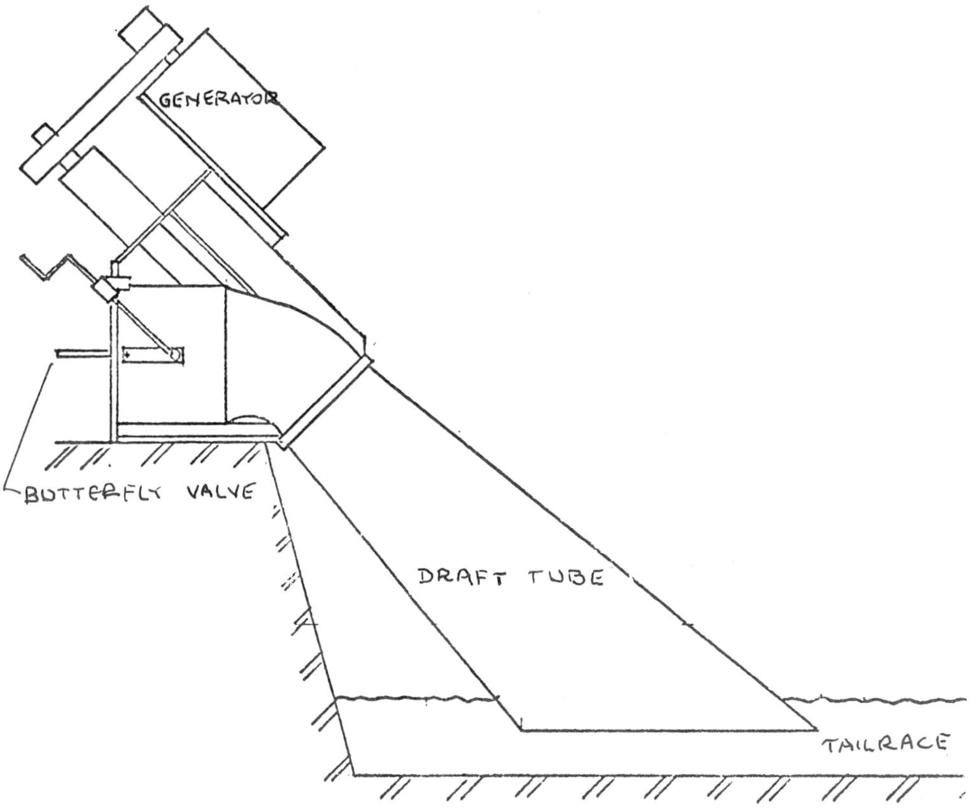

Fig. 4.

REFERENCES

Agnew P.W. (1981). "Appropriate control systems for water turbines". Energy for Rural and Island Communities, In J. Twidell (Ed.), Pergamon Press.

Pfafflin, G.E. (1978). "Tube Turbines". Low Head Hydro, In Gladwell & Warwick (Ed.), Idaho Water Resources Institute, Moscow, Idaho.

Stepanoff, A.J. (1948). Centrifugal and Axial Flow Pumps. John Wiley & Sons, New York.

Micro Hydro in the Everest Region — a Measure Against Deforestation

M. T. Crawford

Department of National Parks and Wildlife Conservation, Katmandu, Nepal, India

ABSTRACT

This project is designed to make the most appropriate alternative energy source to firewood available to rural communities in Nepal. It is designed to promote total indigenous production of micro hydro electric units through development of available technology and expertise and the development of production technology. Low cost, and therefore availability to communities where income is low is ensured by rationalising the design technology and making full use of local available expertise for installation. Siting the small units in or nearby centres of population keeps transmission costs and problems and environmental impact to a minimum. Low initial cost reduces the need for feasibility studies.

Over the year term of the project a minimum 120kW will be installed in 16 small schemes providing electricity for 100 households. This could save 10950 trees annually or 547.5 cubic metres of firewood annually. During this period the personnel trained in installation and maintenance will take the technology developed in the Everest region to other areas of need.

This is believed to be the best available answer, at present, to help halt deforestation. It is relevant to all Third World countries.

KEYWORDS

Community self help; Appropriate low level technology; Development of production technology in an undeveloped country.

INTRODUCTION

Man and Environment

Man is the only animal who can create or destroy the environment. Twenty-five per cent of the world's population faces poverty or starvation. Man's struggle for survival on a hand to mouth basis at subsistence level is a reality in the Third World.

THE EFFECTS OF TOURISM

In the Everest region of Nepal, prior to the opening of Nepal to the West, the Sherpas lived in harmony with their environment. Forest Guards (Shingo-Nawa) controlled the cutting of wood and enforced rigid regulations to conserve the forest. The region became dependant on mountaineering and tourist trade as China overran Tibet and halted all trade with Nepal and India. Previously this trade had provided the income to make the Sherpa communities viable in an area of uncompromising agricultural production problems. The care for wildlife and forest vanished as Western influence crippled Buddhist ideology.

Now approximately 6000 tourists and mountaineers visit the region annually. They and their entourage of guides and porters severely increase the burden on the forest. Everyday the Sherpa wood collectors must go further from the hearth. Every year endangered species of flora and fauna disappear from the area, or forever, as forest habitat is lost. The viability of the Sherpa communities is now threatened by the firewood crisis.

RESPONSIBILITY

Can the responsibility for conservation be expected to come from a country of 5% literacy where man's actions are the result of a fight for survival? Western developed man has not even put a value on ecologically sound environmentally conserving technologies, has taken little responsibility and made few attempts to make good the exploitation of his environment. Much interest has been taken in the conservation of the Everest region. Is the world's environment not our childrens' world heritage and our responsibility?

We in the West should now be aware that the rebuilding of habitat and ecosystems and renewing of the earth's environment can only be achieved by reforestation integrated with agriculture, to increase production, and by population control. The short term for environmental conservation and the energy crisis is Alternative Technology. It is exciting but the only long-term answer on a world wide scale is Reforestation.

THE FUTURE

'One hectare of forest is worth a thousand tranquillisers' (D.B. Crawford, 1981) is perhaps the beginning of real assessment of the value of the quality of the environment.

The next 20 years contain the greatest challenge in realising the value of the environment. A start could be made by quantitatively analysing through cost and benefit studies the real value of ecologically sound energy and food production practices.

Energy for rural communities should create freedom from ecologically unsound environmentally shattering practices.

ENERGY POLICY IN NEPAL

The Food and Agriculture Organisation estimates annual losses in Asia of 5 million hectares of forestry (Anon 1978). The price of 85% self sufficiency in energy production in Nepal is distilled, horrific environmental catastrophe. Life expectancy has doubled and there is no hope of family planning education in a country where tuberculosis is still endemic. In the next 20 years the population may double. Land productivity will be lowered and production land is lost due to deforestation resulting in loss of moisture and erosion.

In Nepal we are developing Alternative technologies to ease the burden on the forest as a source of energy. The impetus for the development of these alternative sources comes from individuals:

'Development and dissemination of new and renewable energy resources in Nepal to a great extent originate in the small workshops of private individuals. It can be noted that in many places trial and error methods bring quick and economical results. What is needed is information and also people with enthusiasm and skill.' (Bachmann, 1981).

'The success of appropriate low level technology as applicable to the Third World depends on a high level of commitment and involvement of the people who make installations possible, and the people the technology is installed for. The success in development of Appropriate technology will only be ensured by a concerted effort of all the people involved in this field.' (M. T. Crawford, 1981).

ALTERNATIVE ENERGY SOURCES FOR NEPAL AND THE EVEREST REGION, TO REPLACE THE USE OF FIREWOOD

Prerequisites include:

1 Low initial cost and low running cost.
2 Low maintenance requirements and cost.
3 Easy installation.
4 Simple technology comprehensible to rural undeveloped people.
5 To supply energy for cooking and heating and to increase agricultural productivity.

1
Burning dung as an alternative reduces the fertiliser value at a time when organic building up of land is required to increase production.

2
The cost of Kerosene in Namche Bazar may be four times the cost in Kathmandu due to portering. The supply is not reliable and basic costs rise prodigiously.

3
The best biomass alternative in Nepal is Biogas which also produces beneficial fertiliser. But, it requires a temperature of 30° C for fermentation; in the Himalayas it therefore requires integration with another alternative source of energy to produce this temperature, so at present it is not cost effective for this area.

4
Until heat storage cookers are further developed, Solar Collectors are suitable only for space and water heating therefore implementation is ineffective.

5
Windpower is not at all known or developed in Nepal.

6
Water Power

There are between 25,000 and 50,000 water mills in Nepal. Water power technology on a micro scale is well comprehended by rural communities. Micro Hydro Electric units can be simplified and through community involvement and the use of non-conventional technology the cost can be kept low. If the technology is simple little maintenance is required. Installation where traditional civil works and local expertise is integrated is simple and rapid. Micro Hydro Electric can supply energy for cooking and heating with the added extra of lighting. The electricity can be used directly to increase agricultural productivity and development by releasing labour from firewood collection and providing irrigation pumping and agricultural processing. More efficient use can be made of a Hydro Power unit by power drive directly to mechanical machinery from a unit which can also be used to produce electricity. Other benefits from development could be setting up of cottage industries and amenity power for village lighting, cultural, religious, education and health centres and drinking water schemes. Such a unit has been designed and is being developed. It could be the beginning of Nepalese self help for the future to utilise her vast water power resource:

Economically feasible Hydro Electric generation potential in the Himalayas is estimated to be 25,000 megawatts.

HYDRO ELECTRIC DEVELOPMENT

In the past large schemes have been proposed and some implemented. These schemes are now thought to be environmentally unsound and too expensive. The Himalayas are very young mountains and as such geologically unstable. Large schemes are necessarily sited on main rivers, subject to glacial melt and monsoon flooding. The civil works often irreparably silt up or are washed away. The potential consumers' villages are usually far away from these river gorges. The cost of transmission is often the most expensive single item in a scheme.

The problem of transportation of required materials and equipment can be almost incomprehensible. One proposed scheme in the Everest region requires 900 tons of cement for the civil works, to be portered for 15 days from the road head. This constitutes 39,600 porter loads and 594,000 man days. A total amount of 1782 cubic metres of wood will be required throughout the access corridor to meet the porters' energy needs for cooking. This constitutes enough wood to supply Namche Bazar for 3 years and 4 months. A scheme in Eastern Nepal requires only the alternator for completion. Completion may have to wait until a larger helicopter is developed.

Feasibility studies undertaken due to the size of investment involved in a Mini Hydro Electric scheme (between 100 and 1000kW), can constitute between 30% and 50% of the total cost.

The cost of feasibility studies has made schemes not feasible.

Ninety per cent of the electricity from these large schemes in the Third World is used for lighting - this only increases the use of firewood.

In the West permissable voltage drop and power loss for domestic use electricity is 2.5%; in Kathmandu, the capital of Nepal, from schemes some 10 years old the power and voltage drop is usually 40% and sometimes 60%. The electricity,which is available to a quarter of the consumers at one time, is only just suitable for standard electric light bulbs;

'The effect these large schemes have on development of the country is as a drop in the Ocean. All the technology and expertise is imported. The indigenously available expertise does not even rise to maintenance.' (M. T. Crawford, 1981).

Micro Hydro Electric schemes can follow the example of the traditional mills built by the people who after all know their environs best. The mills can be sited on small streams nearby or actually in villages. If more power is needed more units can be sited one after another on the way down a stream as with traditional mills.

COMMUNITY INVOLVEMENT

The Sherpa region and people through tourist and mountaineering income and involvement are well developed and finance is available in the community to support the project. The houses which provide the greatest amenities for tourists are the ones using the most wood. They are however the ones with the greatest interest in Alternative energy schemes and the ones with the most income available for purchase.

The project took a demonstration unit to Namche in June this year to assess feasibility. A community advisory committee consisting of members of the community and local government officials was set up. The Sherpa Sagarmatha National Park warden responsible to the Department of National Parks and Wildlife Conservation was chosen as project co-ordinator. Wood and a carpenter were provided by the community - the wooden penstock was built. (Although steel or an HDPVC penstock pipe could have been used, it was decided to pilot the use of wood as a measure to involve the community and reduce transport problems.) The unit was installed and commissioned within one week. Electric lighting and power unit were demonstrated to the people. Six other sites were surveyed. The objectives of the project were moulded to what the community thought and wanted and what was possible in the region.

Twenty-five households initially signed a paper supporting the project and saying they would like electricity in their households and would pay for it. Letters of support for the project were given by the staff of the National Park Head Quarters and the religious leader - the Head Lama, Reinpoche of Thyangboche.

Three households were selected by the community as those using the most fuel and with the greatest interest for the first household installation of electricity and appliances for cooking, heating and water heating from the second pilot unit.

FINANCE AND PROJECTION

The Department of National Parks and Wildlife Conservation as the responsible department of HM Government for Everest Region, Sagarmatha National Park and liaison for the project at present, have agreed to finance the Alternator and Transmission equipment for the Pilot scheme. Approximately £2500 has been made available by an international research centre to fund a socio-economic feasibility study, as relevant to areas of need in Nepal, of the project's second pilot scheme. The subsequent report will be available in November. It will also assess community involvement and the impact micro hydro can have on halting deforestation.

The Department propose that the second scheme should be installed for the Parks Department, Visitors Centre, Park Guards and staff quarters. This would benefit the community by substantially reducing the burden on forest resources and a power diversion system would be installed so that the power, when not used by the Park HQ kitchens, could be used for community amenity such as lighting and water pumping. The Department would finance this 7.5kW scheme totally.

The third scheme of 10kW would then be installed for households where much fuel is used at present in Namche. Again a diversion or load sharing system would be implemented enabling other kitchens to be powered as substantial excess kW hours would be available from a unit of this size.

At this point in Namche a grid system could be implemented with a total 20.5kW available. This could be enough to power 80 kitchens on a kWh basis. Although in an agrarian community the peak need periods of energy for all houses coincide - but this warrants further investigation to make full use of the power available.

Also at this stage it should be possible to establish a community fabrication, maintenance and repairs workshop. This is the level to which decentralising the technology is desirable as efforts should be concentrated in development of production technology requiring centralisation. It may also be desirable to set up the community amenity system, if not already implemented.

ECONOMIC EVALUATION AND POLICY

Training of Nepalese people to implement installations is vital. Foreign experts should not be required in the long term. The technology is simple and training should be rapid with careful selection of personnel and on the job training basis. Nepal should be made self sufficient in the required level of technology for total indigenous production of the units.

The Sirdars/trainees should be selected at community level. They should have a keen interest, be involved with their community well-being, have a practical aptitude and a will to succeed. They should be paid a standard wage for Nepal. In Kathmandu grants and loans for the purchase of parts and payment of expenses and labour should be made to people working in the relevant fields in production technology.

The cost per kW installed to the people is 5000/- NR (Nepalese currency - 21/-Rupees = £1 approximately) at this stage but as a funding body makes loans and the units available to other areas of need and the technology becomes mature, the full cost of 7000/- NR should be paid by the community.

The unit may last for 20 to 25 years but at this stage the unit kWh cost is calculated on a 10 years unit life basis plus maintenance.

An extra 200/- NR should be levied per installed kW per year for a maintenance contract with the workshop in Namche and for permanent maintenance and repairs staff. If 100kW is installed in the region during the project term then 20,000/- NR will be levied per year after the project term. This means that 10,000/- NR will be available for workshop, maintenance and 10,000/- NR for personnel per year. This is considered to be a realistic sum.

This means that after the year term of the project 680/- NR must be levied per year for maintenance per installed kW. One trained person should be retained in the Everest Region. The other trained personnel should be involved in installation and setting up similar project development programmes in other areas.

Cost per kW hour

$$\frac{\text{Unit Cost } 7000 + 10 \text{ years maintenance } 6800 = 13800}{\text{Total kW hours per 1kW unit available in 10 years} = 78642} = 0.17 \text{ NR/kWh}$$

MICRO HYDRO ELECTRIC IN THE EVEREST REGION A MEASURE AGAINST DEFORESTATION

Summary

Deforestation and the firewood/energy crisis are major problems in the Third World. In the last 25 years the area of forest in Nepal has been reduced by half. In the next 20 years the 14,000,000 population of Nepal is expected to double. In the next 15 years the Himalayas could become completely denuded of forest.

At present one household may expend 50% of its time and effort or income on energy requirements. Deforestation brings desertification, loss of agricultural production and land through erosion and loss of moisture.

Erosion of the Himalayas causes silting up of rivers producing severe flooding, famine and disease in the plains of Bangledesh, Pakistan and India.

In the next 20 years between 15% and 20% of all present species of plants and animals may be exterminated due to loss of habitat.

It is essential that an alternative source of energy which is easy to maintain and install and is of low cost be made available for installation now. Although the only long term answer for environmental conservation is reforestation.

The Everest region of Nepal has attracted much international conservation interest. But the paper writing must end and implementation of recommendations and findings must begin.

REFERENCES

1 BACHMANN, A. and SHAKYA, G.R. (1981). Development and Dissemination of Renewable Energy Resources in Nepal . Paper for International Solar Energy Congress, Brighton.

2 CRAWFORD, D.B. (1981). Forestry - Is Conflict Inevitable? Scottish Peat and Land Development Association, Land Use Planning - The Problems of Conflicting Interests.

3 CRAWFORD, M.T. (1981). Micro Hydro Electric in the Everest Region. A Measure Against Deforestation. Report on Project prepared in Nepal for Departments and Development Agencies.

4 HIS MAJESTY'S GOVERNMENT, MINISTRY OF WATER RESOURCES, NEPAL. (1981) Water - the Key to Nepal's Development . Resources Development.

5 UNIDO (1981). Mini Hydro Power Stations. A Manual for Decision Makers.Advance Edition. Development and Transfer of Technology Series.

6 WORLD BANK (1978). Forestry Sector Policy Paper.

The Hydro Heat Pump

A. W. K. MacGregor

Napier College, Colinton Road, Edinburgh EH10 5DT, Scotland, UK

ABSTRACT

Thispaper examines the principles and potential of a heat pump which is driven directly by a water turbine and which uses the flowing water as a heat source. It is concluded that there is a wide range of potential applications for the hydro heat pump in rural and island communities and that the cost-effectiveness should compare favourably with conventional small-scale hydro electric plant.

KEYWORDS

Heat pumps; hydro power

INTRODUCTION

Rapidly rising energy prices have revived interest in an energy device which apparently produces more energy than it consumes - the heat pump. In fact, the heat pump does not break any of the thermodynamic laws. The total amount of energy coming out is precisely equal to the total amount of energy which goes in. The heat pump's economic interest lies in the fact that a relatively small amount of commercial energy can be used to 'pump' a larger amount of low temperature ambient energy up to higher, and therefore more useable, temperatures. Typically, the ratio of useful heat output to the commercial energy input (the Coefficient of Performance) is in the range 2 - 5 .

PRINCIPLES

Most heat pumps are of the mechanical type with the compressor which re-compresses the working fluid (the refrigerant) being driven by an electric motor. The refrigerant is alternately evaporated and condensed in two separate heat exchangers, with the consequent capture and release of its latent heat. Because these two processes are arranged to take place at different pressures, and therefore different temperatures, the input of heat takes place at a lower temperature than the output. Hence the ability of the device to 'pump' heat from a lower to a higher temperature.

THE HYDRO HEAT PUMP

In the hydro heat pump the electric motor of the conventional heat pump is replaced by a water turbine and, by inserting a heat exchanger into the draft tube of the turbine, the water which has driven the turbine is used as the low temperature source of heat. In effect, the idea is to use both the hydraulic and the thermal energy of the stream, in contrast to conventional hydro electric plant which uses only the hydraulic energy.

ADVANTAGES

The particular advantages of the hydro driven heat pump are:

1. The energy potential of a given size of stream is multiplied several times.
2. Since both the water turbine and the heat pump are fluid machines, it should be relatively easy to match their characteristics and to control them. In particular, over-speeding under no-load conditions, a problem in hydro electric plant, should not be difficult to solve.
3. A large mass of moving water, with its high thermal capacity and excellent heat exchange properties, is an ideal heat source for a heat pump. Since the same water drives the compressor and acts as the heat source, the energy quantities are well matched.
4. Since there are no electrical parts, the whole system should be reasonably tolerant of difficult ambient conditions such as spray, wind and rain.

LIMITATIONS

The limitations of the hydro heat pump appear to be :

1. Since heat, not electricity, is produced the output can only be used for heating purposes. However, most of the energy requirements in buildings are for low grade heat so there is a wide potential range of applications. In addition, it is easier to store heat than electricity so there may be less difficulty in meeting a fluctuating load from a constant supply.
2. Heat is more difficult to transport than electricity so there is a limit on how far apart the hydro heat pump and the heat load may be located. Typically, heat outputs of several tens of kilowatts can be economically transferred up to several hundreds of metres. For larger heat outputs, in the order of several hundreds of kilowatts, the economic distance increases to several kilometres. The technology of heat transfer using hot water is now well proventhrough the development of district heating schemes in the Scandinavian countries and elsewhere.
3. Heat output is reduced to zero if the stream totally freezes.

APPLICATIONS

Typical potential applications of the hydro heat pump in rural and island communities are :

a. Space and water heating for individual buildings, groups of buildings (mini district heating schemes) , hospitals , schools and other institutional buildings.

b. Heat for agricultural and horticultural developments. This could include crop drying and heating of glass-houses.

c. Heat for fish farming and fish processing.

d. Heat for drying fuel such as timber or peat.

e. Cooling of buildings and refrigeration of produce in hot climates. This can easily be achieved in the hydro heat pump by reversing the flow of refrigerant so that instead of absorbing heat by cooling the stream, heat is dumped in the stream by heating the water which has passed through the turbine.

EXAMPLE

To illustrate the possibilities of using a hydro heat pump to provide space and water heating, consider an isolated house in the Highlands of Scotland presently heated by oil. The annual heating load for space and water heating might be typically 30,000 kWh. Assuming that there is a small stream nearby, the householder considers installing either a hydro electric plant or a hydro heat pump. To meet an average heating demand of about 4 kW using the hydro electric plant, the stream must have a hydraulic potential power of about 6 kW, for example a flow of 200 l/s and a head of 3m. By contrast, a hydro heat pump with the same heat output only requires a stream with 2 kW hydraulic power potential, for example 70 l/s and a head of 3m, assuming a coefficient of performance of 3. Not only is the cost of the hydro heat pump installation likely to be considerably less than the hydro electric plant, because of its much smaller size, but the probability of having a suitable stream available is much greater.

It should be noted that the climate of the Highlands and Islands of Scotland is particularly well suited to the use of heat pumps since the heating season is distinguished by its length rather than by its severity, giving a high annual load factor. Incidentally, this also applies to solar space heating.

DEVELOPMENT

A proposal to develop and test a hydro heat pump has been made to a manufacturer of small hydro electric plant in the Highlands of Scotland[1]. A suitable rotary compressor to replace the electrical alternator has been located. The characteristics of the compressor and its mechanical drive arrangements seem well suited to the water turbines being presently produced. Preliminary work on the design of the refrigerant-to-water heat exchanger indicates that the size of the exchanger can be quite compact without inducing freezing on the heat exchange surfaces. A suitable location for testing the combined system is now being sought.

REFERENCES

1. Mackellar Engineering, Granton on Spey, Morayshire, Scotland

Tidal Current Power Utilisation Using Moored Turbines

P. R. Cave and K. R. Welsh

Department of Mechanical Engineering, University of Dundee,
Dundee DD1 4HN, UK

ABSTRACT

The potential power output of open turbines installed beneath pontoons moored in tidal currents is discussed. Using the Tay Estuary as an example, the long and short term characteristics of tidal currents are discussed and methods developed for choosing a suitable site and estimating the power output of a moored turbine generator. It is concluded that the choice of site needs careful consideration. The design of a prototype turbine is described and limited test results presented. The desirability of flow acceleration schemes in the design of a turbine is discussed and some proposals presented.

KEYWORDS

Tidal power; tidal currents; tidal characteristics; Tay Estuary; moored turbine generators.

INTRODUCTION

Tides are of gravitational origin and unlike many other renewable sources of energy they are very predictable in magnitude and occurrence. Attempts to tap the energy of tides are usually based on the change in sea level and involve the construction of barrages. This approach is typified by the successful plant at Rance and by the installations proposed for the Severn Estuary and for the Bay of Fundy in Canada. A barrage, while yielding large amounts of energy, involves large scale civil engineering works and puts a high threshold on the minimum scheme that can be proposed. Where smaller amounts of energy are required, an alternative is to tap tidal currents, and a simple way of doing this is to use turbines installed beneath pontoons moored in the tideway. The pontoons align themselves with the flow and one or more would be used depending on the required output. Figure 1 is a schematic of the proposal. Advantages of this approach are that the strongest currents tend to occur in relatively sheltered waters such as estuaries, and also in shallower water near land. There need be no interference with navigation and the environmental impact of such a scheme should be small. When compared with alternatives such as wind power, tidal currents offer predictability and a higher power density, so that a commonly found 1 m/sec current corresponds in power output to a 9 m/sec wind for the same rotor area.

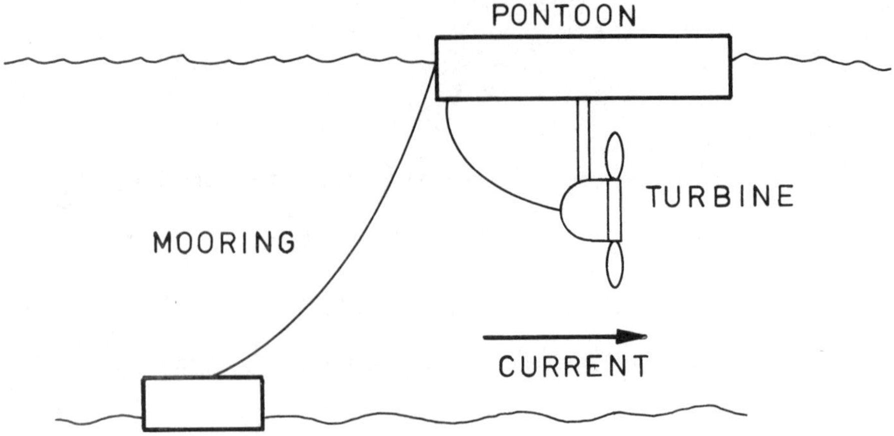

Fig. 1. Moored turbine generator

In order to assess the costs and problems of this approach the authors have chosen the Tay Estuary for investigation. This is a reasonably typical estuary and the methods used should be of general applicability. The tidal characteristics measured, however, are specific to the location.

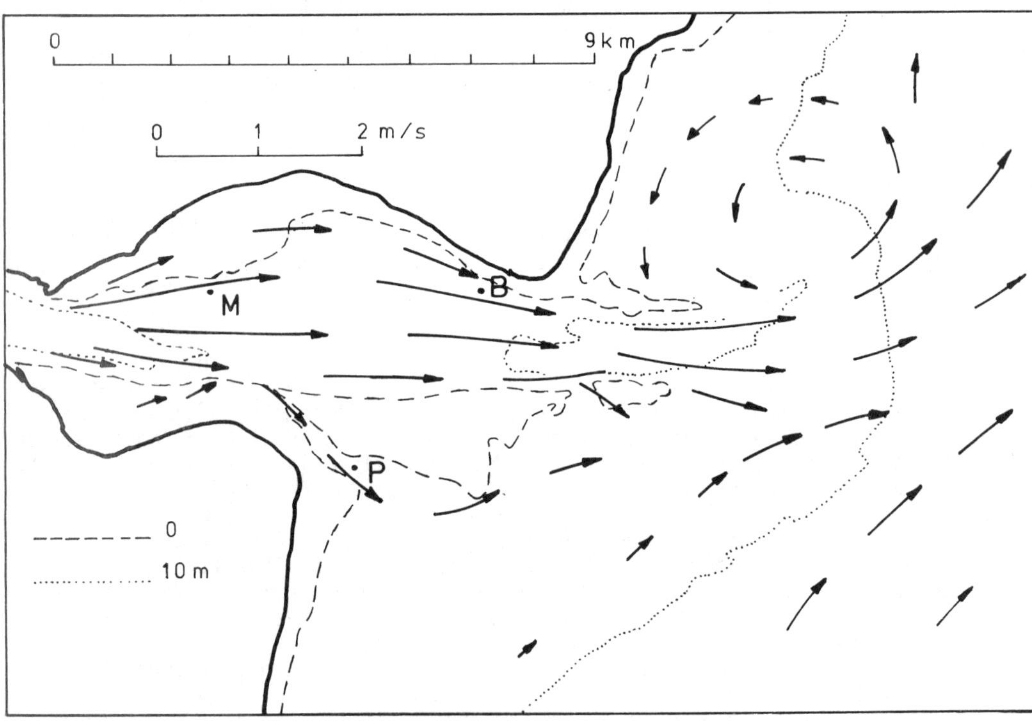

Fig. 2. Tidal currents of the outer Tay Estuary 4 hours after high water Dundee.

TIDAL CHARACTERISTICS OF THE TAY ESTUARY

The tidal characteristics of the Tay Estuary have been studied for a number of reasons including sedimentation and pollution dispersal. Much of this work has been done using the facilities of the Tay Estuary Research Centre (TERC), part of Dundee University, and reported by Charlton (1980,1981). The methods used were direct measurement using radar tracked float drogues and time lapse photography of the Dundee Port Authority tidal model of the Estuary. These measurements are used to produce hourly tidal charts of which Fig. 2 is a typical example. The current vectors shown on the charts are mean values. There are also short term variations most probably due to eddies and turbulence. In the Tay these have two components, a short term (5 second period) variation of up to ±20% and a longer term (5 minute period) variation of up to ±23%. An input varying with these frequencies would be a significant factor to be considered in the mechanical design of a power extrac--tion system.

POWER AVAILABILITY

To select a suitable site for a generator, the set of tidal charts like Fig. 2 are examined for locations where there are large current vectors, small slack periods and which are out of the navigation channel. Three such sites are apparent in the area covered by the chart. These are marked on Fig. 2 and are at Buddon Bay (B), Monifieth Bay (M) and Pool (P). Examination of the Admiralty chart of the area shows that Pool is rather shallow at low water and should be discarded, leaving B and M. Current magnitudes taken from the tidal atlases are plotted in Fig. 3. The curve for M is fairly typical of estuarine flow, with dwells at high and low water. That for B is less typical, the short slack periods being due to some peculiarity of the site.

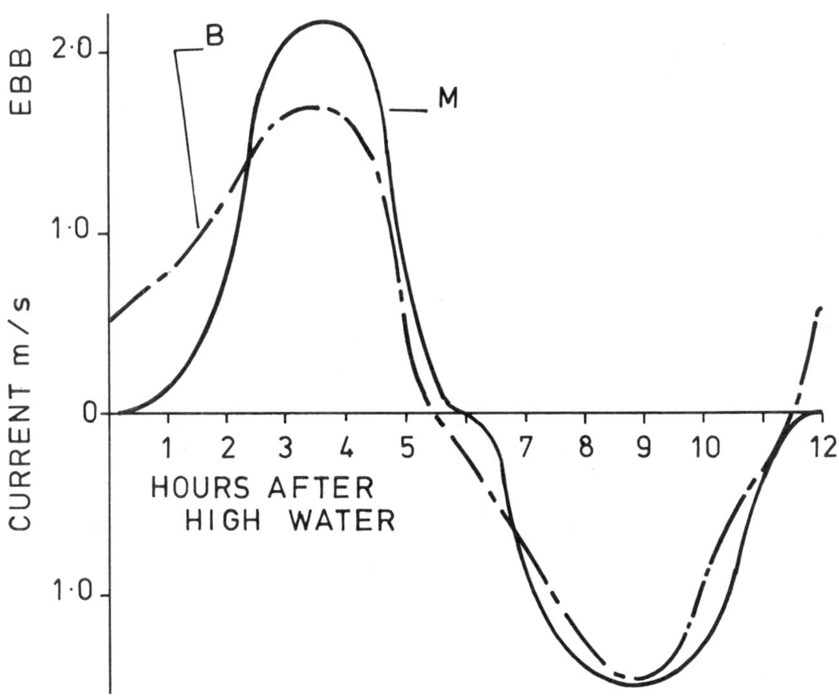

Fig. 3. Tidal currents (springs) at Buddon Bay (B) and Monifieth Bay (M).

Potential power output is given by the conventional equation:

$$P=\tfrac{1}{2}\eta A\rho V^3 \qquad (1)$$

where P is power output in KW, η is overall efficiency, A is rotor presented area, ρ is fluid density and V is the flow velocity. To obtain V, a sinusoidal variation from springs to neaps of 100%-60% is assumed, with 27 tides per 14 days. Rotor presented area is found by assuming a diameter of 4.0 m, which takes into account the low water depth at B and M with an allowance for waves. Density is taken as 1.02 and efficiency, for reasons discussed below, is assumed to be constant at 50%. With these assumptions the curves in Fig. 3 can be integrated over a 14 day period and an average output power calculated, together with a maximum power. Results are given in Table 1.

TABLE 1 Estimated Power Output

	Monifieth Bay (M)	Buddon Bay (B)
Max. current m/sec	2.16	1.70
Mean power KW	4.36	2.88
Max. power KW	32.28	15.74
Ratio max/mean power	7.4	5.5

Choice of Site

The generally higher currents and power levels at M suggest at first that this would be an optimum choice of site. Closer examination of the figures, however leads to a different conclusion. Mean power is a continuous average, and may be taken as an indication of the size of process that can be associated with each unit. If more power is required, more units can be installed. It should be remembered that this is a continuous, 24 hour, average. The maximum power output is an indication of the size and cost of the associated power conversion and storage installation, and can be taken as an indicator of the total cost of the tidal power generator. Of interest, therefore, is the ratio between the mean output, the benefit, and the maximum output, the cost. A low ratio indicates better utilisation of plant and in this respect B would be the better choice. The short slack water periods at B, due to some local peculiarity, most probably cause the difference.

These arguments suggest that the choice of site is likely to be specific to the area being considered and should be considered carefully. The periods of slack water should be taken into account as well as the maximum current.

PROTOTYPE TESTS

In the analysis above, a constant efficiency of 50% was assumed. As a check on this a prototype turbine has been built and tested. It is illustrated in Fig. 4. It is of GRP construction with 8 blades, a hub diameter of 600mm and overall diameter 1250mm. The design was based on the potential flow round a sphere and optimised to give a power gain from the resulting flow acceleration. Blade twist was based on the same velocity distribution. Under test, the turbine was cantilevered out from the side of a TERC research vessel which then anchored in the tideway. Flow velocity was measured with a conventional impeller type instrument. Power was absorbed by a friction brake and power developed was measured by recording torque and rotational speed. Torque was initially measured using spring balances on a band brake, but this arrangement was later replaced by an electronic system based on a

quartz load washer. Typical results are shown in Table 2. These are a selection and show considerable scatter, but in general may be taken as justifying the assumption of 50% overall efficiency. The causes of the scatter are several. Both experimental technique and instrumentation have suffered from restricted testing periods and are in need of development. There is also an optimum rotor speed at each flow speed and this was clearly not achieved in all cases. However, similar results were obtained with two quite different ways of measuring torque and thus the results are presen--ted in support of the assumption made above.

TABLE 2 Test Results

Flow vel. m/sec	Rotor spd. rev/min	Power avail. W	Power msd. W	Efficiency %
0.98	45	584	397	68
1.08	31	788	151	19
1.18	31	1028	495	48
1.29	33	1330	257	19
1.44	63	1868	1160	62

DESIGN OPTIMISATION AND MODEL TESTS

In equation 1 the velocity cubed term makes very attractive the possibility of artificially increasing the flow velocity, for example with a venturi or reducer. If flow could be effectively exhausted there would be a small power gain and the addition of a shroud to the rotor would help to eliminate tip losses and improve efficiency. In addition, the intake to a venturi could be made rectangular to intercept more flow and could be shaped to compensate for the velocity profile of the stream, giving better, more controlled flow through the rotor, improving efficiency and power output. From a mechanical point of view, a smaller rotor would be preferable and a shroud would make it less easily damaged. The venturi could be incorporated into the pontoon and the pontoon, rotor and venturi would then form a single unit of high mechanical integrity.

These ideas are being explored in the laboratory. The systems shown in Fig. 5 have been built for flume test and there are preliminary indications that the venturi proposal is worthwhile.

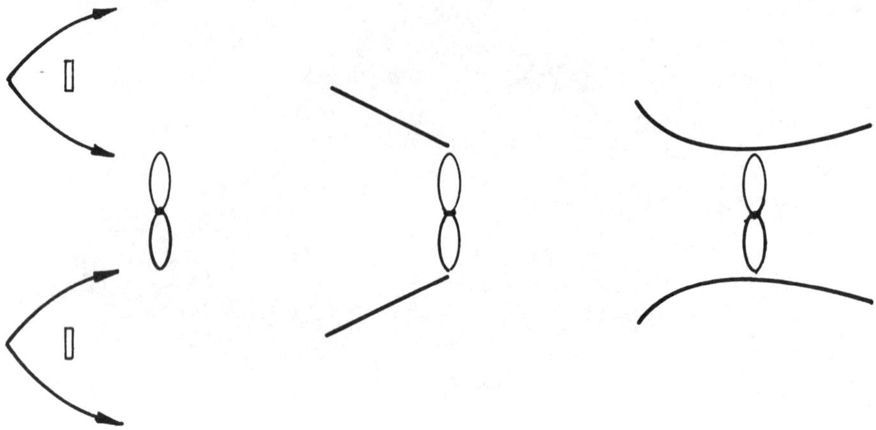

Fig. 5. Flow acceleration schemes.

CONCLUSIONS

The arguments above have shown that tidal currents can be a useful source of power. The site needs to be chosen carefully and long slack periods avoided. In terms of power extraction, a moored pontoon supporting a shrouded rotor with some form of flow acceleration system seems to offer the most promise.

ACKNOWLEDGEMENTS

The Authors wish to thank Dr. J.A. Charlton for his advice and the staff of the Tay Estuary Research Centre and of the Mechanical Engineering workshops for their assistance.

REFERENCES

Charlton, J.A. (1980). The tidal circulation and flushing capability of the outer Tay Estuary. Proc. Roy. Soc. Edinb., 78B, 33-46

Charlton, J.A. (1981). Tidal Atlas of the Outer Tay Estuary. Tay Estuary Research Centre Report No. 5, University of Dundee.

Admiralty Chart No. 1481.(1976). River Tay. Hydrographer of the Navy.

TOPIC E
Biofuels

The Integration of Energy Production into Tropical Agricultural Systems

B. Pound and T. R. Preston

CENIP/CEAGANA, Aptd 1256, Santo Domingo, Dominican Republic

ABSTRACT

The integration of crop production, animal production and energy generation in the tropics is discussed. The fractionation of high-yielding tropical crops such as sugar cane and Leucaena leucocephala into feed and fibre components allows best use to be made of resources, and a model is proposed for a three hectare farm producing 100 Kwh of electrical energy/day.

A simple charcoal-making process for low density crop residues is described, together with experiences gained with a variety of simple biogas digesters. The role of gasification at the farm and village levels is discussed.

KEYWORDS

Agricultural systems; energy; biogas; producer gas; charcoal; biomass crops.

INTRODUCTION

In tropical - as in temperate - countries the cost of taking centralized power supplies to isolated, rural communities is often disproportionately high. In developing countries a high proportion of the population lives in rural areas. In the Dominican Republic 50% live in the countryside, and an estimated 80% of these do not have access to centralized power supplies. Even in energy rich countries such as Venezuela and Mexico there are large areas without power, which is necessary for the development and comfort of their communities. When centralized power supplies are available they may not be sufficiently reliable for energy dependant enterprises such as dairy farms, where power may be needed for machine milking, milk cooling, water pumping, forage chopping and domestic lighting.

Decentralized power supplies, at the village or even at the farm level, are therefore necessary, and it is suggested that this objective can be attained by the integration of animal and crop production with energy generation.

Specialized versus "Dual-Purpose" Crops

Tropical countries have the advantage of high insolation which, coupled with high average temperatures, gives high biomass production per unit area. Fast growing trees such as Eucalyptus can yield 200 GJ/ha/year. Tropical grasses such as Pennisetum purpureum have given up to 1150 GJ/ha/year (Alexander, 1978). These "energy crops" are grown for one specific purpose - that of energy generation. Similarly crops grown for alcohol production (e.g. sugar cane, cassava and maize) use the primary product (sugar, starch and grain respectively) for the distillation process.

An alternative approach is to use the primary products as human or animal food or for industry, and the residual products for energy generation as shown in Fig. 1.

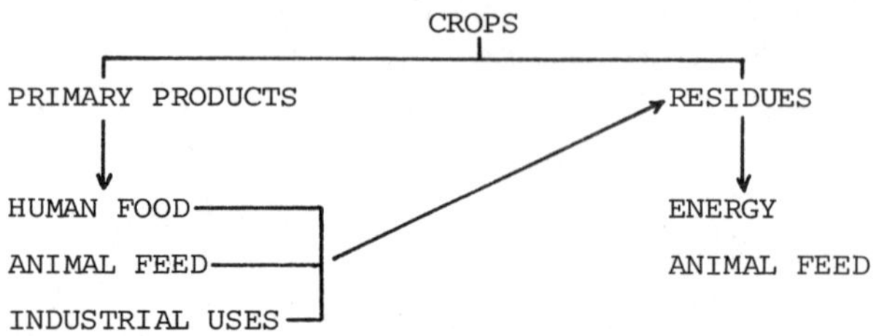

Fig. 1. Priority uses for crop products

Two dual purpose crops that we have been evaluating in the humid, lowland tropics are Leucaena leucocephala and sugar cane. The tropical, leguminous, shrubby tree Leucaena leucocephala separates easily into leaves, which are used as a protein and vitamin source in animal feeds, and wood for fuel. Sugar cane can also be fractionated into juice and fibre. This of course is the process used in a large scale sugar mill, but it can also be achieved by small 3 or 5-roller crushers at the farm or village level when the juice can be used for animal feed or crude sugar production, and the fibre for energy production through gasification to producer gas or anaerobic fermentation to biogas.

Integrated Farm Model

A simple model of a farm incorporating the ideas outlined above is shown in Fig. 2.

The three hectares are split into two of sugar cane and one of leucaena. The sugar cane is fractionated through a 3-roller mill with 50% extraction rate to give 300 litres of juice and 300 kg 40% dry matter pressed stalk. The juice, together with the cane tops and leucaena leaf meal form the basal diet for 10 cows and their calves or 5 sows and their piglets. The fibrous residues, which amount to 155 kg of dry matter/day can be processed in a number of ways. Vapourisation is probably uneconomic on this scale (Breag and Chittenden, 1979), but gasification, charcoal production and biogas production could be used. The output from gasification of a 100 Kwh/day is calculated on the basis that 1 kg DM produces 2.3m^3 producer gas. Each m^3 of gas has a calorific value of 5.5 MJ, giving 1961 MJ daily. At an overall conversion efficiency for the engine/generator unit of 18% this would give approximately 100 Kwh electricity/day.

Through anaerobic digestion the output of electrical power is less, amounting to only 31 Kwh/day. This assumes the use of the cow slurry (approx. 35 kg DM/day)

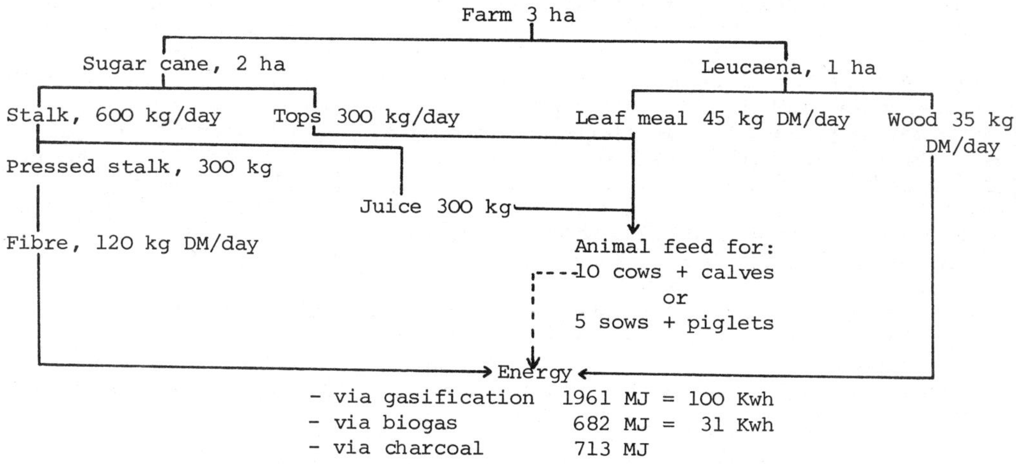

Fig. 2. Integrated farm model

together with the 120 kg pressed cane stalk. Under experimental conditions this mixture has given 0.2m^3 biogas/kg total solids. (Pound, unpublished data). Thus 31m^3 biogas should be produced daily. Our experience with both dual-fuel (diesel/biogas) and spark ignition biogas motor/generators has indicated a conversion of 1 Kwh/m^3 biogas, giving 31 Kwh/day if the biogas is used in this way. More efficient use of the energy available in the gas is possible through direct combustion of the gas for cooking when a conversion efficiency of 65% is possible using a Lorena-type stove.

The dried cane and leucaena residues could otherwise be converted to charcoal. The final charcoal has a calorific value similar to that of biogas. The charcoal has the advantage that any surplus could be sold as a cash earner.

This example demonstrates how an integrated crop, animal and energy production unit can not only be self-sufficient in energy, but also provide a surplus, which could be pooled for community use.

Simple Energy Technologies

Several simple technologies are being evaluated in the Dominican Republic for the sort of system outlined above.

Charcoal. The use of low density crop residues such as pressed sugar cane stalk for charcoal production requires different conditions from those employed in the carbonisation of wood. A major difference is in the air supply to the oven. Low density material combusts so fast that virtually no restriction of the air supply is necessary. An experimental scale oven was made by Ffoulkes, Elliott and Preston (1980) which consisted of a 200 litre oil drum with ten 1 cm holes drilled in it. This is filled with 15 kg of dried, pressed cane stalk of about 95% DM, and lighted at the top. After 11-18 mins. the whole of the burning material glows red. At this point water is sprayed onto it with a hose pipe to stop the burn. The yield of fragmented charcoal is 2.3 kg, with a calorific value of 30 KJ/g. To convert this into a saleable product it has to be compressed into briquettes. The process is relatively inefficient as the combustible, volatile components are lost in the burn. However the process is simple, rapid and effective.

Biogas. This technology has the great advantage that it can be made as complex or as simple as the situation demands. An advantage of the tropical climate is that unless very high yields of gas are needed then heating of the digester is unnecessary. A number of different types of digester have been constructed at CEAGANA in the Dominican Republic. The first was a 90m^3 digester with a mild steel gas holder floating directly in the slurry provided by 64 cattle. This has been working for 6 years, but is now showing signs of serious corrosion. Two digesters of the Chinese rectangular design were constructed. The walls were made from concrete blocks and the flat roof from poured concrete. All inside surfaces were given 4 coats of plaster. Despite this, and further efforts to seal them with sealants, biogas has continued to leak from the gas storage space.

These two types of digester mentioned above, in addition to the limitations of their construction materials, also need specialized artisans for their construction and installation. In the light of experience with these digesters and from talking to farmers, a list of points to be borne in mind for tropical small-farm situations was drawn up as detailed in Table 1.

TABLE 1 Priority Considerations for Small-farm Digesters

1. The digester must have low initial cost.
2. Installation must be simple.
3. Management must be simple.
4. Operating labour requirements must be low.
5. The digester must reliably produce a satisfactory amount of gas over an extended period.
6. The digester must be built into the farming system as a whole.

Plastic and P.V.C. Digesters. Bearing these points in mind a programme to evaluate plastics and P.V.C. as construction materials was initiated. As a first step a 10m^3 (slurry volume) "Red-Mud P.V.C." digester was purchased from Taiwan (Union Industrial Research Laboratories). Its cost, including all installation costs, was £445, which was less than half the cost of the concrete digesters of the same size - and it worked without problems. Installation was simple, and with a single daily charge of 6% total solids cattle slurry, and a retention time of 40 days at 26°C, biogas production has averaged 40% of slurry volume for several months. This is sufficient to run a 3 burner stove for 7 hours/day, which is considered sufficient for a farming family. A detailed report on this digester has been published (Pound, Bordas and Preston, 1981).

The next step was to construct locally a digester of similar design using local materials and workshops. This was done using ordinary 0.6mm P.V.C. sheet and P.V.C. welding equipment available locally. The digester has been working for several months, and a source of U.V. stabilized P.V.C. with extra plasticizer is being sought for this, and two further designs described below.

Batch digesters for crop residues require that the digester be opened between batches to remove the spent charge and introduce the fresh one. Thus the tube like design of the Taiwanese is not suitable for this application, and a very simple design using only two sheets of plastic was evaluated - as depicted in Fig. 3. The lower sheet lines the hole and the upper sheet covers the fermenting material and also acts as the gas storage space. The sheets are sealed with a water seal. This was tested with a mixture of 67% pressed cane stalk, 29% cattle slurry (DM basis) and 4% urea (to raise pH and lower C:N ratio) which experimentally has given 0.2m^3 biogas/kg total solids.

An even simpler design of digester is presently under evaluation which we believe to be superior to the one described above. It is shown diagramatically in Fig. 4 and

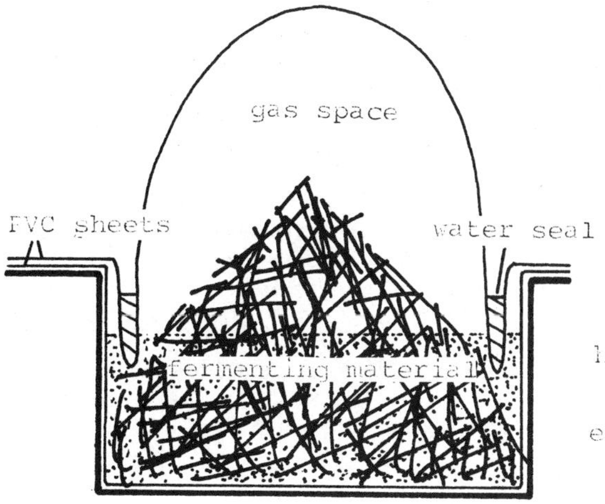

Fig. 3 "Water-seal" digester

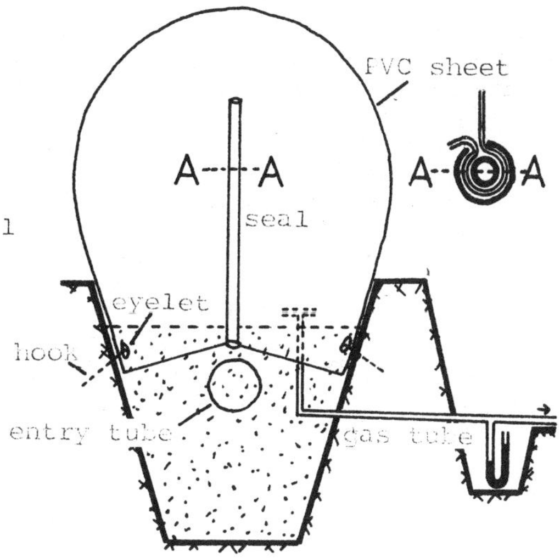

Fig. 4 "Tarpaulin" digester

consists of an unlined trench into which a plastic tarpaulin has been anchored by means of bent reinforcing rods hammered into the earth walls. The tarpaulin is easily removable and this design can be used for either batch or continuous operation. Flexible, light-stable P.V.C. would be a better material than tarpaulin as the flexibility reduces pressure fluctuations with gas production and use which is important when one is using the gas for cooking. The cost of this type of digester is very low. A 10m^3 (slurry volume) digester is estimated at £215 (£60 for P.V.C., sealing and eyelets; £60 for labour at developing country rates of £5/manday; £50 for inlet and outlet cisterns; £25 for gas tubing and connections and £20 for rope and reinforcing rods). If the trench had to be lined with ferrocement, this price would be increased to £300 - still substantially cheaper than the Taiwanese design.

It has been shown that biogas technology can be simple and it must be emphasized that the process is not only useful as a fuel producer, but also as a treatment process for residues to reduce their contamination load, whilst retaining the majority of their nutrients. The effluent is useful as an organic fertilizer and soil conditioner or as a supplement to animal diets. Recycling of nutrients in this way reduces the amount of inputs that have to be bought in from outside the farm. We are therefore nearing the situation in which the farm is self-sufficient in energy, produces the majority of the feeds necessary for its livestock, and also optimises the recycling of nutrients.

Gasification. The generation of electrical power through gasification is more efficient in terms of the power generated/kg of dry raw material. It is however a more complex technology even in its simplest form than anaerobic digestion, and requires a moderate level of skill and attention to operate effectively and safely. Producer gas has a lower calorific value than biogas (5.5 MJ/m^3 as compared to 22.0 MJ/m^3) and cannot be burned directly in stoves or lights. Initial investment is high with a 25 Kw unit (including motor/generator) costing around £15,000 installed. The range of raw materials is limited to fibrous residues of 5 - 20% moisture content and normally to a particle size of 1 - 8 cm. The place of this technology will probably be at the level of the small industry which has a steady production of fibrous residues (e.g. coconut husks or rice hulls), or at the level of the small community, with its members contributing agricultural residues such as maize cobs or pressed sugar cane stalk.

CONCLUSIONS

1. A three hectare integrated farm unit, by growing dual-purpose crops can be self-sufficient in energy and grow nearly all the feed necessary for its livestock.
2. To be acceptable to small farmers in Latin America biogas digesters must be cheap, simple and reliable. P.V.C. and plastics are suggested as the most appropriate construction materials.
3. Gasification, although efficient, is not as versatile nor as simple as anaerobic digestion - and its major role will probably be at the small industry or community level, where adequate raw materials and supervision are available.

REFERENCES

Alexander, A. A. (1978). In N. Quezada (Ed.), Caribbean Energy Consultation. Instituto Superior de Agricultura, Santiago, Dominican Republic.

Breag, G. R., and A. E. Chittenden (1979). Producer gas: Its potential and application in developing countries. Report No. G130 Tropical Products Institute, 56/63 Grays' Inn Rd., London WC1X 8LU.

Ffoulkes, D., R. Elliott and T. R. Preston (1980). The Potential for Charcoal Production from pressed sugar cane stalk. Trop. An. Prod., 5, 135-139.

Pound, B., F. Bordas and T. R. Preston (1981). Characterisation of production and function of a 15m^3 Red-Mud P.V.C. biogas digester. Trop. An. Prod., 6, 146-153.

Union Industrial Research Laboratories. Instruction manual for the installation of waste digesters; Union Industrial Research Laboratories, 1021, Kuang Fu Rd., Hsinchu, Taiwan 300.

The Use of Wood for Industrial and Domestic Energy in Rural Areas of Britain and the Transfer of Wood-Energy Technology to Developing Countries

A. I. Fraser

International Forest Science Consultancy, 21 Biggar Road, Silverburn, Penicuik, Midlothian EH26 9LQ, UK

ABSTRACT

The energy properties of wood and charcoal are described because of the confusion that exists through the use of improper or inappropriate units. Large wood resources have been created in predominantly rural parts of Britain, which could supply energy for domestic and industrial purposes. Development of appropriate harvesting systems will be necessary, but use of small roundwood branches and roots would bring additional revenue to forest owners, while minimising competition with traditional wood processors. The energy in wood can be extracted in many ways and the production of intermediate energy carriers such as charcoal, charcoal-oil mixtures, gas and methanol can offer a number of technical advantages. A wide range of industrial energy requirements could be met by using the most appropriate energy conversion route, and much of the experience gained would be relevant to developing countries.

KEYWORDS

Energy, wood, rural communities, industrial energy, domestic energy.

INTRODUCTION

Most energy consumers in Britain have a choice of several types of fuel and patterns of consumption of the different fuels have varied over the years as relative prices have changed. During the past decade oil has changed from being one of the cheapest forms of fuel to being almost the most expensive, while the opposite has happened with wood.

One important factor which has a bearing on the cost of energy for the end-user is transport and distribution so that rural communities and rurally based industries are at a serious economic disadvantage compared with town dwellers and urban based industries. Improvements in distribution systems for all the main industrial fuels mainly benefit the large single consumer or large groups of small consumers in one locality, while the more remote and scattered communities in rural areas are left with the inefficient and more costly labour intensive methods of distribution.

These economic disadvantages for rural communities extend to all other aspects of the local economy as goods which are not produced locally and must therefore be imported have the cost of transport added, while all produce not consumed locally must be transported long distances to the market. If rural communities are to survive and prosper it will be vital that they increase their level of self-sufficiency in order to reduce the economic disadvantage of their relative remoteness. Since energy costs make up more than 10 per cent of domestic expenditure, and may account for a similar or higher proportion of industrial manufacturing costs, it is clearly important that every effort should be made to reduce them. For a rural family which has a high degree of self-sufficiency in food, energy may be the highest single item of cash expenditure.

Agriculture is generally the most important rural industry and the level of farm income will have a major influence on the strength of the local economy. Most of the energy used in primary agriculture is in the form of fuel for machinery and cannot easily be replaced by other types of fuel. However, crop drying can be done with a variety of fuels and lends itself well to using agricultural or forest residues. Food processing is also a heavy energy consumer. It also adds considerably to the value of food and if done in rural areas can bring substantial economic benefits to the rural communities.

Many of these food processing industries, such as canning, milk processing, meat curing, milling and jam manufacturing are located in semi-rural areas and utilise fossil fuels for their energy. Some could be converted to use biomass fuels which would bring the added benefits of extra rural employment in the harvesting and processing of the energy supply.

As Britain's forest plantations reach maturity and timber processing industries develop to convert the output, supplies of wood residues are likely to increase. Most forest industries are likely to be located near the forest resources, which means that they will tend to be in the predominantly rural parts of the country. Sawmill residues in particular could be an important source of energy, either for the industries themselves or for local communities, and are likely to show a good return to the sawmills, unless some large pulp or particleboard plant is situated nearby.

Another important component of the rural economy is tourism which increases local demand for energy for heating of hotels and domestic accommodation.

As more specialised leisure and recreational complexes are developed they will also add to energy requirements. If the energy has to be imported, the cost of these facilities is increased for the tourist, leaving less to spend locally on other things and more of the income derived from the tourists is lost to the urban areas which supply the energy.

The location of accessible coal deposits has had a big influence on industrial development in Britain in the past so that most of the remaining rural areas are in regions with no indigenous reserves of fossil fuels. On the other hand, the large afforestation programme of the past fifty years has been in the rural parts of the country and so has (unintentionally) created a source of energy close to many rural communities.

Plantation grown trees require to be planted fairly close together and thinned regularly to achieve maximum yield, but there are few markets for the small sized trees that are produced early in the life of the crop. These early thinnings represent not only an important energy resource for the rural communities, but also a source of employment and a means of strengthening the rural economy.

In many developing countries a majority of the population are rural dwellers and wood is the main source of domestic energy. Fuelwood plantations are being established on an increasing scale in many countries and there is a growing need for new technology to enable the energy to be extracted efficiently and converted to more convenient forms so that people can enjoy some of the conveniences of life which most people in developed countries take for granted. The development of techniques for converting the energy stored in the wood to more convenient forms such as charcoal, gas and heat in Britain will thus indirectly bring benefits to developing countries, as well as providing opportunities for creating export-oriented industries which can be located in rural areas.

ENERGY PROPERTIES OF WOOD AND CHARCOAL

Wood, like oil and coal, is not a consistent commodity with standard or uniform properties and the forest and timber industry has its own units of measurement which bear no relation to energy content. One of the difficulties encountered when considering wood as a fuel is obtaining an accurate definition of its energy properties in a form which enables comparison with other fuels.

When a tree is felled, the wood will have a moisture content of 50 to 60 per cent of its fresh weight (wet basis), or 100 to 120 per cent of the dry wood fibre (dry basis), that is about half the fresh weight of wood is water. As the water has no energy value, it is necessary to determine first the dry fibre weight of the wood before the energy content can be estimated.

Most measurements in the forest industry are based on volume and wood is generally sold by volume (the cubic metre) but because freshly felled hardwoods have a density of 850 to 950 kg/m^3 it is common to refer to a cubic metre as a "ton". This can be very misleading for the uninitiated owner of a wood-burning stove who generally has no idea of how much energy is being purchased.

The energy content of cellulose fibre varies very little from one species to another so that dry fibre density is the property which has the greatest influence on energy content. Some resinous conifers may contain a proportion of high energy aromatics which increase the energy content but for practical purposes a value of 19 MJ/kg equivalent to 5 360 KWh per tonne can be taken for the heat value of bone dry wood.

The water in the wood not only has no heat value and so reduces the net heat value of the unit weight of moist wood, but it also has a specific heat capacity and a latent heat of evaporation which must be supplied from the energy in the wood when the temperature is raised to a level where combustion can take place, which is equivalent to about 2.5 MJ/kg of water. The energy content of wood therefore falls off rapidly as moisture content increases. Table 1 gives some of the more important heat values associated with wood and bi-products derived from wood during conversion.

TABLE 1 Heat Values of Wood in Various States and Bi-Products Derived from Wood

Item	Moisture content (dry basis)	Energy content		
Combustion		MJ/kg	(MJ/m^3)	kWh/kg
Freshly felled wood	100-120%	8.3		2.30
Air dry wood - Hardwood	15- 20%)	16.4)	(10.7)	4.56
- Softwood	)	)	(7.4)	
Bone dry wood	8- 12%	19.3		5.36
Charcoal per kg	5- 10%	29.0		8.05
Wood gas per kg	-	7.0	(5.6)	1.94
Pyrolisis				
Charcoal per kg wood)	5- 10%	7.2		2.01
Recoverable heat per kg wood)		5.6		1.55
Wood gas per kg wood	-	6.6		1.83

The items listed under "combustion" represent the fuels as they would be used for conventional burning, while those listed under "pyrolisis" represent the products recoverable from the conversion of 1 kg of air dry wood by a pyrolitic reaction. The figures in brackets in the third column illustrate the difference between hardwoods and softwoods on a volume basis. The use of the last column enables wood to be directly compared on a cost basis with other fuels.

The principle attributes of wood as a fuel were given by Fraser (1980) but are worth repeating in summary for the sake of completeness:

(a) Renewable provided the forest resource is so managed.
(b) Easy to store - though bulky.
(c) Can be converted to other forms such as heat, charcoal, gas or electricity for more convenient use.
(d) Production and harvesting systems can be low cost and small-scale or large-scale as appropriate.
(e) Fuelwood plantations have important conservation, environmental and aesthetic benefits.
(f) Energy may be extracted from re-cycled wood as well as roundwood, and trees may be divided up so that non-industrial parts only are used for energy.
(g) Ash content is low (1.5 to 3.0 per cent) and has a high potash content and so can be recycled as a fertiliser.

Charcoal is obtained from wood by heating it in the absence of air to about 400°C. During the reaction a number of volatile substances are given off which can either be discharged into the atmosphere, condensed and separated into the various constituents, or burnt to raise heat. The charcoal contains about 50 per cent of the energy in the original wood and about 25 per cent of the weight. The recoverable heat from the reaction, combined with the heat released from combustion of the volatiles and gases, can yield a further 30 per cent of the original heat value in the wood so that the overall thermal efficiency of the pyrolisis reaction can be about 80 per cent (Thoresen, 1980).

Table 2 gives typical values for the more important physical and chemical properties of charcoal.

TABLE 2 Physical and Chemical Properties of Charcoal

Property	Value
Density	220 kg/m³ (softwood) - 320/kg/m³ (hardwood)
Bulk density	120 kg/m³ (softwood) - 200 kg/m³
Heat value	29 MJ/kg
Fixed carbon	75 - 80 per cent
Volatiles	10 - 15 per cent
Moisture content	5 per cent
Ash content	5 per cent

Charcoal is widely used as a domestic fuel (utility in most developing countries, luxury barbecues in most developed countries) but the biggest single use is for iron and steel smelting where it is valued because of its low sulphur content. It is also an extremely useful energy carrier as it has consistent properties, especially when compared with wood and it is also very reactive. Recently considerable interest has been shown in Charcoal-Oil Mixtures which can be made with 30 per cent or possibly more by weight of charcoal in suspension in a heavy fuel oil. The energy value of the oil per unit weight is reduced slightly but it enables a substantial proportion of charcoal to be burnt in oil-fired equipment and so saving the cost of adapting existing equipment to use a new fuel. This technique could have a number of applications in Britain but more so in certain developing countries which import large quantities of fuel oil for power generation or mineral treatment and could substitute a significant proportion with domestically produced charcoal. A 30 per cent charcoal-70 per cent fuel oil mixture would have a heat value of about 39.5 MJ/kg compared with 44 MJ/kg for pure oil and at present UK industrial fuel oil prices of £110 per tonne, would give the charcoal a value of £74 per tonne on an energy equivalent basis.

HARVESTING SYSTEMS AND FUEL PREPARATION

The solar energy which is trapped and stored by trees is distributed between the main structural components - the stem, the branches and the roots. Traditional forest management, especially in developed countries, has concentrated on the production of industrial wood (stems) and little interest is taken in the branches and foliage (crown) or the root system. In developing countries these latter parts of the tree are usually the main source of rural domestic energy.

Studies by Fraser (1969) on conifers in the UK have shown that the proportion of the total weight of a tree found in each main part is in the following range:

Stem	- base to 7 cm diameter	- 59-39.0 per cent
Crown	- stem<7 cm and branches and leaves	- 23-37.5 per cent
Root	- roots to about 1 cm diameter	- 18-23.5 per cent

where the first figure is the proportion found on fertile favourable sites and the second on poorly drained infertile areas. The latter are more representative of most coniferous plantations in Britain. Thus 50 to 60 per cent of the total biomass is likely to be in the crown and roots. A similar distribution might be expected in fast grown hardwood coppice crops, although as these are most likely to be grown on reasonably good sites, a higher proportion in the stem can probably be expected (perhaps 50 to 60 per cent).

Open grown hardwood trees or multi-stemmed unmanaged woodland trees have not been studied to determine the proportion of the biomass in the various parts, but volume tables which are prepared for forest management purposes generally refer only to the volume in a single stem, to the point where the major branches of the crown arise. Few studies have been made of the biomass in root systems.

It is apparent from this that traditional harvesting techniques designed to extract only the industrial stem wood will leave a very high proportion of the biomass in the forest. Under British conditions it would appear that the yield of woody biomass could almost equal the output of industrial wood by taking the crowns and roots of the trees currently being harvested. In addition to this there is a large volume of small trees which should be removed in thinnings to improve the residual crop, for which no industrial market exists.

The transfer of woody material from the forest or woodland to the end-user should be considered as a system which is designed to have as high an overall efficiency as possible. Several distinct components can be distinguished, for each of which there are generally several possible methods and several possible positions in the sequence. These components can be described as follows:

- stump operations
- in-forest transport
- out-forest transport
- wood preparation
- wood drying
- wood energy conversion process.

At the present time in Britain the main use of wood for energy is for domestic heating and no real distribution system exists. In fact most firewood merchants operate on a casual basis and have little or no special equipment. Some of the larger forestry companies, whose main task is forest management or timber processing, may handle 'firewood' as a sideline but there are as yet few, if any, examples of what might be called a 'fuelwood harvesting system'.

The nearest equivalent to an industrial use of wood for energy is a pulpmill which requires large volumes (100 000 to 500 000 m^3/an) of fairly uniform small roundwood. A few exist in Britain though numbers have declined substantially in the recent past. Pulpmills utilise wood chips and the roundwood is delivered in standard lengths for chipping at the mill. However, the wood is better wet for the pulping process so that drying is not part of the harvesting system.

In Scandinavia, demand for wood for both pulping and energy has outstripped the supply of traditional roundwood and additional supplies of wood fibre are being sought from branches and roots (Bendz, 1980). The former lend themselves very readily to being chipped in the forest by small mobile chipping machines and the chips are generally easier to handle than tree tops and branches. The roots can either be extracted by uprooting the whole tree or they can be grubbed up by a special machine in a separate operation. The Scandinavians have so far opted for the second approach. After grubbing up, the roots must be left to weather so that all soil can be washed off before taking them for processing. However, in Scotland a high proportion of the conifer plantations are on peat, which might be incorporated with the woody material as a fuel.

A felled or uprooted tree may be divided into various components at the stump, or moved intact to a roadside landing. The conversion may be complete into the form required by the end-user such as billets or chips, or it may be a partial conversion to some more convenient form for long haul transport out of the forest.

Although equipment is available for burning wet wood, the energy value is so much lower per unit weight or volume of wood that drying is almost essential for wood-fuel. Little research has been done into techniques for air drying in Britain so it is not possible to say whether drying can be achieved more quickly by leaving the trees more or less intact or by reducing the trees to some smaller sizes. Studies by Plumptre of drying various timbers in solar kilns in South Britain suggest that in summer small sizes would take from three months for softwoods to seven months for hardwoods to dry from green to about 12 per cent. In the air about double the time is required but it will depend to a large extent on how the wood is exposed to sun and wind and how much re-wetting by rainfall takes place. Wood is a bulky material so that a considerable area of land is required to stack a 6-12 month supply of wood. Relatively low cost solar kilns (Hardie and Plumptre, 1979) reduce the drying time and a combination of initial air drying and finishing off the last two to three weeks in the solar kiln may prove effective.

DIAGRAM: WOOD ENERGY HARVESTING SYSTEM: THE INTER-RELATIONSHIP BETWEEN THE VARIOUS OPERATIONS THAT ARE REQUIRED BETWEEN THE TREE STUMP AND THE END-USER

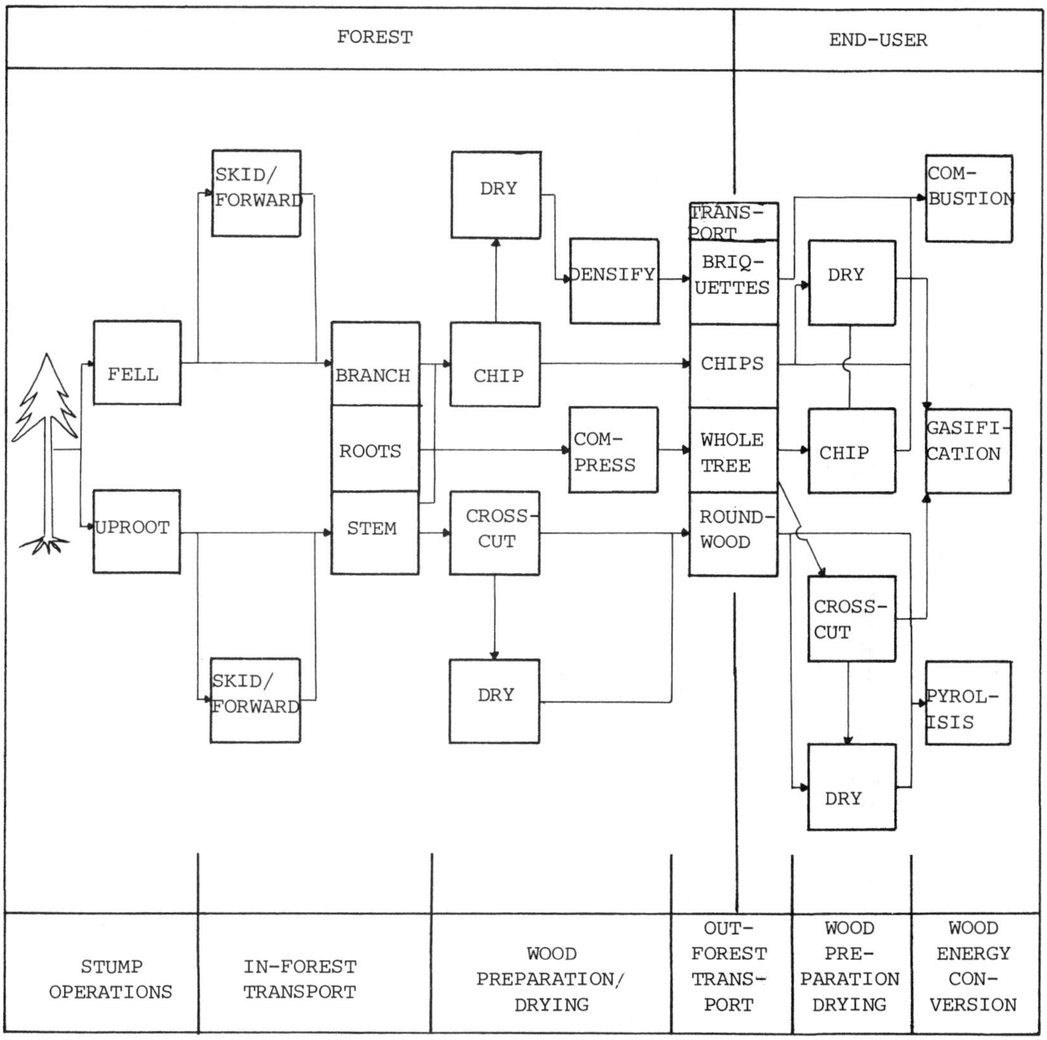

Transportation is generally the most costly part of the harvesting system because of the low bulk density of wood, but there is considerable variation in bulk density according to the form of the wood when it is transported. Unless the wood is densified by briquetting wood chips, the greater bulk density would be achieved by squaring solid wood, while chipping wood reduces bulk density very considerably. Table 3 illustrates the range of the energy content which can be transported by a standard truck with maximum capacity of 25 tonnes or 75 m³, whichever is reached first, with the wood in different forms.

TABLE 3 Bulk Density and Energy Content per Truck Load According to the Form of the Wood

	SOFTWOOD		HARDWOOD	
FORM OF WOOD	Volume/ 1 000 kWh (m^3 gross)	Energy content kWh per 25 t/ 75 m^3/truck	Volume/ 1 000 kWh (m^3 gross)	Energy content kWh per 25 t/ 75 m^3/truck
Squared wood - green	1.0	57 500	0.6	57 500
- air dry	0.5	114 000	0.3	114 000
Stacked roundwood				
- green	1.6	57 500	1.1	57 500
- air dry	0.8	114 000	0.55	114 000
Chips - green	2.0	37 500	1.4	57 500
- air dry	1.0	76 150	0.7	107 000

The table shows that the transport efficiency is more sensitive to the effect of the form in which it is transported for softwood than it is for hardwood, and also that pre-drying wood has a very substantial effect on the amount of energy that can be transported per load, and hence the cost per unit of usable energy: green softwood chips are likely to cost three times as much as dry round hardwood per unit of energy transported.

The average annual energy requirement for space and water heating for a British household is about 20 000 kWh, which would require a gross storage volume ranging between about 10 m³ for stacked hardwood to 40 m³ for green softwood chips. The former would involve a stack about 6 m long x 1 m wide x 1.7 m high, which is not unreasonable for many country gardens but would be unrealistic for most town houses. Green chips should not be stored for more than 2-3 months in order to avoid a build-up of fungi which liberate large quantities of spores that can be a health hazard. However, about thee-quarters of the total energy requirement will be during the winter six months so that storage capacity for about one-quarter of the total volume, or about 10 m³, is required. The shape and location of a wood chip store is critical if automatic stoking equipment is to be used. Although a wide variety of automatic stokers is now available, their cost is such that they are unlikely to be justified for installations with a heat output of much less than 30 kW, which precludes their use for most of the smaller domestic requirements, but they are well suited to the needs of large houses, farms, commercial premises or institutions such as colleges and hospitals.

Softwoods represent the biggest proportion of the available wood resources in Britain so that any large-scale industrial application for wood energy is likely to use softwood branches, roots and small roundwood from thinnings. Transport cost will therefore be a critical factor in determining economic viability. Chips or small pieces made by a hammer mill (hogged fuel) are probably the most convenient for automatic stoking, but will be costly to transport. A solution to this problem may be the formation of briquettes from dried chips which increases the density of the wood and hence the energy content per unit volume. Most briquetting machines require the wood to be pre-dried to about 8 to 10 per cent mc, but with softwoods it is generally not necessary to use a binder. The briquettes can be formed to a standard size and shape and so can be readily handled by automatic equipment. Equipment has not yet been developed to make briquettes in the forest but a mobile briquetting plant, burning forest residues to raise steam for pre-drying, would appear to be a desirable prerequisite for the industrial use of wood energy.

The various conversion routes which can be used to extract the stored energy in the wood are discussed in the next section but clearly the form of the woodfuel will be largely determined by the conversion process being used and the scale of operations. These in turn will determine which combination of the various options discussed above is the most efficient and economic in the particular circumstances.

CONVERSION TECHNOLOGY

Wood is chemically stored solar energy which can be released in various ways and converted into more useful forms of energy. In an industrial society most energy is ultimately used as heat, mechanical or electrical energy. Wood can be converted to any of these but it may be desirable to use some other form of carrier to store the energy until the moment it is required, or to transfer it in the final stages of conversion. Solid fuel is not appropriate for internal combustion engines.

The following diagram illustrates the main conversion processes and carriers which can be used to convert the stored energy in wood into more useful forms:

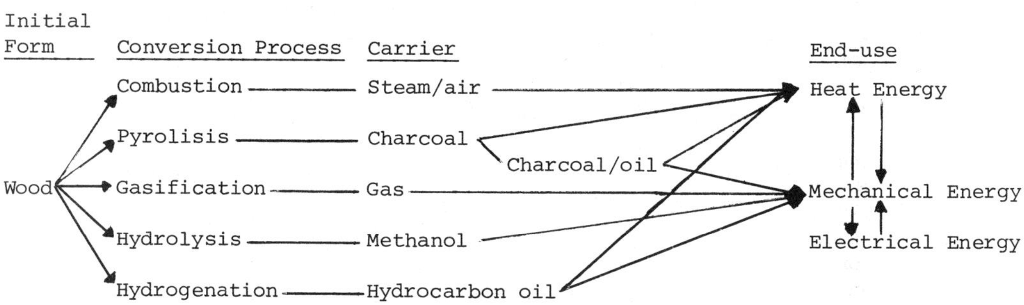

Combustion When wood is heated with an abundant supply of air it dissociates to carbon dioxide and water vapour and releases energy. If there is insufficient air, the reaction will not be completed and carbon monoxide and hydrogen will be produced. When these gases are discharged into the atmosphere a considerable amount of the original energy will be lost. It is therefore very important to ensure that combustion is complete, and in properly designed plant for burning wood there are primary and secondary air supplies to achieve this. The secondary air is injected above the main combustion zone in order to react with the gases.

There is a very large number of different types of appliance commercially available for the combustion of wood, ranging from small 1 to 2 kW units suitable for warming a room or heating small amounts of water, to very large industrial furnaces with a heat output of 2 MW or more.

The thermal efficiency of an appliance is expressed as the quantity of useful energy recovered as a proportion of the total amount of energy in the original fuel. About 20 to 25 per cent of the energy in wood being burnt is required to maintain the temperature of the flue so that water vapour does not condense. Further losses may occur if only partial combustion takes place and volatile substances or unburnt gases are discharged up the flue. Thus wood burning appliances normally show efficiencies in the range 50 to 80 per cent with dry wood, depending on design. The efficiency achieved can have a large effect on the cost of usable heat.

The presence of free water in the wood not only reduces the heat value of the fuel as described earlier, but also reduces the temperature of the reaction and increases the amount of condensation in the flue, leading to very low thermal efficiencies. The problem has been partly overcome by the use of a refractory lined combustion chamber, and a forced air supply which raises the temperature to around 1 000^oC. This allows green wood chips to be burnt and maintains flue gas temperatures sufficiently high to avoid condensation problems. If the heat value of dry woodfuel is taken as the input, the real thermal efficiency is only about 30 to 40 per cent, though considerable time and expense is saved through not having to pre-dry the fuel.

Small stoves and boilers must generally be hand fed because the cost of automatic stoking equipment is high. An appliance should therefore be selected which has a large enough firebox to hold 10 to 12 hours of fuel if it is to burn overnight. This can be calculated from Table 3 which indicates, for example, that a 10 kW central heating boiler to burn for 10 hrs (100 kWh) on air dry round hardwood would require 0.055 m^3 (55 000 cm^2), which is equivalent to a firebox of about 35 x 35 x 50 cm.

Larger installations with a heat requirement of more than about 30 kW can justify the cost of chip storage and automatic stoking facilities which means that the unit can be left unattended and programmed to provide whatever heat regime is required. Such arrangements are particularly well suited to commercial and industrial applications.

One variant of the green wood chip burner is to have a small refractory lined chip preburner attached to the front of an existing gas or oil fired boiler which is used to burn the hot gases emitted from the preburner. Control of the air supply to the preburner regulares the composition of the gases. This avoids the need to change an existing central heating system and allows conversion to woodfuel at a relatively low capital cost.

<u>Pyrolisis</u> When wood is heated in the absence of air to about 400^oC an exothermic reaction takes place in which most of the cellulose is converted to carbon and water vapour, together with a number of volatile organic compounds, and gases which are given off. The volatiles may be condensed and recovered or they can be burnt with the gases in a second stage to add to any heat that is recovered from the reactor. The carbon is recovered as charcoal which is a more convenient form of fuel than the original wood.

The combined heat value of the charcoal and the heat recovered from the reactor and gas combustion can be as much as 80 per cent of the original heat value in the wood (Thoresen, 1980).

1 000 kg	air dry wood contains:	4 400 kWh
converts to 250 kg	charcoal containing:	2 015 kWh
plus	recoverable heat	1 555 kWh
		870 kWh is lost

There are at present two basic ways of making charcoal with several variants of each. One uses kilns in which the reaction takes place in the presence of a restricted supply of air so that part of the wood charge is burnt within the pyrolisis reactor to initiate the reaction. The other uses retorts in which the reaction takes place in the complete absence of air and heat is either applied externally or by circulating hot gases through the charge.

Little charcoal is manufactured in Britain at the present time though imports have been rising at about 7 per cent per annum over the last decade. In many developing countries charcoal is an important source of domestic and industrial fuel and is preferred to wood because of its higher heat value per unit weight, its more consistent properties, especially low moisture content, and the freedom from smoke when it burns because most of the volatiles are removed in the manufacture.

Charcoal has a similar calorific value to industrial grades of coal but has the advantage of having a lower ash content (1 to 3 per cent) and a very low sulphur content compared with most grades of coal. It therefore has potential as an industrial fuel. The current price of imported charcoal, cif UK port, is about £160 per tonne, which is about double that of coal and means that it only finds a market as a luxury product for barbecues and various horticultural products. One of the main reasons for the high cost of charcoal is the low recovery (25 per cent by weight) and the loss of about 50 per cent of the original heat value of the wood.

The value of charcoal as an industrial fuel depends on the price of the alternative fuels that it might replace. During the pyrolisis reaction heat can be recovered both from the reactor and from burning the gases (Breag and Harker, 1979) and this heat might be sold in the form of hot water or steam for industrial purposes. A combined heat and power plant in Southern England currently sells heat to an adjacent industrial complex at £0.30 per Therm, which is equivalent to £0.01 per kWh. Coal currently sells to industry at about £45.00 per tonne (£0.0056 per kWh). Recent tests have indicated that it will be technically feasible to make charcoal-oil mixtures with up to 30 per cent milled charcoal in a heavy residual oil. The heat value of the mixture is about 39.5 MJ/kg compared with 44 MJ/kg for the pure oil. Coal-oil mixtures have recently been subjected to a wide range of tests under commercial conditions and performed satisfactorily (Wall, 1981) and charcoal-oil mixtures seem to offer a number of advantages, especially the lower ash content compared with coal. A pyrolisis plant could therefore show the following gross revenue per tonne of air dry wood utilised:

1 555 kWh Heat	@ £0.01 /kW		- £15.55
2 015 kWh Charcoal	@ £0.0056/kWh	(coal substitute)	- £11.28
			£26.83
or 1 555 kWh Heat	@ £0.01 /kWh		- £15.55
2 015 kWh Charcoal	@ £0.0091/kWh	(charcoal/oil mixture)	- £18.50
			£34.05

(The charcoal/oil mixture would have to sell at £100.00 per tonne to compare on a unit energy basis with pure oil.)

The manufacturing costs (excluding the wood cost) are difficult to estimate for a general situation because they are very much influenced by local factors. However, a figure of £12.00 per air dry tonne of wood raw material is in line with current charcoal manufacturing costs in Britain, and would leave a residual value for the wood of £14.80 to £22.00 per tonne, which is comparable with current industrial wood prices.

This analysis illustrates the importance of being able to recover and sell the heat as well as the charcoal and also shows the advantage of making a synthetic liquid fuel as opposed to a solid fuel. It also shows that charcoal could be substituted for either coal or partially substituted for oil without the need to change existing combustion equipment. The latter point may well have considerable appeal to those developing countries which import large amounts of fuel oil for mineral processing such as Copper and Bauxite and which have adequate local resources of low value wood residues which could be converted into charcoal.

Gasification Techniques for converting wood or charcoal to a usable gas have been known for a long time and were intensively developed in several European countries before and during World War II. Interest in the technology declined after the war when abundant supplies of natural gas and oil were available at a low price. In the past five years, as the cost of hydrocarbon fuels has risen, interest in the gasification of biomass has been renewed.

There are a number of processes which may be used to convert wood to gas, and several plant designs which can be used for each process. Most of the processes and plant designs which are currently available commercially are fairly small scale.

The simplest and most widely used process involves heating wood or charcoal to about 400°C in a closed container and drawing air through the charge. The result is a gas with a low heat value of about 5.6 MJ/m³, containing about 30 per cent CO and 11 per cent H_2: the balance is mainly Nitrogen and CO_2 which has no energy value (Kissin, 1942). Yield is about 2.5 m³ of gas per 1 kg of wood, giving a thermal efficiency of about 75 per cent.

A variation of this process has been developed in North America and involves the use of oxygen rather than air. This results in a medium heat value gas with about 13.5 MJ/m³ and the proportion of CO and H_2 are increased to 60 and 28 per cent respectively (Van Swaaij, 1980: Reed, 1980).

If the wood or char is heated to about 1 000°C, steam may be used instead of air or oxygen to produce what has sometimes been called water gas. The resultant mixture has a similar composition and heat value to the oxygen gasification process (Van Swaaij, 1980).

A fourth process involves a further reaction with the CO:H_2 mixture under pressure in the presence of a catalyst which results in the production of methane, with a high heat value of about 31.5 MJ/m³, more nearly comparable to 'natural gas' (38 MJ/m³). This process has been developed for the gasification of coal and is not yet commercially available for wood.

The principle advantage of gasification compared with direct combustion or pyrolisis is that the gas can be used in an internal combustion engine to produce either mechanical energy or electrical energy.

Interest in the technology has revived recently and a number of manufacturers now offer gasifiers ranging in size from about 20 kW to about 2 MW output. One with an output of 250 kW has been supplied for the island of Bora Bora in the South Pacific for the generation of electricity from coconut husks. At least one manufacturer offers diesel engines specially modified to burn wood gas for the generation of electricity with outputs of up to 400 kW.

The thermal efficiency for the gasification process is claimed to be as high as 94 per cent (Roth, 1978) and the efficiency of the conversion to electricity by diesel generator will be about 33 per cent, giving an overall efficiency of about 31 per cent.

Overall efficiency could be increased by applying the principals of a combined heat and power plant (CHP) so that waste heat from the generator is recovered. This would raise the efficiency of gas utilisation to about 75 per cent and overall efficiency to about 65 per cent.

A gasifier and CHP plant would therefore produce the energy outputs given as follows, together with their current commercial value:

1 000 kg air dry wood	contains	4 400 kWh		
converts to		1 510 kWh	low grade heat @ £0.01/kWh	= £15.10
		1 380 kWh	electricity @ £0.025/"	= £34.50
		1 550 kWh	lost	£49.60

<u>Hydrolisis</u> When wood cellulose is treated with an acid or an enzyme, it is hydrolysed to glucose which can then be converted to alcohol by fermentation. During World War II a number of 'wood sugar' plants were operated in Europe and N. America using both concentrated and dilute acids. The most successful process used dilute sulphuric acid at a temperature of about 150°C (Di Novo et al, 1978) and gave a yield of 45 to 50 per cent by weight of sugar, and of up to 200 kg alcohol per tonne of dry wood. The ethyl alcohol has a heat value of about 28 MJ/kg which means an overall thermal efficiency of about 35 per cent in converting the energy in the wood raw material. Development work has been re-started on a number of processes, especially on enzymatic hydrolysis. This technique has one particular advantage in that the process is more efficient with moist wood, so reducing the problem of pre-drying, but it has not yet been commercially proved.

The economic viability of this process will, in the short-term, be very dependent on local factors, particularly on the abundance of low cost wood residues and the price of alternative fuels. In the longer term viability will depend on the relative premium value attached to liquid fuels as opposed to solid fuels and on improvements in the yield under commercial operating conditions.

<u>Hydrogenation</u> This process involves reacting hydrogen with organic material at elevated temperature and pressure to produce a gaseous or liquid fuel with a high proportion of hydrocarbons that give the product a high heat value, in the range 24 to 41 MJ/kg depending on the proportion of oxygenated compounds.

According to di Novo et al, 1978, laboratory trials of the process show an oil yield of 30 to 40 per cent by weight. In the prototype reactor the gases which are given off are flared, while additional energy is supplied to preheat the feedstock, and in the form of hydrogen for the reaction. Overall thermal efficiency for the process is therefore not known.

A similar technique is used in the Fisher Tropsch process, developed by Sasoil in South Africa for the liquifaction of coal. The coal is first converted to Carbon Monoxide and Hydrogen and then these two gases are heated in the presence of catalysts to produce hydrocarbons, water and carbon dioxide. The resultant liquid fuel has a higher heat value than the oils obtained by pyrolisis and may have economic advantages in the future in reducing the transport cost of wood-based fuels.

WOOD ENERGY IN DEVELOPING COUNTRIES

Most developing countries have predominantly rural populations, despite the widespread migration to cities of the past few decades. Few of these countries have large reserves of fossil fuels, but those that do either export most of it to earn foreign exchange or use it for their own industry.

Arnold and Jongma (1978) discuss the role of wood as a source of energy in developing countries and quote energy consumption figures ranging between 1 200 kWh per caput per annum in Bihar, India to 9 700 kWh per caput per annum in the Bolivian Andes. This compares with an average of about 8 200 kWh per caput per annum for 'domestic' energy consumption in the UK.

This widespread use of wood is mainly because it is the only fuel available locally and can either be collected or bartered for by people with very low incomes. Growing populations mean that energy needs are rising, but generally the forest is being destroyed both to clear land for agriculture and by overcutting for fuelwood. In many of the more densely populated parts of the world, and especially where forests do not regenerate easily, there is now a very serious crisis because villagers must spend more time gathering less fuel and destruction of trees means loss of shelter, shade and sometimes valuable supplementary food from fruits and nuts.

There are two approaches which are being pursued in tackling this problem: one involves taking steps to increase the supply of wood, and the other involves improving the efficiency with which the energy is utilised so that the same benefits can be obtained from less fuel. Joseph et al (1981) have examined a large number of cooking stoves used in developing countries and report efficiencies ranging from 6 to about 25 per cent when burning wood. In practice the efficiencies are likely to be lower than this because of heat lost during warming up and after cooking is finished. Liquid and gaseous fuels are not only more convenient to use but they are also more efficient as they can be turned off instantly when not needed and the heat can be directed where it is needed.

The use of charcoal is a step in this direction because despite the loss of about 50 per cent of the energy in the wood during carbonisation, a much higher proportion of the energy in the charcoal can be utilised effectively and if heat or gases are recovered, as described earlier, the process can be even more efficient.

Many developing countries have substantial reserves of minerals such as Copper, Nickel and Bauxite which need large amounts of energy for drying and refining. This is generally obtained by burning heavy fuel oil but its rising cost is stimulating interest in alternative fuels and especially in wood where it is available locally.

Although social and cultural conditions in developing countries are different from those in rural parts of Britain, many of the problems are similar and can be overcome by the same techniques. In the very poorest developing countries more efficient equipment is badly needed but in the more advanced developing countries with a substantial monetary sector, there is a growing need for more convenient types of fuel. Those countries which do not have reserves of fossil fuels but still have forests, or have the potential to grow fuelwood plantations, will need technology for converting their biomass fuel to liquid or gaseous fuels or electricity. Thus all the conversion routes described above are likely to find a wide range of applications in developing countries.

REFERENCES

Arnold, J.F.M. and J. Jongma, (1978). Fuelwood and charcoal in developing countries. Unasylva, 29, (118), 2-9, Rome.

Blowers, M.J., D.E. Fussy and R. Liversedge, (1980). Evaluation of Char-oil slurry combustion in a commercial boiler. Report 1/80 Energy Planning Unit, Dept. Minerals and Energy, Papua New Guinea.

Breag, G.R. and A.P. Harker, (1979). The utilisation of waste heat produced during the manufacture of coconut shell charcoal for the centralised production of copra. Rep. Trop. Prod. Inst., G.127.

Chereminisinoff, N.P., (1980). Wood for energy production. Ann Arbor Science: Ann Arbor.

di Novo, S.T., W.E. Ballantyne, (1978). Preliminary environmental assessment of biomass conversion to synthetic fuels. Ind. Environmental Res. Lab., Ohio.

Fraser, A.I., (1969). Some investigations of the relative development of crown, stem and root of trees growing on shallow peat soils in Britain. Suo, 20, 1-8.

Fraser, A.I., (1980). Wood as a source of energy for rural and island communities in Energy for Rural and Island Communities, Ed. J. Twidell, Pergamon: Oxford.

Hall, D.R., (1981). The handling and combustion of stable Coal/Fuel Oil dispersions. Proc. 3rd Int. Symposium on Coal/oil mixture combustion, Orlando, Florida.

Hardie, A.D.K. and R.A. Plumptre, (1979). A solar heat dryer for timber. Appropriate Technology, Vol.6, No.1, 4-6.

Joseph, S. and Y. Shanahan, (1980). Compendium of tested stove designs. ITDG: London.

Kissin, I., (1942). Gas producers for motor vehicles and their operation with forest fuels. Imperial Forestry Bureau, Oxford. Tech.Commn.1.

Levelton, B.H. and D.V. O'Connor, (1978). An evaluation of wood-waste energy conversion systems. Environment Canada: Western For. Prod. Lab.: Vancouver.

Reed, T.B., (1980). Biomass gasification: yesterday, today and tomorrow. Proc. Conference on Energy Generation and Co-generation from Wood. Atlanta, Georgia, USA. Feb. 1980. For. Prod. Res. Soc.

Roth, H.E., (1978). Possible applications of stationary wood-gas generators system Roth for energy production. Proc. ECE Symposium on Energy questions of Wood Industry, Udine, Italy.

Thoresen, P.E., (1980). Charcoal a possible energy carrier for wood. Proc. International Forestry Energy Meeting, Jonkoping, Sweden.

Van Swaaij, W.P.M., (1980). Gasification - The process and the technology in Proc. Energy from Biomass Conference. Ed. Palz, N., Chartier, P. and Hall, D.O. Applied Science Publishers: London.

A Study of the Feasibility of Coppice for Energy as a Lowland Farm Enterprise

B. J. Unwin and C. P. Mitchell

Forestry Department, Aberdeen University, St Machar Drive,
Aberdeen AB9 2UU, UK

ABSTRACT

An area in South Worcestershire was surveyed to assess the feasibility of coppice for energy as a lowland farm enterprise. Three per cent of land in the area was suitable for growing coppice without interference with existing farm enterprises. Productivities of between 5 and 9 dt/ha/year on 10 year cutting cycles could be achieved. The estimated internal rate of return ranged from 0 to 5.3 per cent. Farm labour was judged to be available on 10 farms. Twelve of the 25 farms surveyed could produce sufficient wood to provide all domestic heating. Of these, 8 could also provide some farm energy. Recommendations for implementing farm forestry for energy are made.

KEYWORDS

Wood fuel; coppice; farm forestry.

INTRODUCTION

Recently there has been much interest shown in large-scale production of wood for energy particularly in Eire, Finland and Sweden where large expances of cut-over peat provide an excellent site for growing trees. However, in Britain the use of small areas of non-productive land for energy plantations on farms has been postulated. Morgan and Bather (1980) estimated that 10 per cent (1.12 Mha) of lowland Britain is unused or under-used, Pearce (1980) considered that 2.5 per cent (300,000 ha) of farmland including small woods is suited to tree production and Mitchell and Matthews (1978) estimated that 500,000 ha was potentially available for growing trees for energy in the lowlands.

These estimates relate to national figures and do not consider local availability of land for growing coppiced trees for energy for on-farm use, its cost of production, availability of labour or the attitudes of farmers to coppice as a farm enterprise. The work reported here examines these aspects in relation to a specific area of lowland Britain.

STUDY AREA

The study area (2,355 ha) is situated in the lower Seven-Avon lowlands of south Worcestershire, between the Malvern hills and the Cotswold scarp and has a population of 500. Agriculture is the only industry within the study area, in which, including owners, about 90 people are directly employed. The area contained 15 farms in entirety, more than half of a further 8 and part of 5 more.

The area has a Land Use Capability of Grade 3 due to surface wetness and difficulties of cultivation of the marl-derived soils, producing a pattern of predominantly stock and dairy farms. However, the trend in the area has been increased drainage leading to an increase in cereal area and a decrease in sheep.

Rainfall in the area is low (less than 700 mm) but medium to fine textured soils which cover more than 95 per cent of the area provide adequate moisture to deep-rooted crops and trees until late-July and early August.

The whole of the area was surveyed in 1980 to ascertain the number of sites and total amount of land which could be used for growing trees for energy without interfering with agricultural production. This land was, for the purpose of this study, called 'available land'. The survey resulted in a total of 142 potential sites. As a result of subsequent interviews with 23 of the farmers 36 sites were rejected as unsuitable, but 20 more were suggested bringing the total approved by farmers for planting coppice to 126 sites (69.41 ha).

Most of the sites were less than 1 ha in size and all of the larger sites were existing woodland. Indeed, 54 per cent of the 'available land' is at present under broadleaved trees in small woods or isolated groups. About 5 per cent has a cover of conifers, 28 per cent scrub and 13 per cent pasture on steep banks and in awkward corners.

'Available land' amounts to 3 per cent of the study area. This represents 3 ha on an 'average' farm in the area (99 ha). However, 8 farms have less than 1 ha and 7 more than 3 ha. There appeared to be no relation between type of farming and amount of 'available land' on a farm. 'Available land' (and farm enterprise) appear to be related to physiography and past land management.

COPPICE PRODUCTION

Of the 23 farmers interviewed 19 were in favour of the idea of planting coppice for on-farm use. In order to assess the feasibility of such an enterprise each of the 126 sites were examined in detail and estimates of coppice production, costs of operations and labour requirements were made.

The 126 sites were classified by their physiography in order to ascribe a yield and costs. Four site types were identified:-

1. Lower slope - soil moisture receiving and sheltered.
2. Mid-slope - soil moisture run-off approximately equals the amount of water received from upper slopes.
3. Top of slope, ridge tops - soil moisture shedding and exposed.
4. Wet - hollows with nearly flat or depression topography.

The object of growing trees on 'available land' is the maximization of wood production for on-farm use where the major constraint is a requirement for manual harvesting which requires a minimum cutting cycle of 10 to 12 years. This cutting cycle in turn determines the choice of species and planting density. The tree species prescribed for farm coppice in the different site types and their optimum productivities appear in Table 1.

TABLE 1 Coppice Productivity: 2.0 - 2.2 m Square Spacing

Species	Site Type	Yield (Dry t/ha/yr) Maiden	Coppice
Poplar clones	1	8	10
Willow clones	4	8	10
Nothofagus spp	2	7	9
Sweet chestnut	2	5 ?	6
Sycamore	3	4.5	6

Poplar and willow clones should have productivities of 8 dt/ha/year before coppicing (maiden) and 10 dt/ha/year thereafter on a cutting cycle of 10 years. These are 'working' yields for good sites and on heavily shaded or poorer sites yields will be much lower.

On 51 per cent of the area of 'available land' (26 sites) a coppice system under existing standards is proposed. Here the productivity of coppice might be 5 to 7 dt/ha/year on a 10 year cutting cycle.

If all the 'available land' were established as coppice then the estimated productivity from the 69.4 ha would be 5.9 dt/ha/year. The range for the farms was 0.5 to 42 dt/year (Table 2).

The coppice will be felled by chainsaw and chipped with a tractor-mounted chipper. In this way all above-ground woody material is utilized. Production of chips enables easy bulk handling, storage and automatic stoking of boilers.

ECONOMIC RETURN

The economic return of growing coppice for energy was assessed using the costing method of revenue minus variable costs plus assignable fixed costs including labour. The value of fuelwood is unknown because it has no established market price. Therefore, the opportunity cost of pulpwood was taken as the revenue for wood.. The local roadside price for pulpwood in 1980 was £9.25 per green tonne which equates with £18.50 per dry tonne.

Costs and yields for one maiden and two 10 year coppice cutting cycles (30 year rotation) were estimated for each site independently giving a cash flow and discounted flow for each farm. A rate of return of 0 to 5.3 per cent is encouraging for such small areas with high establishment costs, and should increase as energy prices rise. The average costs and yields for coppice production over the study area would give an annual net revenue of £38.63 per hectare.

LABOUR AVAILABILITY

When the standard-man days for the existing enterprises on each farm were examined in relation to existing labour force and seasonal requirements a considerable surplus of labour was found on 10 farms. This could be used for working coppice.

TABLE 2 Potential Wood Production on Farms in Study Area

Farm	No. of Sites	'Available Land' (ha)	Annual Yield (dt/farm)
A	7	3.94	22.5
B	6	2.31	18.5
C	7	2.54	17.6
D	13	5.08	31.9
E	6	3.74	20.5
F	4	3.37	22.4
G	6	4.40	25.2
H	2	0.22	1.8
J	5	1.23	4.4
K	4	2.51	20.3
L	6	1.62	7.4
M	1	0.30	2.4
P	1	0.40	1.6
Q	2	0.94	8.5
R	2	0.06	0.5
S	5	1.28	7.3
T	2	0.15	1.4
U	5	2.13	10.0
V	8	5.74	41.9
W	5	1.89	12.0
X	2	0.13	1.0
Y	4	5.35	30.1
Z	2	0.26	2.2
Others	21	19.82	97.2
TOTAL	126	69.41	408.5

COPPICE ENERGY

These coppice systems were designed with a view to on-farm consumption of energy. Unfortunately, we do not know the actual energy consumption farms in the area but estimates can be made. If we assume a range of energy demand for domestic consumption per farm house of 50 GJ/year to 260 GJ/year then this represents a range of of 4.4 to 19 dt/year of wood assuming a combustion efficiency of 72 per cent for air-dried wood compared to 85 per cent for oil (Tillman, 1978). Eight farms produce less than 4.6 dt/year and 8 have an estimated production greater than 20 dt/year. An intensive stock farm might require 700 - 1500 GJ/year for static installations. This represents 60 - 130 dt/year of wood which clearly cannot be achieved by coppice on 'available land' alone. Coppice for energy could therefore make 4 farms self-sufficient in energy for 1 house (10 - 20 dt/year) and 8 farms self-sufficient for 2 houses (20 dt/year).

CONCLUSION AND RECOMMENDATIONS

The results of this study indicate that growing coppice for on-farm use of energy on 'available land' by farmers is a feasible proposition. What, however, is lacking in the system is a means whereby the farmer can be made aware of and be advised how to capitalize on this overlooked resource.

Farmers must be interested and not thwarted by lack of immediate information. Therefore, to establish coppice for energy as a farm enterprise advisors are required who are knowledgeable in forestry as well as farming and backed-up by a simple grant scheme.

REFERENCES

Mitchell, C. P. and J. D. Matthews (1978). The Potential of Forest Biomass as a source of energy. Aberdeen University, Report to Energy Technology Support Unit, Harwell.

Morgan, K. E. and M. Bather (1980). Energy - making the most of it. Paper II National Power Farming Conference - Survival through the '80's.

Pearce, M. L. (1980). Coppice Trees as Energy Crops. Forestry Commission, Report to Energy Technology Support Unit, Harwell.

Tillman, D. A. (1978). Wood as an Energy Resource. Academic Press. New York.

Biomass Potential in Thailand

M. Slesser* and C. Sintunawa**

*Energy Studies Unit, University of Strathclyde, Glasgow, Scotland, UK
**Mahidol University, Bangkok, Thailand

ABSTRACT

Thailand is a country which imports energy and exports food. Rising energy prices place an increasing burden on balance of payments and even threatens the energy security of the country. The paper explores, by means of a dynamic simulation model, the potential advantage of switching food producing land to biomass production for subsequent alcohol manufacture as an energy import substitute. It is found that the results, while favourable to energy security, may not favour balance of payments. However these results are very sensitive to assumptions about future energy and rice prices and the potential net energy of biomass systems.

KEYWORDS

Biomass fuel, Food intensification, Energy for rice intensification, Rice paid energy import, Net energy from surplus sugar cane, Energy cost balance, Rice stock, Net energy, land diversification.

INTRODUCTION

Thailand, like many other countries of the Third World, is faced with an energy crisis, for its energy supply depends on increasingly expensive imported oil. Since 1973, when the crude oil price increased fourfold, the progress of economic development of Thailand has been severely affected. This situation forces Thailand to explore other sources of energy and means of alleviating dependence on imported oil.

Bio-energy sources have been proposed as one way of combating the effect of rising oil prices and impending oil scarcity. While such means may not be open to the highly developed and densely populated countries, they do seem, on the face of it, to be feasible for the less developed countries, especially those that are already food exporters, such as Thailand.

In general, the development of bio-energy sources means a shift from land used for food production to land to be used for energy production. For most, if not all, developing countries, existing biomass fuels are being used at the limit of their renewability, or far beyond it. The analytical problem facing one,then is, in essence, a very simple one. If food producing land is conceded to the production of energy, then on the remaining land the output of food must be intensified. It is well known that to intensify food production carries with it an energy penalty, so the simple question is : How far can one switch land from food production to energy production, before the energy gains from bio-energy are balanced by the additional energy needed to produce food more intensively upon the remaining food producing land?

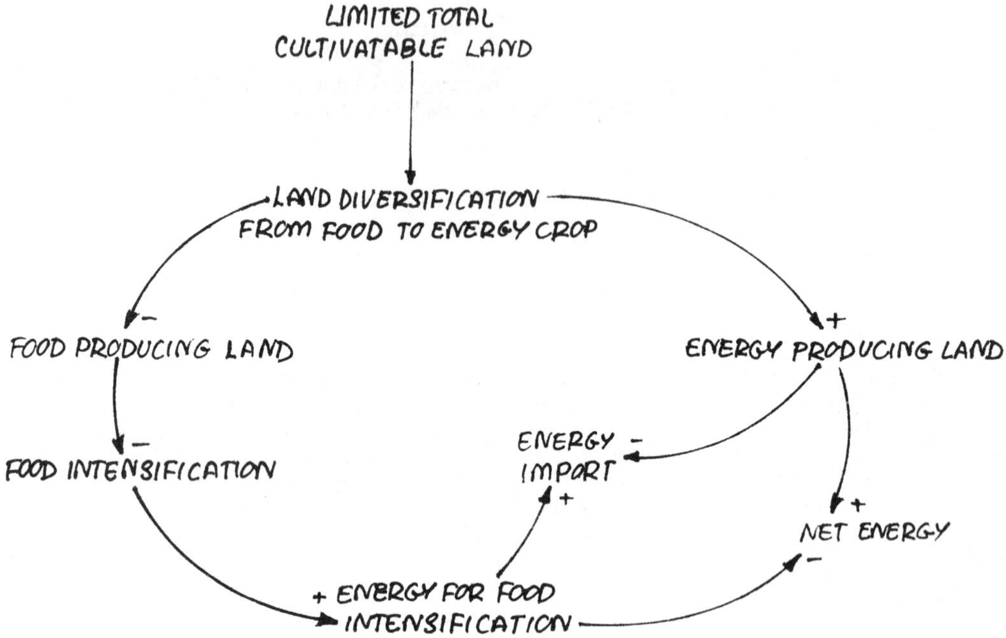

FIGURE 1. CAUSAL RELATIONSHIP OF ENERGY-FOOD SYSTEM.

In the case of Thailand, as for any food exporter, there are at least two policies which could be adopted. In Thailand's case food exports pay for oil imports. One could therefore choose to intensify food production and export more food, thereby having more money to buy energy. This policy leaves the country at the mercy of external markets, and offers no development towards national energy security. Another policy would be to maintain food exports, while intensifying food production within the country to meet the declining availability of land, as food producing land is shifted to energy producing land.

This paper explores the second policy, considering only one fuel type : ethanol via sugar cane. Figure 2 depicts the causal relationships of such a policy. These in turn have been turned into system dynamics langugage so that a dynamic mathematical model may be constructed to reveal change through time. The model is operated using system dynamics.

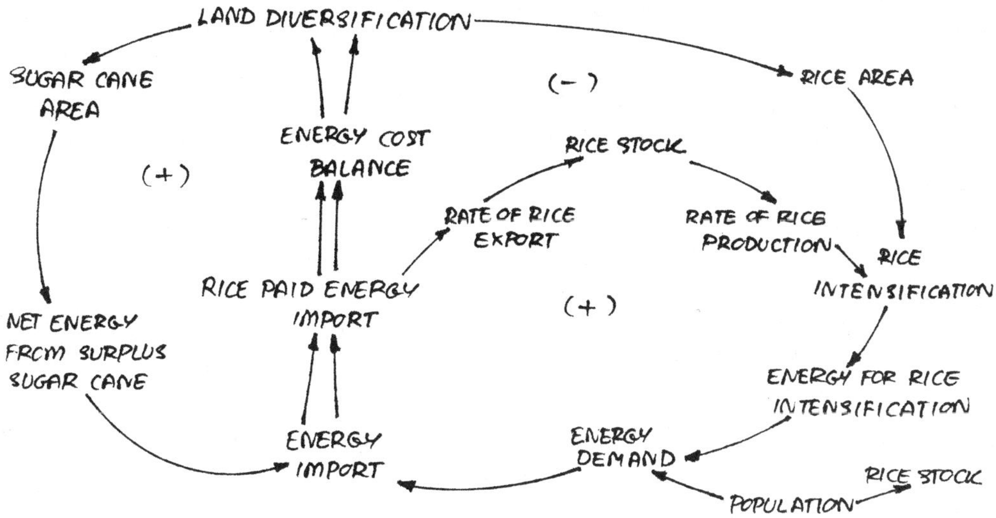

FIGURE 2. CAUSAL LOOP DIAGRAM OF ENERGY AND FOOD SYSTEM.

DATA

There are several key pieces of data necessary for such a model, beyond a description of the country in terms of agricultural and economic statistics of today. Firstly it is necessary to know what is the net energy that may be obtained from a bio-energy system per hectare of the land so used. Table 1 gives data for a number of possible Thailand crops harvested and turned into ethanol. (1, 2). The net energy per hectare is obtained by considering all the energy needed to grow the crops, transport them, process them, and yield, finally, commercial grade ethanol. Sugar cane offers the best net energy (27.2 GJ/ha.yr).

Secondly it is necessary to know the relationship between intensification of a crop and the energy requirement of the added inputs to intensification. The data should be of the form input intensity (GJ $ha^{-1} yr^{-1}$) versus output intensity (net energy) (GJ $ha^{-1} yr^{-1}$) according to Slesser, Lewis and Edwardson (3). This data is not known with any accuracy for Thailand, and is to be the subject of detailed study in Thailand in the forthcoming year. However we can make our model operational by using already developed global plots (3) to get a feel for the problem and the resulting policy options.

It is of course, also necessary to know how the Thai population will grow through time, what will be the future cost of oil, what non-agricultural energy demand there will be, what will be the price obtained for Thai food exports.

There are no reliable answers to any of these questions. On the long haul, one would expect a relationship between energy prices and food prices as is indeed formed in thailand. Figure 3 depicts how Thailand has been affected in recent years. However, one cannot affirm with any certainty that such a 'gearing' will persist.

TABLE 1 Net Energy (MJ) of Various Major Crops in Thailand Converted to 1 kg of Ethanol.

Major Inputs (MJ energy to produce)	Rice	Cassava	Maize	Sorghum	Sugar Cane
Starch	24	25	15	18	-
Fermentable sugar	-	-	-	-	6.5
Chemicals for Fermentation	0.9	0.9	0.9	0.9	0.6
Water	0.4	0.4	0.4	0.4	0.3
Plant Construction	1.2	1.2	1.2	1.2	0.5
Energy Consumption	30.5	30.5	30.5	30.5	16
Total energy input per kg ethanol (MJ/kg)	57	58	48	51	24
Energy in ethanol MJ/kg)	29.7	29.7	29.7	29.7	29.7
Net Energy MJ/kg)	-27.3	-28.3	-18.3	-21.3	5.7
Net Energy available per hectare (GJ/ha yr)	-9.7	-13.0	-7.0	-.49	27.2

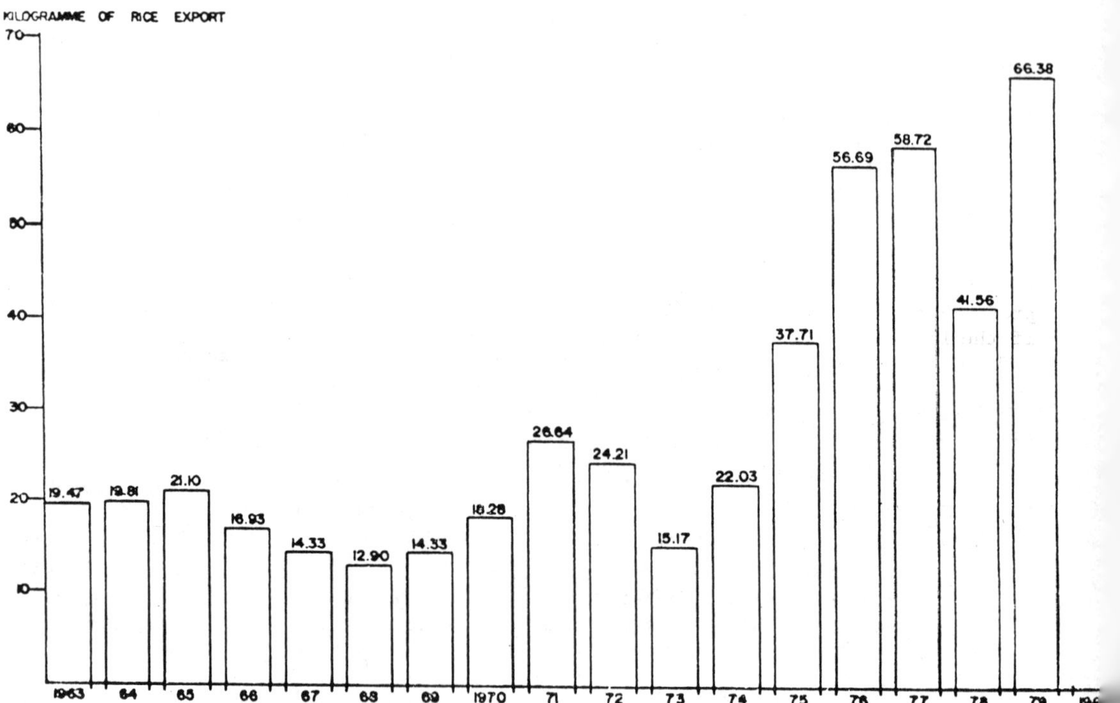

Figure 3 – KILOGRAMMES OF RICE EXPORT EQUIVALENT TO ONE BARREL OF CRUDE OIL IMPORT

RESULTS

Population and prices are constant

Bearing these factors in mind, the model is set up to examine first of all, the outcome for zero population growth, and no change in energy import or food export prices. Land is switched from food production to bioenergy production in units of area equivalent to the optimum size of bioenergy manufacturing (crop to ethanol) units. In Thailand today this means units at 20,000 ha (200,000 litres/day distillery). We have, after exploration, decided to switch land from rice production to sugar cane production at the rate of 100,000 ha/year (in keeping with the creation of five new processing units per year). Figure 4 reveals that such a policy is good. Food exports are maintained, the balance of payments improves. Thai energy production from renewable sources rise from 2.5 to 8% of energy demand, while oil imports drop 9%.

Population increases : prices are constant.

We now allow the population to increase at the current rate of 2.3% annually. Figure 5 depicts that outcome. Land has been diverted to produce more bio-energy and the balance of payments improves progressively until the highest potential of rice intensification has been reached after which the payments begin to decrease rapidly due to falling rice export. It also shows the more land diverted the less energy will be imported, though energy demand increases proportionally with population increase, i.e. population has a direct influence on energy demand.

Population and prices increase

Figure 6 shows how the outcome is affected if oil prices rise 15% a year, and rice prices rise only 9% per year. The outcome shows that Thailand (on the basis of assumed increase in oil and rice value) has to increase rice intensification rapidly in order to pay for the increasing value of energy imports. Balance of payments improve till rice intensification reaches the highest level and then drops rapidly afterwards. The larger the land transferred to bio-energy production the less energy Thailand has to import thereby becoming less dependent on energy imports.

CONCLUSION

The present rate of population growth on the limited land area will cause increasing pressure on both energy and food systems. If oil prices increase at a faster rate than rice prices, the only way in which Thailand can increase bio energy production is to intensify both bio-energy and food crops. In Thailand's case if the balance of payments is the main policy goal it is better to export rice and import energy. If energy security is the main policy goal, transferring land to bio-energy will provide national energy security but at some loss to balance of payments. These tentative conclusions are basedon uncertain projections of future oil and rice prices, and on trial data. Further studies particularly to provide on site data may alter these perceptions.

REFERENCES

1, Sintunawa C. (1978). Power Alcohol from Agricultural and Agro-industrial products : An Appropriate Measure to develop Energy Sources for Agricultural Developing Country, Mahidol University, Thesis.

2 Slesser M. and Lewis C. (1979) Biological Energy Resources E & FN Spon, London.

3. Slesser M. Lewis C. and Edwardson W : (1977) Energy System Analysis for food policy. Food Policy, 2, 123.

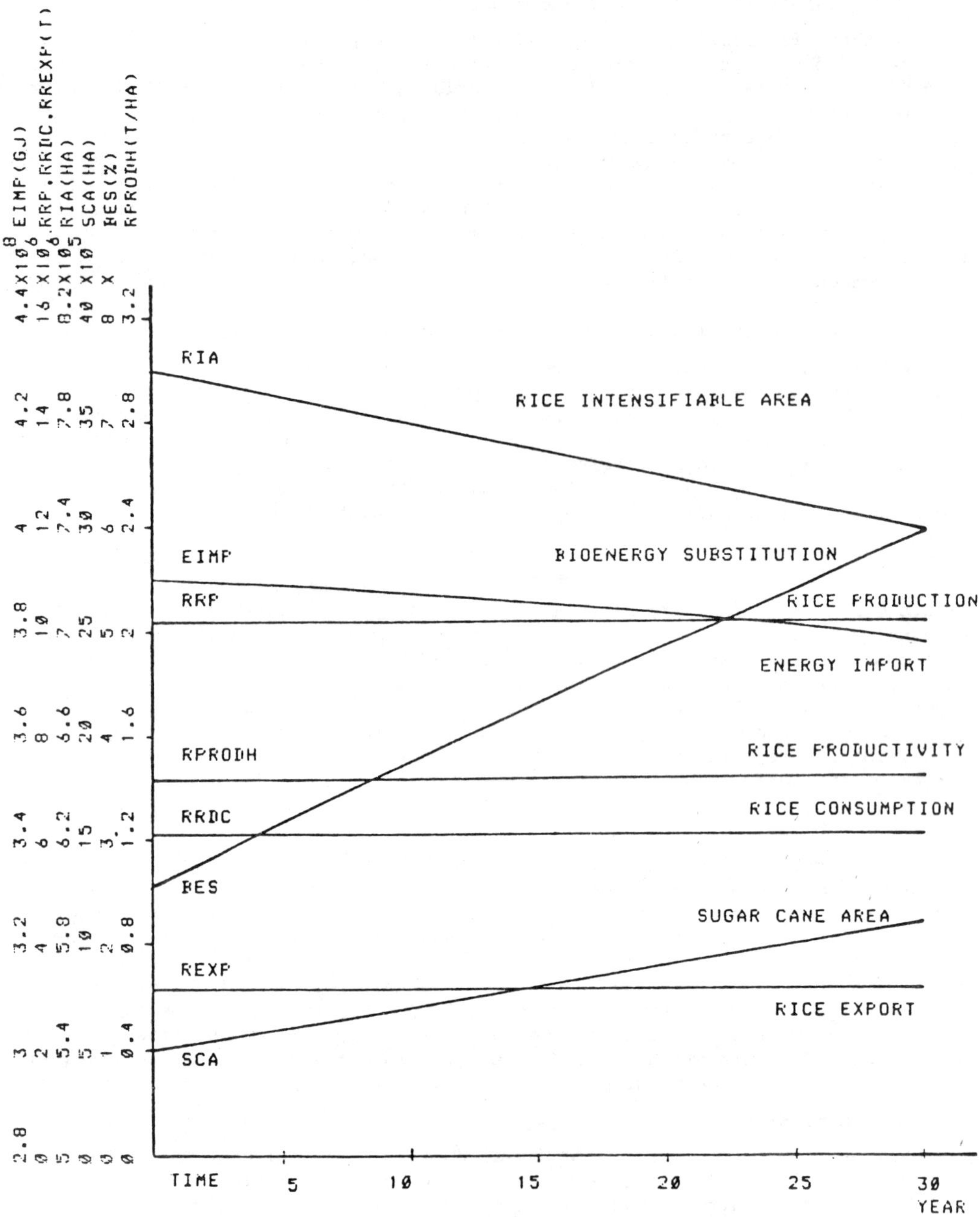

FIGURE 4.EVOLUTION OF SYSTEM:RICE-SUGAR CANE
POPULATION AND PRICES ARE CONSTANT

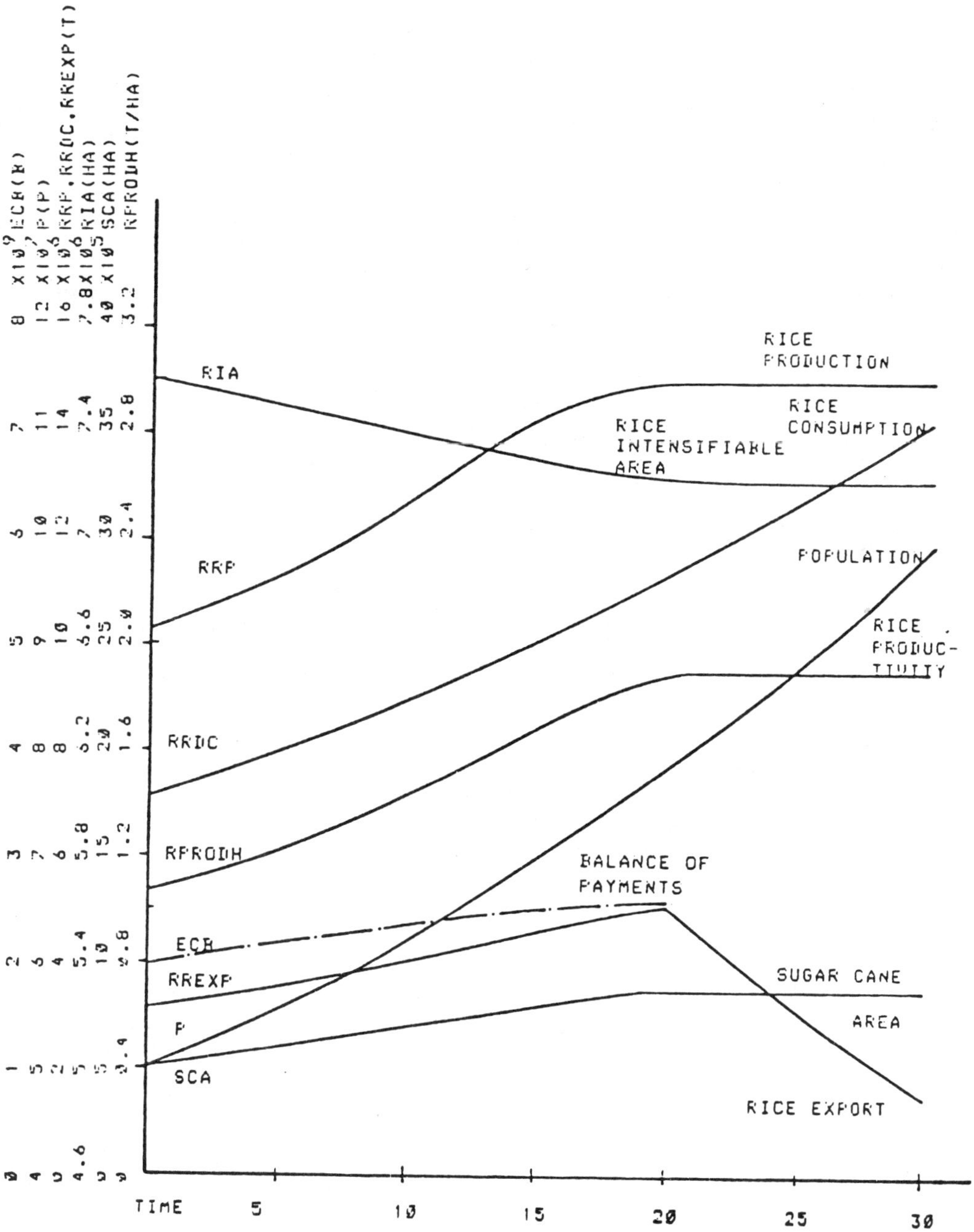

FIGURE 5.EVOLUTION OF SYSTEM:RICE-SUGAR CANE
POPULATION INCREASES:PRICES ARE CONSTANT

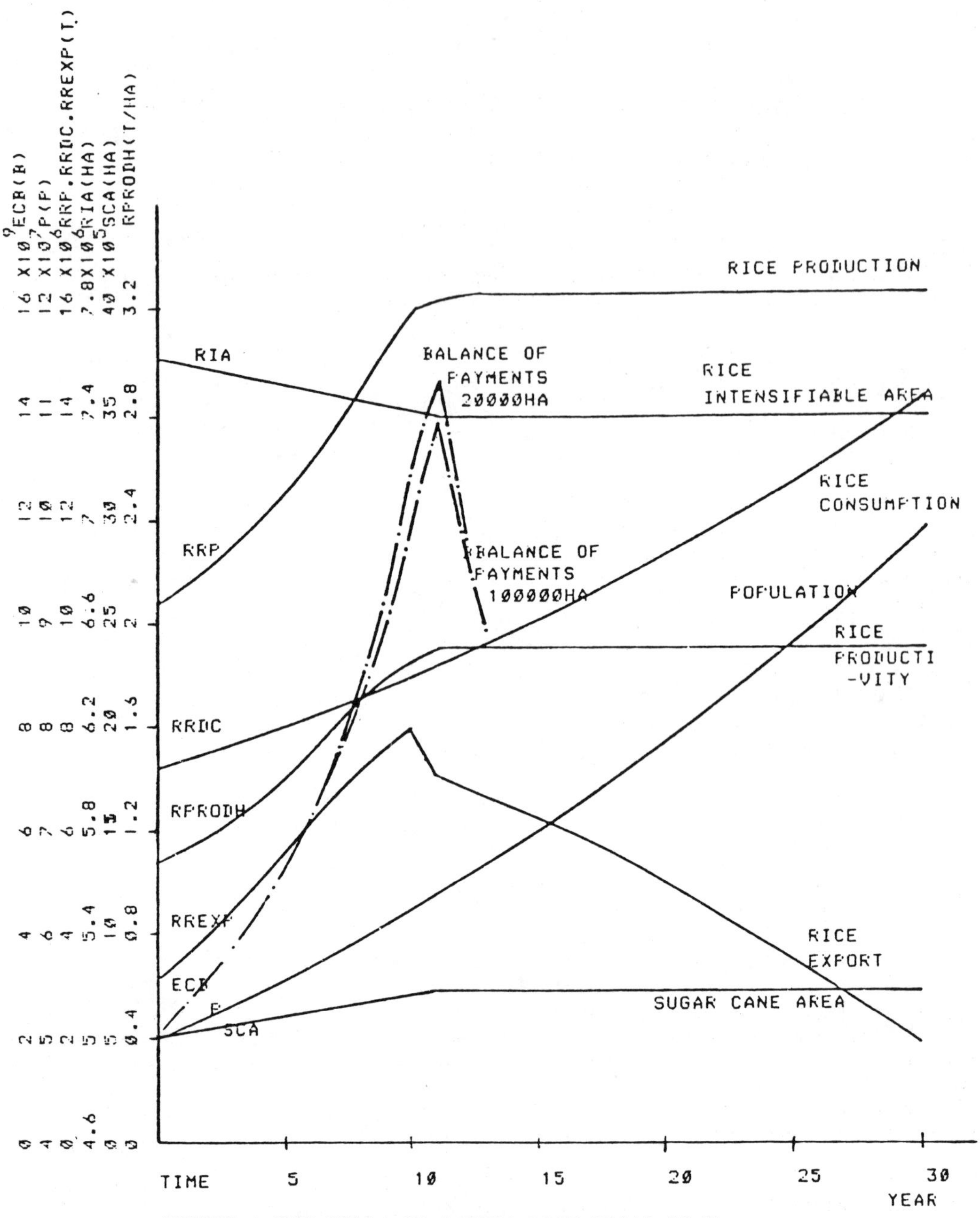

FIGURE 6.EVOLUTION OF SYSTEM:RICE-SUGAR CANE POPULATION AND PRICES INCREASE

Rural Energy Strategy — Alternative Applications in Rural Pakistan

J. L. Gaddy* and I. H. Usmani**

*Department of Chemical Engineering, University of Arkansas, Fayetteville, Arkansas 72701, USA
**United Nations, New York, NY, USA

ABSTRACT

About one quarter of the world's population lives in rural areas in remote regions of the world. Conventional forms of energy are usually not available in these areas. The use of alternative sources of energy is the only hope that these regions have of modernizing their societies. Agriculture is the primary means of support and agricultural residues are available as an energy resource. These residues can be converted into biogas by anaerobic digestion. Biogas can be used as a source of heat for cooking and also to generate electricity. This paper describes the design of biogas units for remote villages in Pakistan. Systems to supply energy for a population of about 500 cost around $200,000 and deliver biogas for about $5-8 per million BTU and electricity for $.03-.05 per kwh. Considering the increased income from irrigation of crop land, the cost of these systems could be recovered in about five years.

KEYWORDS

Biomass; Rural Energy; Anaerobic Disgestion; Agricultural Residues.

INTRODUCTION

Approximately 20 percent of the world's population live in small rural communities in Asia, Africa and Latin America (Usmani, 1978). In general, these communities do not have access to fossil fuels or electrical power distribution systems, deriving their small energy consumption from the land: wood and animals. As a result of the lack of conventional energy sources, the economic development and quality of life of these people is behind the rest of the world.

With rapidly increasing cost and scarcity of energy, there is little hope that conventional energy sources can be extended to serve remote areas. Most of these tropical regions have an abundance of sunshine, biomass, wind, or, in some cases, hydropower. Utilizing recently developed technology, these alternative energy sources could be used to furnish heat and electricity for remote areas. Over the long term, these technologies should be more economical for such locations, since they are based upon available and renewable resources.

Recognizing the needs in these remote areas and the potential of employing new technologies to fill these needs, the Governing Council of the United Nations requested: "the Executive Director to accord high priority to the establishment in some of the typical rural areas of the countries of Asia, Africa and Latin America, in cooperation with the governments of the countries concerned and such agencies within and outside the United Nations system as may be considered appropriate, of a few demonstration centers harnessing, individually or in combination, the renewable resources of the energy locally available."

The purpose of this paper is to describe the feasibility of utilizing biomass to supply energy for rural villages in Pakistan. The requirements for heat and electricity in a typical village are presented. The design of a unit to produce biogas and electricity from biomass is developed. The capital cost and economics of this installation are discussed.

BIOMASS ENERGY RESOURCES

The biosphere annually assimilates 10^{18} kcal of energy as biomass, or about 15 times the world's current energy consumption. Biomass, in the form of native vegetation, agricultural residues or energy crops could supply a significant fraction of the world's energy needs today.

In remote regions, present energy needs are small. These people depend upon agriculture as a primary means of support and agricultural residues are available. These regions are often tropical and biomass is plentiful. In more arid areas, ground water is available for irrigation to increase crop yields, if energy were available to raise and distribute the water.

Biomass Conversion

Agricultural residues could be burned directly to supply heat, but heat makes up only a small part of the energy needs of these villages. Biomass can also be converted into synthesis gas and charcoal by pyrolysis at high temperatures. This process produces a gas that can be burned to produce mechanical or electrical energy. The pyrolytic conversion process requires a large input of energy and the resulting efficiencies are low. More importantly, the nutrients in the residue are largely destroyed at high temperature. Consequently, this process is not particularly attractive for agricultural residues, since the land would be depleted of nutrients and minerals.

Biomass can be converted into methane biologically at normal temperatures. Agricultural residues have been found to be readily digestible in the University of Arkansas laboratories, giving high yields of methane and good overall energy efficiencies. The biological conversion process converts carbon in the biomass into gas and leaves the nitrogen and other minerals in a form suitable for reapplication to the land as fertilizer. Furthermore, the biological process is reasonably simple to operate and would be more appropriate for unskilled technicians.

The mixture of methane and carbon dioxide produced can be used for cooking and to drive engines to power machinery or generate electricity. Engines must be derated when operating with biogas, and resultant efficiencies are poor (as is also the case with syngas by pyrolysis). Nevertheless, biogas represents a versatile form of energy that is compatible with the agricultural environment. Biomass can be stored easily and electricity generated from biogas could be integrated with photovoltaics and wind energy to provide an economical energy system.

BIOGAS SYSTEM DESIGN

The components of the biogas system are the reactors to convert the biomass, gas storage and distribution and electrical generation. The primary criterion for design of a biogas system for rural application is that of simplicity, dependability and ease of operation. The system should be economical in terms of both capital and operating cost. The size of the biogas unit will depend upon the heating and electricity requirements of the village. The basic design concepts are described and a specific design presented.

Reactors

Two types of reactors could be used for conversion of crop residues into methane: batch or continuous. Batch reactors, while less efficient, are much easier to operate and are preferred for smaller sizes. However, to meet the requirements of an entire village with batch reactors would require impractical reactor volumes. To reduce the complexity of continuous reactors and eliminate the need for continuous pumping of feed and effluent, a semi-continuous operation is proposed. This system is fed only once per day (or less frequently if desirable).

Data for the design of reactors digesting corn residues have been obtained in the University of Arkansas laboratories. These data, obtained in 10 ℓ reactors, are summarized in Table 1 (Foutch and Gaddy, 1981).

TABLE 1 Results of Laboratory Corn Stover Digestion

Retention Time (Days)	Gas Production (Liters/Day)	C_{in}	C_{out}	Reaction Rate (gmoles/L-day)	Conversion (Percent)
20	6.366	2.904	1.769	0.0568	39.1
20	8.815	3.630	2.059	0.0786	43.3
30	6.635	3.630	1.710	0.0642	52.9
40	6.052	3.630	1.472	0.0540	59.5
40	5.944	3.630	1.418	0.0553	60.9
100	2.889	3.630	1.054	0.0258	71.0

The feed concentration for these runs was 5-10 percent solids. Methane concentration averaged 55 percent. Conversions of 50-70 percent were obtained over the range of retentions. The data show a range of methane yields from 2.4 CF/lb at 20 day retention to 4.3 CF/lb at 100 day retention. Figure 1 is a plot of reaction rate and carbon concentration, yielding a first order kinetic constant (slope) of .045/day. The kinetic constant can be used to determine the reactor volume for various biogas production rates. The reactors for this system are cylindrical steel tanks mounted vertically. Gas space of 25 percent of the liquid volume is necessary. Several reactors operating in parallel give good operating flexibility. For example, six 10 ft. high tanks would produce 1 x 10^6 CF methane per year. The reactors are designed for operation with 10 percent (weight) solids, periodic agitation and constant temperature of 95°F. Operation without agitation is possible, but at reduced performance, and is not recommended for digestion of agricultural residues.

Reactor Agitation

Periodic agitation of the reactors is essential to good conversion of the biomass and to avoid solids or scum accumulation in the reactors. Agitation for a few minutes several times a day is adequate to promote good digestion. Agitation could be provided in three ways: mechanical, liquid circulation or gas recirculation.

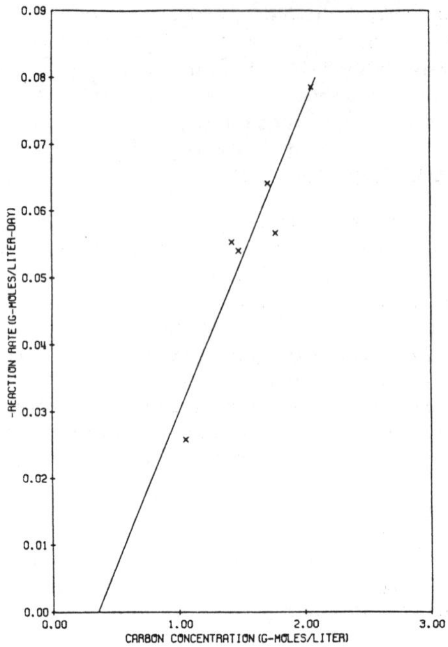

Fig. 1 Reaction rate versus corn stover carbon concentration.

Adequate agitation can be accomplished by periodically recirculating a portion of the product gas into the bottom of the reactors. This method requires a gas compressor, and this equipment will be installed for distribution of gas in the village. Therefore, the only additional equipment needed are gas distributors in the bottom of each reactor. This type of agitation can be done automatically and should be simple and trouble-free. A gas rate of .65 CF/min - ft^2 is suggested for moderate agitation (Perry, 1973).

Reactor Heating

It is essential that the reactors be maintained at a constant temperature. Two temperature ranges are possible: mesophilic - about 95°F, or thermophilic - about 130°F. The reaction rates are reported to be slightly higher in the thermophilic range. However, the additional energy required to keep the system at these higher temperature cannot be justified by the improvement in yield. Therefore, an operating temperature of 95°F is recommended.

Heat must be added to maintain the system at 95°F and to warm the fresh water fed with the biomass. This heat could be supplied from a portion of the methane produced. However, substantial quantities of waste heat are available from the exhaust gases from the engines driving the electrical generators or the compressor. This heat is more than enough to supply the requirements and should be used.

Several options are available for supplying heat to the reactors. These options include circulating warm water or hot gases, pre-heating the make-up water, electrical heaters and solar panels. Another alternative, and the simplest and most economical method, is to enclose the reactors and generators in an insulated building and to use the exhaust heat to warm the building with a space heater. This concept is the simple and easy to operate, but would require a rather large building for these installations. An expensive type of building is not required and would cost no more than all-weather insulation for the tanks. Water heating can also be easily accomplished with a storage tank installed in the building.

Biomass Preparation and Reactor Feeding

The agricultural residues must be ground and mixed with 9 parts of water, and this mixture heated to 95°F before introduction into the reactors. The residue should be reduced in size to 1/8-1/4" in size for good digestion. A standard forage chopper powered by a gas or electric motor is suggested for this service. The ground residues can be blown into the reactors or conveyed manually with a cart. Warm water would be introduced by gravity flow.

Prior to feeding of biomass, effluent from each reactor is withdrawn into a drain outside the building. The drain is sloped for flow into a holding pond or directly into the irrigation drains for delivery to crop land.

Gas Storage and Distribution

Distribution of gas throughout a village would require short transmission lines in most cases. Two inch piping with a gas pressure of 30 psig is adequate for this service.

The compressor must furnish gas for agitation as well as for delivery to the village. For the systems under consideration, agitation is the greater requirement and fixes the design and a 5 hp compressor is required. Low pressure gas storage is available in the gas space in the reactors. Additional low pressure storage may be provided in PVC bags, but high pressure storage for about one half day requirement should also be provided.

A schematic of the biogas and electricity operation system is given in Fig. 2.

APPLICATIONS IN PAKISTAN

Seventy percent of Pakistan's approximately 70 million people live in rural areas and depend upon agriculture for their livelihood. The principal crops are corn and wheat with rice, cane, peas, cotton, and vegetables grown in smaller amounts. Buffalo, cattle and goats are kept to furnish meat and milk.

The rural population lives predominantly in small villages. About 40,000 such villages in Pakistan have no electricity. Energy for cooking is supplied from cow dung, wood and twigs from surrounding forests or crop residues. Camels and donkeys are used for transportation. Table 2 gives a summary of the data for a typical village.

Energy Requirements

The village of Tara Garh consists of permanent mud and tile houses. Energy is needed in this village for lighting, cooking, refrigeration, irrigation and drinking water pumping. The energy consumption factors (Usmani, 1978) for these requirements are given in Table 3.

Applying these factors to the requirements of Tara Garh gives the resulting energy usage as shown in Table 4. The lighting requirements include houses, streets, school, hospitals and public buildings. Drinking water pumping energy is based upon the ground water depth given in Table 2.

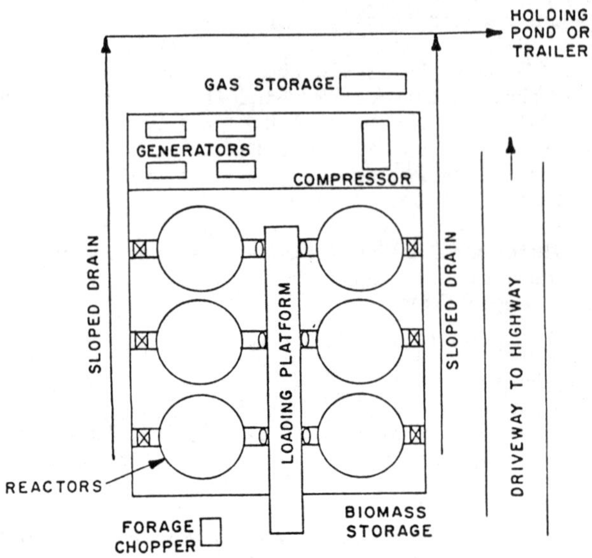

Fig. 2 Schematic of biomass conversion system (plan).

TABLE 2 Statistics for Village of Tara Garh

Item	Data
Population	700
Houses	70
Animals	530
Cultivated Land (acres)	781
Ground Water Depth (ft)	66

TABLE 3 Energy Consumption Factors

Usage	Quantity
Cooking	2 mm Btu/person-yr.
Lighting	150 wh/day-house
	45 w/house
Street Lighting	25 w/8 houses
Drinking Water	20 ℓ/person-day
	40 ℓ/cow-day
Irrigation Water	50 m^3/day-ha

TABLE 4 Annual Energy Usage in Tara Garh

Energy Farm	Amount
Cooking (10^6 BTU)	1400
Lighting (kwh)	4470
(kw)	3.6
Drinking Water (m^3)	35
(kwh)	1080
(kw)	.3

Pakistan Biomass Resources

The villages of Pakistan maintain large buffalo and cattle populations. However, these animals are not confined, except at night in some instances, and collection of residues would be impractical. The villages do have considerable acreage under cultivation and crop residues from corn and wheat crops could be used as a source of energy.

Estimates of crop yields ranged from 6-18 maunds (82 pounds) of wheat per acre and 8-30 maunds of corn per acre. The average wheat yield is 6 mds/acre and corn is 8 mds/acre. Residue quantities were given as 1.33 pounds residue per pound grain, on the average. Using these data, .33 tons of wheat straw and .44 tons corn stover are produced per acre. These yields are considerably below the U. S. averages of 1.5 tons wheat straw per acre and 2 tons corn stover per acre. However, these data are for unirrigated Barani regions of Pakistan, where yields are very low.

In rural Pakistan, crop residues are used as animal feed and cooking, and it was reported that no residues were unused or available for energy production. Other reports (Mubayi and Chatterjee, 1979) have speculated that countries like India could supply substantial portions of their energy from unused crop residues. Therefore, if crop residues are to serve as an energy resource, additional crops must be produced.

Experiments in Pakistan have shown that nearly a ten-fold increase in wheat yield and a five-fold increase in corn yield are possible with irrigation. Adequate underground water for irrigation is available at most locations; however, in most cases, energy is not available for irrigation.

Biomass becomes a very important resource if used to supply energy for irrigation. When the potential for increasing crop yield and income are considered, irrigation becomes a primary objective of the rural energy centers in Pakistan. It has been conservatively estimated (Gaddy, 1979) that 20 inches of irrigation water should produce a wheat yield of 2.5 metric tons per hectare and that 15 inches of irrigation water should easily permit a corn yield of 3 metric tons per hectare. If two crops are produced, corn in summer and wheat in winter, the residue available from irrigated corn and wheat would amount to 3.3 tons per acre, based on the previous biomass yields. Some of this biomass will be required for animal feed, at present .8 tons per acre. Allowing for wastage, an amount of 2 tons per acre should be available for energy conversion.

DESIGN OF TARA GAHR SYSTEM

Table 4 gives the cooking requirements as 1.4×10^9 BTU/yr or 1.4×10^6 CF methane/yr. The electricity requirements for lighting and pumping drinking water are 5550 kwh/yr. Regular gasoline engine generators can be operated with biogas. However, the units must be derated because of the reduced heating value of the fuel, and the efficiency ranges from 16 to 35 percent depending upon the size of the unit (Onan, 1977). A generator to produce 4 kw (5550 kwh) operating at an efficiency of 20 percent would consume 85,000 CF methane/yr. Therefore, a total of 1.485 x 10^6 CF methane is required annually for the basic village energy needs.

Based upon a biomass conversion of 50 percent, the production of 1.485 CF methane would require 240 tons of crop residues. This extra amount of biomass could be produced from 120 acres of irrigated land. However, additional energy for irrigation must be supplied. To furnish the 35 in. of irrigation water at Tara Gahr would result in a net biomass yield of 55 percent, i.e., 45 percent of the additional 2 tons of biomass/acre is needed to furnish the energy for pumping. This efficiency is based upon a 65 percent pumping efficiency, a 67 percent field application efficiency and a 20 percent efficiency for conversion of biogas into electricity.

Applying the yield factor to the acreage requirement, a total of 218 acres must be irrigated at Tara Gahr (98 acres supply the biomass for irrigation energy). Pumping of the water for irrigation would consume 80,000 kwh/yr and the total annual electricity requirement is 85,550 kwh. A total of 1.3 x 10^6 CF methane are needed to generate this amount of electricity. Table 5 summarizes the energy requirements. Therefore, a biogas unit with a capacity of 2.8 x 10^6 CF methane/yr must be installed.

TABLE 5 Total Energy Requirements at Tara Gahr

Energy Requirement	Annual Amount Electricity (kwh)	Annual Amount Methane (10^6 CF)
Cooking	--	1,400
Lighting and Drinking Water	5,550	85
Irrigation Water	80,000	1,300
Total	85,550	2,785

System Economics

The design for the system at Tara Gahr consists of six reactors 14 ft. in diameter by 14 ft. high. A 32.5 kw generator must be provided for operation at a 30 percent load factor. An insulated building 40 x 60 ft. must be provided. The capital cost for this equipment is given in Table 6. The total cost of the installation is $200,000 (1981), including detailed design, training of operators and a 30 percent contingency.

Economics have little meaning in energy supply for remote areas without conventional energy sources. However, based upon 15 percent capital charges, the cost of gas is $7/$10^6$ BTU and the cost of electricity is $.045/kwh. These costs are considered quite reasonable by a comparison with present fossil fuel energy prices.

BENEFITS OF RURAL ENERGY CENTERS

The benefits of the rural energy centers include an enhanced quality of life as a result of the availability of electricity and gas. Such items as refrigeration, lights, television, appliances etc. will improve the health, education and occupational skills of the peoples of these remote areas. Gas for cooking will eliminate the necessity for cutting wood and vegetation for cooking, thus halting further erosion and degredation of the soil.

In addition to these advantages, there are definite cost benefits to be derived from rural energy centers in Pakistan. The primary benefit is through the additional income from increased crop production by irrigation. If it is conservatively estimated that an additional 10 maund wheat ($140/ton) per acre and 20 maund corn ($160/ton) per acre are produced, an increased income of $188 per acre irrigated would result. At

Tara Gahr, irrigation of 218 acres would return $41,000 per year, a substantial economic benefit for this village. Of course, irrigation of the remaining 550 acres could also be readily achieved.

TABLE 6 Capital Cost and Economics of Village Energy Production

Cost Item	Amount, $
A. Capital Cost	
Reactors	$ 25,000
Building	25,000
Piping	20,000
Forage Chopper	5,000
Compressor	5,000
Gas Storage	5,000
Utilities Distribution	25,000
Generator	25,000
Design & Training	30,000
Contingency	35,000
Total	$200,000
B. Operating Cost	
Methane	$7/10^6 BTU
Electricity	$.045/kwh

SUMMARY

The concept of rural energy centers in Pakistan and elsewhere is technically and economically feasible. Adequate biomass resources exist to furnish the energy needs in such villages. Without irrigation, biomass resources may be minimal; however, if energy is made available for irrigation, a fourfold increase in biomass and grain could result. Only a fraction of this additional biomass is needed to provide the energy for irrigation, leaving a substantial amount of biomass as an energy resource for village needs.

Conversion of crop residues into methane by anaerobic digestion is an efficient means of utilizing these resources and would preserve the minerals and nutrients for recycling back to the soil as fertilizer. Capital and operating costs for producing biogas and electricity from biogas are considered reasonable at these remote locations and are competitive with fossil fuel energy costs in urban areas.

REFERENCES

Foutch, G. L., G. C. Magruder and J. L. Gaddy (1981). Production of ethanol and methane from corn stover. Agricultural Energy, ASAE Pub. 4-81.

Gaddy, J. L. (Nov. 1979). Feasibility of utilization of biomass as an energy resource in rural pakistan. UNDP Report.

Mubayi, V., J. Lee and R. Chatterjee (Sept. 1979). The potential of biomass conversion in meeting the energy needs of the rural populations of developing countries. Presented at ACS Meeting, Washington D.C.

Perry, R. H. (1973). Chemical Engineers Handbook, Fourth Ed., McGraw-Hill, New York, NY.

Usmani, I. H. (June 1979). Energy banks for small villages. Bulletin of Atomic Scientists.

Usmani, I. H. (1978). Rural electrification: an alternative for the third world. Nat. Res. Forum, 2, 271-277.

Biogas — the State of the Art in the UK

P. J. Meynell

British Anaerobic and Biomass Association Ltd., The White House, Little Bedwyn, Marlborough, Wiltshire, UK

ABSTRACT

Anaerobic digestion can play an important part in local energy supplies, rather than nationally. Planning of the system and fitting production of the biogas to varying demand is vital to ensure viability; understanding of the process and its requirements are necessary at this stage. This paper summarises these requirements and the influence they have had upon designs in the U.K. The economics of anaerobic digestion is complex because of the benefits which contribute more or less to the overall economics depending on the situation. Management and monitoring during operation are discussed, but usually the digestion itself is fairly stable; it is the ancillary equipment and poor management which have caused most digester "down time".

KEYWORDS

Biogas; anaerobic digestion; U.K. experience; economics; management.

INTRODUCTION

Of all the biomass sources of energy, biogas production is arguably the most viable at the present time, in terms of the minimum size, the technology and the possibility of fitting to current agricultural practice without the food versus energy conflict. Although biogas was once hailed by some as the panacea solution to the energy crisis, it is now generally recognised as having a small but significant contribution to make in terms of the U.K. national energy supply. If all the available material - sewage sludges, animal wastes, organic industrial wastes, landfill sites - were digested anaerobically to produce biogas, the yearly production would not be more than about 5% of our total energy requirement.

This statistic, however, entirely misses the point, because energy is not used by a nation, it is used by people and machines in specific places. The importance of biogas is not its five per cent contribution of the nation's energy, but its possible contribution of most of the energy needs of a farm, or community. The more isolated this farm or community, the more attractive biogas production becomes, because it confers a degree of independence from costly imported fuels and electricity.

Biogas can be produced with very simple technology as evidenced in China and India and the modifications made to the process in the U.K. and Europe are essentially ones to increase efficiency, to minimise wastage of energy, and to reduce the labour requirement. The process remains the same while the costs and values placed on the various benefits change; it is these which affect the viability. To some extent the minimum size is determined by the demand for energy in a given situation, rather than the amount of material available for digestion. Along with other aspects of the planning process, the matching of constant biogas production with a fluctuating energy demand is considered vital to the viability of the process in a given situation.

Every situation is different and should be treated as such, but the basic process remains the same with different approaches to design and ancillary equipment to adapt it. An understanding of the process is necessary to see what implications this has had upon the different designs, and an attempt is made to illustrate this with the U.K. digesters, currently available.

PROCESS REQUIREMENTS

The biochemical process of anaerobic digestion occurs naturally in muds and polluted waters where air or oxygen is excluded. This then is the primary criteria and all designs have an airtight digester tank which is usually heated (especially in cool and temperate climates) to the optimum of 30°-35°C. If the process is working well it will maintain itself at a neutral or slightly alkaline pH and only when unusually acid wastes are digested or when the methane producing bacteria are inhibited will pH adjustment be necessary.

The process can be broken down into three distinct phases involving different sets of bacteria, each set using the waste products of metabolism of the previous set. (Meynell 1981, Stafford 1980). These three phases are:-

(1) The Hydrolytic phase in which organic matter is broken down to smaller molecules, such as amino acids, simple sugars and long chain fatty acids. This phase may in fact take place outside the bacterial cells; it is a phase of preparation of the organic matter so that it can be absorbed by the next stage. It is important because for many materials it is the rate limiting step and the one which holds most promise for enzyme enhancers etc. Certain intransigent materials may need pretreatment before digestion to speed up the process, but generally for animal wastes this is not necessary.

(2) The Acetogenic phase is the phase in which the smaller organic molecules are broken down further to form volatile fatty acids (VFA) of which acetic acid predominates. The VFA are the end products of metabolism of this set of bacteria and, as with alcohol production by yeasts, the acids would eventually poison the acid formers were it not for the fact that the methane producers of the next phase remove it. The acid formers are less sensitive than the methane producers and so under some conditions, they will continue to produce acid which is not then removed.

(3) Methangenic phase. The methanogens are the most sensitive bacteria. They convert the VFA to methane and carbon dioxide which, being gases, escape from the liquid system; they therefore do not remain to poison the process.

The picture which emerges of the process is one of interdependent phases, a balanced ecosystem which will respond to deal with minor variations in the feedstock and environmental conditions; bacteria populations in a dynamic state of change according to the conditions. A particular waste will develop its own balance of bacterial flora adapted to that waste. Thus some piggery wastes contain a very high ammonia content (over 2000 mg/l); provided the ammonia content does not change too radically or too fast, the bacteria will be able to function satisfactorily under conditions which would certainly inhibit a digester unused to those conditions. In general the whole process is stable provided it is fed regularly; however it is probable that there have been more failures of digestion due to inadvertent changes in the feedstock than any other cause.

ENGINEERING REQUIREMENTS

The digester design must ensure that conditions within it are maintained within the range of the process requirements. Secondly the design must have an eye to net energy efficiency; without this much of the point of the process is lost (though not all of sewage sludge digesters whose prime purpose is sludge treatment rather than energy). It is probably easiest to consider the engineering requirements in terms of the functions of the components which are generally considered to be necessary by the U.K. industry.

(1) Digester tank. Depending on the waste, this should be sized to give 10-20 days hydraulic or sludge retention time. Newer types of digester, such as the upward flow anaerobic sludge blanket and anaerobic contact processes for dealing with wastes with low solids contents, may have much shorter retention times. Digester tanks show a remarkable variation ranging from the concrete tank in the ground in which the flow is essentially horizontal to the vertical vitreous enamel tanks standing above ground. The former system is used in the design by Helix Multiprofessional Services, whilst the latter is used by Hamworthy Engineering, Farm Gas, Farrow Irrigation amonst others. Some tanks are completely enclosed and air tight, whilst others incorporate the gas holding facility on top.

(2) Loading and outlet systems. Loading is often the most problematic stage for farm wastes are notoriously variable causing complications with pumping etc. The actual design of the loading system will depend upon the lie of the land, the collection system of the waste and the design of the digester. A farmer in Wales has designed his own system pushing the slurry up a ramp and into his horizontal digester with a tractor; the Helix designed digester at Bore Place is gravity fed aided by scrapers in channels, whilst the vertical digesters are pump fed from a collection sump. In general the outlet system incorporates a simple weir overflow, with spent slurry being displaced as fresh slurry flows in.

(3) Heating system. In temperate and cool climates, heating is essential to maintain the temperature at the process optimum. (N.B. This may not be the biochemical optimum, since the extra energy required to heat the digester to that temperature may not be justified in terms of net energy gain). Heating systems most often use the gas in a water boiler with heat exchangers to transfer the heat to the slurry. However, the heat exchangers may be situated outside the digester as in the Farrow Kilbees design, the design principle being that nothing should be in the digester to cause spaces etc, whilst most other designs have the heat exchangers in the digester. Often these are combined with the mixing system e.g. the paques design which incorporates mixing through a heated double walled draft-tube. As a general rule the heat required to maintain the temperature is about $\frac{1}{4}$ to $\frac{1}{3}$ of the gas produced. This may be reduced by good energy management and reclaiming otherwise waste heat from the outgoing slurry and from the exhaust and cooling waters if an engine/generator uses the biogas. In all digesters insulation is vital for net energy production. Underground digesters use the ground as the main insulation, but above ground systems are generally insulated with several inches of polyurethane foam sprayed inside or outside the digester tank.

(4) Mixing systems. In the U.K. most designs are completely or partially mixed. Mixing is essential for the high rate process i.e. 10-20 days c.f. 30-60 days of non mixed systems. Mixing is necessary to keep the bacteria in contact with the substrate and so maintain optimum activity; it is also necessary to reduce the incidence of scum formation and sedimentation. Scum forms very

readily with floating materials, especially fibres (pig's hairs etc) forming a thick mat within several weeks of inadequate mixing. Mixing, in U.K. designs, is most commonly done by gas recirculation, although mechanical stirring and draft tube mixing are also used. Gas recirculation involves pumping some of the gas back through the digester in such a way to break up any crust which might form. Mixing is not continuous, but specific mixing regimes would have to be designed from experience with a particular digester and the waste it treats.

(5) Gas collection and storage systems. In current practice gas is used as it is produced without much more than half a day's gas storage. Biogas is not easily compressed; it may take more energy to do this than the biogas produced. Gas storage may take place above the digesting slurry as in the Helix and Farrow/ Kilbees designs. The former involves a novel approach involving interlocking modular gas caps which float on the slurry, whilst the latter has a flexible, nylon reinforced rubber membrane which inflates as gas is produced and sinks down onto the slurry as it is used up. Other designs have separate gas holders, usually incorporating the principle of the bell-over-water gasometer.

(6) Gas use. It is vital to the overall economics of the system that as much of the gas is used as possible. In practice, especially if electricity is generated, one would be lucky to get more than 80% usage. Electricity generation is an obvious choice because of the versatility of electricity and there are a number of engine manufacturers in the U.K. making or adapting engines for this purpose e.g. Fords, Reading Industrial Engines, Applied Energy Systems and Manns of Saxham. Electricity generation is perhaps more viable on the larger installations, but it does add considerably to the expense and to the complexity of the control systems. The other main use is for heating e.g. domestic for the smaller installations, heating of animal houses e.g. farrowing units, etc. Because of the impurities in the gas - hydrogen sulphide especially, the gas should not be burnt directly inside, but rather in a vented system, e.g. AGA cooker, central heating system or radiant tube heaters.

(7) Control system. The controls required will depend upon the size and complexity of the system chosen and the degree of automation. One of the main differences between the U.K. designs and the simpler designs of China and India is the need to reduce the labour input in the more sophisticated designs for the Western market; hence more automation and integration of the different systems. Controls generally show the workings of the various mechanical or engineering parts, rather than of the biochemical processes. However, probes for temperature and pH of the slurry may be incorporated as may some monitor of the calorific value of the gas, especially on the larger installation.

ECONOMICS

The economics of anaerobic digestion is a very vexed subject, mainly because, of all the benefits, only the biogas can be costed in economic terms; some of the benefits e.g. odour control are very important in certain instances but negligible in others. Every situation has to be treated on its own merits, and using the value judgements of the prospective user. In an isolated situation, the subjective importance attached to the degree of self sufficiency in energy may well be much greater than in a situation where fuel and energy supplies are readily available.

Anaerobic digesters are not cheap pieces of equipment. Although manufacturers are learning how and where to make savings in digester design, the costs still go up with the increasing costs of fossil fuels; thus, the competitiveness of biogas as a fuel always seems to be one step away in spite of the costs of conventional fuels. Table 1 shows the range of costs for three different sizes of digester taken from commercial budget prices.

Table 1. Range of capital costs for Anaerobic digestion.

To treat manure from	Biogas utilised by:	
	heating	electricity generation
5000 pigs	£28000 - 47000	£40000 - £60000
1000 pigs	10000 - 26000	not considered viable
100 cows	13000 - 26000 (some consider size not viable)	not considered viable

Costs vary widely according to the design and complexity of the site and operation, and use of the gas. It can be seen that electricity generation can add £10,000 to the cost of a digester. (Grundey 1980, BABA 1981).

The capital cost may not be that large in terms of other pieces of equipment used on a farm, but when it comes to decision making, the farmer has up till now generally put anaerobic digestion fairly low on his list of priorities. One of the main reasons being that the payback period is often too long when assessed in terms of energy alone.

Table 2. Comparative schedule of performance for three farm animals.

INPUT		COW	PIG	LAYING CHICKENS
1. Weight/Unit of Animal	kg/no	520kg	68kg	1000 no.
2. Total Effluent/Day	kg/day	41.0	4.5	115
3. Percentage Total Solids	%	11.	9.5%	24%
4. Total Solids	kg/day	4.51	0.43	27.6
5. Effluent to be treated assuming dilution to 6% by washing down, etc.	kg/day	75kg	7.2	460
6. Retention time	days	20	16	20
7. Tank volume	kg	1500	115	9200
OUTPUT				
Gas produced 0.26m3(3)/kgTS Cows .3m3/kgTS Pigs .4m3/kgTS Poultry.				
Thermal Energy	kwhr/day	7.0	0.8	68.0
Solids	kg	3.5	0.27	21.4
Crude protein 50%TS	kg	1.7	0.12	10.5
Liquid to be irrigated or post treated assuming separation	kg/day	72.5	7.0	450

1. Figures assume that best pollution control is required.
2. Generally 25% of heat produced will be required at 0°C to maintain the digester at 35°C. Assuming 30% heat reclaim.
3. These figures will vary according to the particular requirements of the operation.

(1) Biogas production.
Table 2 shows a comparative schedule of some parameters used in assessing

digester size and gas production from three farm animals (BABA, 1980). Figure 1 shows the energy output to be expected from different numbers of animals (Farm Gas 1980). Usually the biogas or the electricity generated is given the same value as the fuel or electricity it replaces and hence is fairly easy to cost. The problem, however, arises since it is unlikely that all the gas will be used to the maximum - even the CEGB has a problem with meeting fluctuating loads with production of electricity, and on a farm situation one would be lucky to use more than 80% of the electricity produced.

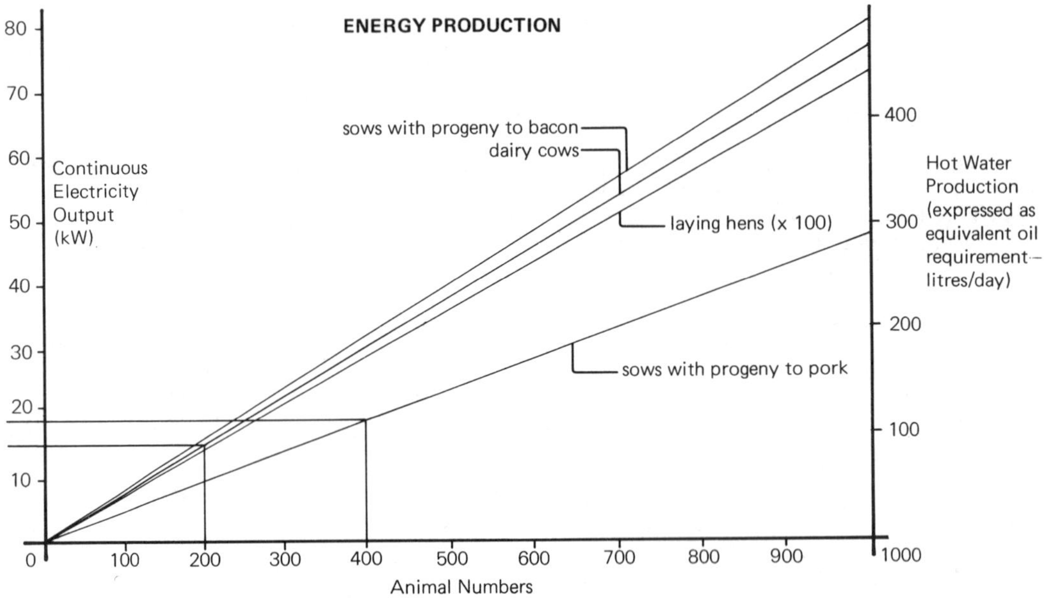

FIG. 1 Energy derived from numbers of animals.

Figure 2 compares the energy value at 8000 hours use/year and 6000 hours use/year before depreciation and other costs of the equipment. This makes the point very well that use of the gas must be maximised if the investment is to make economic sense. It also shows that electricity generation does not make sense for 1000 pigs or 100 cows, but that heating with biogas might. For 100 cows the energy value for heating after subtraction of depreciation might be in the range of £1000 - 1500 per year. On this accounting, with a capital cost of £13,000 - £20,000 the pay back period would be at least ten years at present day fuel prices. This is too long for most accountants to countenance and hence the need for the other benefits and the value judgements.

On the other hand with larger installations, e.g. for 5000 pigs, the energy value after subtracting depreciation might be about £5000 per year for heating or electricity generation. This would pay back the capital cost of the digester used for heating in 6 years and used for generation in under 10 years. Economics of scale are beginning to make themselves felt, but still the payback is too long. (BABA, 1981).

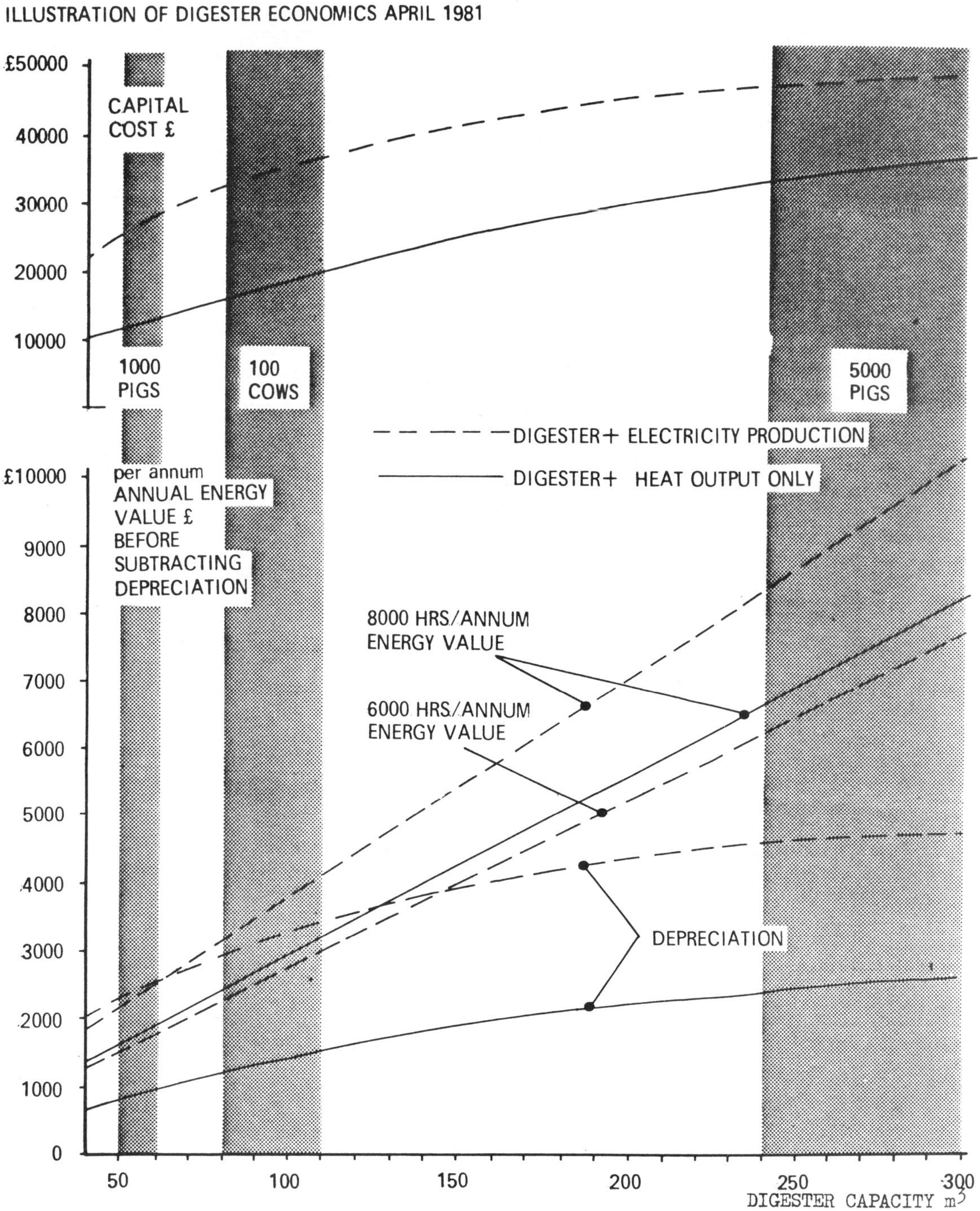

FIG. 2. Digester economics and capacity

(2) Pollution control.

In the U.K. pollution control is perhaps the other most attractive benefit of anaerobic digestion, although perhaps less so for an isolated community. Anaerobic digestion has been used on sewage works for many years as a means of pollution control and the gas is considered a by-product. On farms, especially the intensive livestock units, pollution and odour control are becoming increasingly important. Many such farms do not have enough land to spread their wastes or when they do they get complaints from people round about. In extreme cases they can be closed down. In this sort of situation a digester can be costed against alternative means of pollution and odour control. Since digestion is a method which actually produces a useful gas, it may well have the edge on other methods even though they may be cheaper in the first instance.

(3) Value of the sludge.

The other important benefit is the sludge. It must be realised that anaerobic digestion does not remove the problem of final disposal of the sludge, it merely conditions it so that handling and disposal are easier. But disposal is perhaps a negative way of looking at it since the sludge has a real value of its own.

Considered as a fertiliser, digested sludge does not contain any more nutrients than undigested sludge - if allowed to stand it may in fact contain less nitrogen. However the nutrients are more readily available to plants and the sludge, being more easily handled, can be pumped for regular irrigation, rather than stored for two or three times yearly disposal. Animal and plant pathogens are destroyed during digestion and the problems of tainting and "burning" of grass are considerably reduced. In the U.K. most of the animal wastes are returned to the land, so the fertiliser value of digested sludge may not be so important, but the greater ease of handling may well be a costable benefit to the process.

In other parts of the world the fertiliser value is the most important aspect e.g. in India where the cow dung would otherwise be burnt and the nutrients lost to the ground. This change of emphasis illustrates how different conditions and requirements throws a different light on the costings.

More recently a lot of interest has been paid to the use of the digested slurry as a source of single cell protein for refeeding to animals. Although some way off widespread practice, the separation of bacterial solids leaving nutrient rich liquor for irrigation, seems very attractive. Some costings have put the value of the bacterial solids at seven times the value of the gas. (Stafford, D. Personal Communication).

MANAGEMENT AND MONITORING

One factor which affects the costings, both in terms of capital cost and manning costs is the management. You can either pay for increased monitoring appliances which decrease the need for human intervention, or you can spend a little bit more time in regular maintenance etc. Digestion is a relatively stable and simple process and the easiest way to make sure that it is working well is to see if gas production is normal and if it burns well.

Provided that the digester is well treated and fed regularly with a slurry of similar strength and consistency it is unlikely that the process will go wrong. Most process failures have arisen from changes in the feedstock. If the process does not often fail, the commonest cause for digester "down time" has been the ancillary equipment, including monitoring equipment. There is something to be said for simplicity in a digester; complexity adds to the likelihood of something going wrong. It also adds to the cost.

Monitoring equipment apart from the burn test - or calorific value if more accurate information is required - should include a pH meter, temperature indicator and gas meter. Simple facilities for measuring the solids content of the slurry are useful and, if more information is required, analysis for volatile fatty acids etc. If the digested slurry is to be used as an animal feed, further facilities for measuring its nutritional status might be necessary.

Management and maintenace of the plant also take up time and have to be costed. It may be that an average of five or six hours per week is necessary. If electricity is being generated this may be more, and time would be required for engine servicing etc. Poor management is one of the main causes of digester failure, on the simple plants of India and China as with the more sophisticated plants. The latter also suffer from what is known as "finger trouble", but perhaps this is because of the novelty of having a digester to play with.

CONCLUSION

This all too brief picture of the requirements and factors influencing the economics of anaerobic digestion has highlighted a few of the problems which are being worked upon by the industry. There are not that many digesters installed in the U.K. yet, perhaps because availability and costs of fuel are still reasonable and because pollution control measures are not being pressed, and perhaps because of the economic recession. The paper has concentrated on animal wastes because most commercial digesters have been designed for this purpose, but research on other feedstocks such as weeds, crops and industrial wastes is going on. Biogas from a combination of sewage, animal and vegetable wastes holds a promise of a localised energy source for rural and island communities but, as with all alternative energy sources, the practice is not quite as easy as it first appears.

REFERENCES

BABA 1980. The Digest No. 2., British Anaerobic and Biogas Association

BABA 1981. The Digest No. 4., British Anaerobic and Biogas Association

Farm Gas 1980. Commercial Literature.

Meynell, P. J. 1981. Methane - Planning a Digester 2nd Ed. Prism Press, Dorchester.

Stafford, D. and others. 1980. Methane Production from Waste Organic Matter CRC Press, Florida.

Grundey, K. 1980. Tackling Farm Wastes. Farming Press, Ipswich.

Fuel Alcohol from Sweet Potatoes

B. M. Whitehurst

Texasgulf Chemicals Company, Aurora, North Carolina, USA

ABSTRACT

The production of fuel alcohol from farm waste offers the American farmer a way of reducing costs or improving income. Sweet potatoes are a substantial crop in the southeastern United States. North Carolina farmers planted 38,000 acres of sweet potatoes in 1979. It is estimated that 100 to 150 bushels per acre of sweet potatoes are wasted each year because the cannery will only accept a specific size. This paper discusses the waste sweet potato as a source of fuel alcohol. The economics, alcohol yield, process conditions and process design are described.

KEYWORDS

Sweet potatoes, alcohol, yield, enzymes, farm operation, storability, economics.

INTRODUCTION

The production of fuel alcohol from farm waste offers agriculture a way of reducing cost or improving income. In addition, the otherwise lost protein value from the waste can be recovered as food for livestock. Sweet potatoes are a substantial crop in the Southeastern United States as well as in the warmer climates around the world. In the United States about 1/3 of crop can be lost as culls in some years because the produce and cannery market can only accept specific sizes.

Data shows that more alcohol per acre can be produced from sweet potatoes than from corn or the other cereal grains. The dried by-product remaining after fermentaion is higher in protein than distillers grain. In North Carolina, the sweet potato can produce about twice as much alcohol per acre as corn when compared on average crop yields. The concept of using jumbos and culls to profitably produce alcohol in farm scale units has been demonstrated by Mr. Carl Smith of Wilson, North Carolina. He has a 6 gal./hr.

alcohol unit in operation.

The sweet potato is used primarily as a food crop. Normally it is baked or boiled, but recently its use for "french fries" has been encouraged. The sweet potato has been used to produce alcohol in Japan for more than a century. It was an important raw material for the ethanol used by the early chemical industry in Japan.[1] In the United States, the sweet potato was used 50 to 75 years ago to produce starch for textile finishing. It was equal to other starches, however it was not as economical. The motivation to produce starch from the sweet potato was to use the unmarketable grades such as jumbos and culls. The very same motivation is behind the movement today to convert the jumbo and cull to alcohol. Alcohol is more valuable than the starch, and simpler to produce in a small operation.

Research done by Keitt (1909)[2] showed that the sweet potato had potential for alcohol production. Production was related to the variety, starch content and yield per acre.

The sweet potatoes grown for the food market, however, tend to be the lower starch varieties. Therefore, alcohol production from these varieties is lower. It appears that to have a viable alcohol industry from sweet potatoes we must raise the high starch variety specifically for alcohol, and use the alcohol facility to dispose of the jumbos and culls from the edible varieties. This procedure would maximize farm income on a land use basis.

The present study involves the more popular commercial varieties and some experimental high starch varieties. The data is shown in the following table.

Alcohol Yield Data by Variety

Variety	Dry Matter %(3)	Alcohol Yield, Gal. Anhydrous (Freshly Dug)	Total Potato Yield CWT/A	Alcohol Yield Gal/A
Porto Rico, Red	26.2	1.41 gal/CWT	---	---
Jewel, Yellow	23.8	1.07	297	318
Centennial, Yellow	25.9	1.23	365	450
[3]Pelican Processor, White	26.7	1.54	---	---
[3]Rojo Blanco, White	24.9	1.40	254	355
[3]Whitestar, White	26.4	1.48	---	---
[3]Vogel, White	28.2	1.65	---	---
White Hayman	---	1.93	125	241

[1] Personal communications with M. Tashiro of Toya-Soda Chemical Company.

[2] Edmond, J. B., Ph.D., Mississippi State University, 1971, the AVI Publishing Company, Inc., Sweet Potatoes: Production Process Marketing, Major Feed and Food Crops in Agriculture and Food Services.

[3] Collins, Dr. Wanda, Dry Matter Analysis, Horticultural Research, North Carolina State University, Raleigh, North Carolina.

The American taste usually prefers the red or yellow varieties because they are juicier and sweeter when baked than are the white varieties which are starchy in taste.

The alcohol yield drops about 25% after 10 months' storage. This rate of decrease is not linear with time, but rather flat during the cold months, with a rapid decrease in the warm months. Data is shown in the following table.

Effect of Storage Time on Alcohol Production

	Freshly Dug	10 Months Old
Jewel	1.07 gal/CWT	0.77 gal/CWT
Porto Rico	1.41	1.07

In the production of alcohol from starch, the enzyme, alpha-amylase, is normally added to the cooking step. Sweet potatoes, depending on the variety, require different amounts of alpha-amylase. With some varieties, the enzyme could be omitted without seriously affecting yield.

Effect of Alpha-Amylase Addition on Yield

Freshly Dug	No Alpha-Amylase, Gal Ethanol/CWT	0.05% Alpha-Amylase, Gal Ethanol/CWT
Centennial	1.16	1.23
Porto Rico	1.34	1.41
Jewel	1.04	1.07
10 Months Old		
Porto Rico	0.19	1.07

ECONOMICS OF FUEL ALCOHOL

Definitive economics cannot be presented because situations differ widely based on yield per acre and the number of crops per year. In the United States one of the serious problems with the sweet potato crop is the jumbos and culls. They can be a disposal problem on any given year. In North Carolina in 1979, about 30% of the crop was unsaleable for various reasons. Starting with a material of zero value, and using an on-farm alcohol plant, it is possible to convert this waste to a product of value. The alcohol produced can be used as a fuel to replace No. 2 fuel oil for heating, or to operate a tractor or truck.

Using alcohol as a substitute for heating oil on the farm is the simplest of all approaches. Equipment changes can be made for less than $25. The fuel is clean burning, and the alcohol content of the fuel need be no higher than 180 to 190 proof (90 to 95%).

Using fuel alcohol for heating applications, such as corn drying or tobacco curing has a greater economic potential than its use as a motor fuel (in gasohol). Quality control is simpler, and the expense of vehicle conversion is avoided. The production of anhydrous ethanol for motor fuel (gasohol) is expensive and difficult for an on-farm situation.

Mr. Carl Smith of Wilson, North Carolina, has a 6 gallon per hour on-farm unit operating with sweet potato culls, waste corn, and apples. Mr. Smith raises about 350 acres of sweet potatoes. He uses waste wood from his packing crate production facility to fuel the cooking and distillation processes.

ACTUAL ON-FARM SYSTEM

Twenty bushels of sweet potatoes, approximately 1,000 pounds, are placed in a wood fired open top cooker with the addition of approximately 875 pounds of water. See The Schematic Flowsheet. The mixture is brought to a boil and allowed to boil until the sweet potatoes are cooked (about 1 hour at the boiling point). The temperature should then be lowered to 80°C, and 0.05% alpha-amylase added. Alpha-amylase may not be needed for some varieties of fresh sweet potatoes. While the mixture is still at 80°C, the potatoes should be pulverized into a mush. This can be done by hand with garden tools or with a beater powered by a one-half inch drill. The mush should then be cooled to the fermentation temperature of 30°C by pumping the mush through a heat exchanger back into the cook tank with an air operated sludge pump. After cooling, the mush is pumped to the fermentation tank. Fermentation usually requires four to five days.

After fermentation, the solution is heated to the boiling point and the alcohol stripped from it. This procedure is carried out in the fermentation tank. This tank is equipped with a steam sparger to assist the heating from the direct burning of wood. This technique allows for the large heat load during the heat up period. A small boiler can then be used at considerable saving. The vapor is then stripped off through an inclined pipe which provides reflux to increase the alcohol content of the crude product. After passing through the reflux zone, the uncondensed vapor is condensed to crude alcohol of approximately 55% or 110 proof.

From the crude alcohol tank, the solution is pumped into a wood fired steam assisted reboiler where the alcohol is refined via a packed column to 190 proof. A six gallon per hour refining column is packed with 1/2 inch ceramic saddles, and is 12 feet high. The reflux ratio of this system is three to one. Previous experience has indicated this ratio is effective, and gives a consistently high quality product. The reboiler can be wood fired directly. In addition it will have a steam coil to provide for better heat control and quality control. A wood fired boiler provides steam.

After the alcohol has been stripped from the mush, the slurry would be pumped into a centrifuge or a vacuum filter and dewatered. The wet cake, which contains about 12% solids, can be dried or fed directly to livestock. The protein level of the wet cake is approximately 5%. After drying, the protein level increases to over 30%.

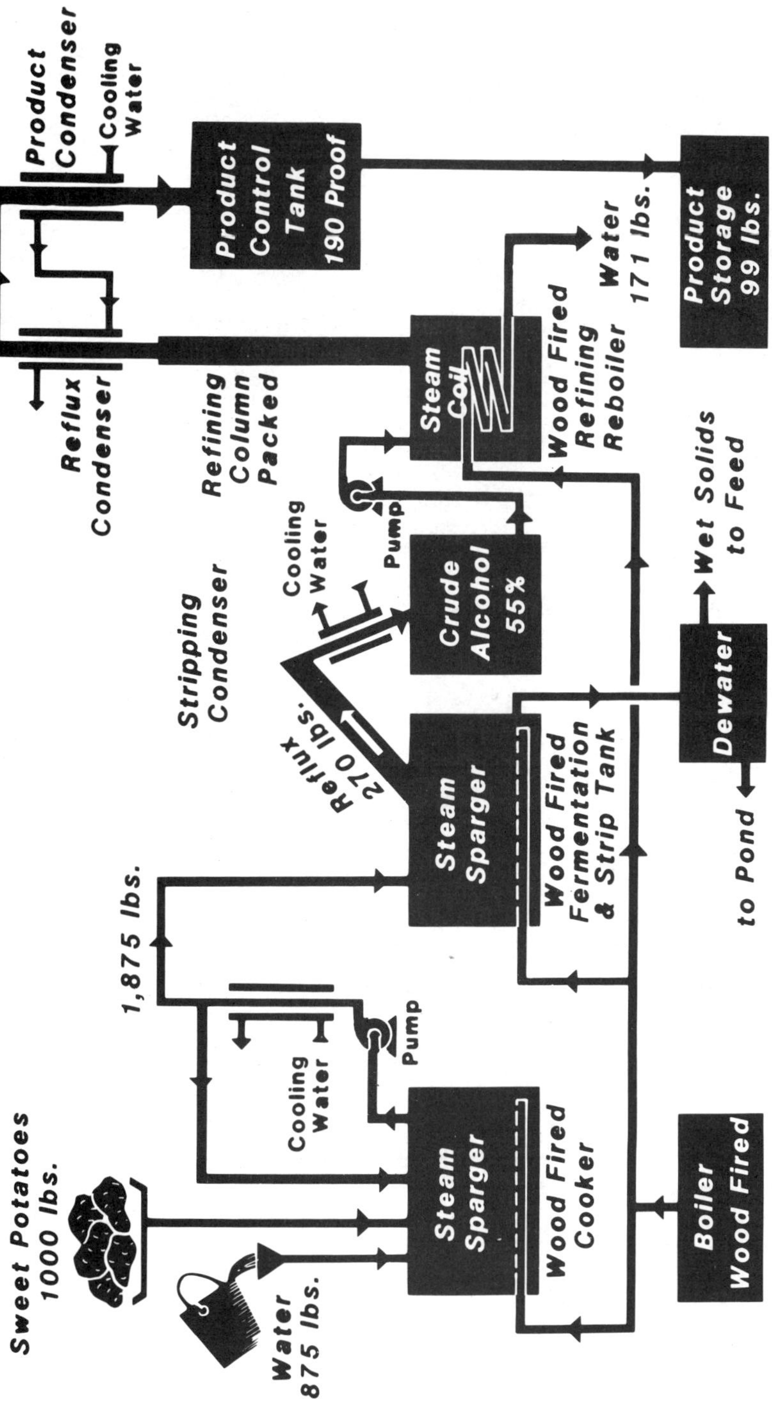

Figure 1: Schematic Flowsheet

Ethanol Production in Kenya — an Economic Perspective

G. W. Barnard and D. M. Kanyi

Department of Chemical Engineering and Chemical Technology, Imperial College, London SW7, UK

ABSTRACT

In this paper the options for the production of fuel ethanol in Kenya are discussed. Trade-offs between economies of scale and transportation costs are considered.[1] An analysis is presented of the relative merits of producing ethanol, on foreign exchange grounds, as opposed to exporting sugar or molasses directly. Relationships are derived for the breakeven ratios of the world market prices for sugar/oil and molasses/oil at which ethanol production is preferable in foreign exchange terms. Viewed in this way, ethanol production is not the most effective use for sugarcane or molasses, given current world price ratios.

KEYWORDS

Kenya, ethanol, sugarcane, molasses, economics, foreign exchange.

INTRODUCTION

In recent years, the Kenyan foreign exchange problem has been worsened by the rising price of crude oil. As much as 35% of her total import bill is currently accounted for by oil imports (Stansell, 1981). This has led the Government to consider possible alternative energy sources, one of which is the production of fuel ethanol from sugarcane and molasses.

A variety of approaches are possible for the production of ethanol in Kenya, involving different choices of feedstock, plant size, plant location, and ethanol use. There are also important economic decisions to be made regarding the optimum use of Kenya's sugarcane and molasses resources in view of the fact the sugar and molasses can themselves be exported directly to earn foreign exchange.

BAKCGROUND

Before looking at the various strategies Kenya can adopt for the development of an ethanol programme, it is relevant to consider the structure of her existing

---

[1] All prices mentioned in this paper refer to U.S. dollars

sugar industry. At present there are seven major sugar mills in Kenya, six of which are located in relatively close proximity in Western and Nyanza Provinces. Current sugar production is in the order of 400,000 tonnes per annum, although actual installed capacity is considerably higher (Kanyi, 1980). Production of sugar at the moment slightly exceeds the demand in the country.

As a by-product from sugar production, approximately 130,000 tonnes of molasses are currently produced each year (Kanyi, 1980). Internal demand is roughly 40,000 tonnes per year. The majority of the remaining 90,000 tonnes is presently exported, but could potentially be diverted for use as a feedstock for ethanol production. There is also considerable scope for expanding cane growing capacity.

The current consumption of gasoline in Kenya is in the order of 418 million litres per annum. Assuming a 20% ethanol blend, the potential demand for ethanol would be approximately 86 million litres per year (Earthscan, 1981).

SYSTEM CHOICE

In view of the availability of potential feedstocks for ethanol production, the most obvious system options are listed as follows:

Option A Production of ethanol from molasses using small-scale facilities at individual sugar plants.

Option B Production of ethanol at a single large-scale facility using molasses transported from nearby sugar plants.

Option C Production of ethanol directly from sugarcane, grown in newly established plantations.

For the two molasses-based options, if a total of, say, 90,000 tonnes of molasses were devoted to ethanol production, this would provide enough ethanol to blend gasoline up to a level of roughly 6%. This assumes current levels of gasoline consumption, and a conversion yield of 278 litres of ethanol per tonne of molasses. For the cane-based case, planting an additional 14,000 hectares with cane would provide enough ethanol for a 20% blend (assuming a cane yield of 50 tonnes/hectare/year, and a conversion yield of 70 litres of ethanol per tonne of cane). Such an area would correspond to a 20% increase in the total area under cane.

Discriminating between these various system alternatives is a complex process, involving a wide range of social, economic and political considerations. Selecting the optimum scale for conversion facilities, for example, will require a detailed analysis of the trade-offs between the economies of scale with large scale facilities and the increased feedstock transportation costs.

Estimates of the capital cost of ethanol production facilities vary widely (Sheehan, 1978). One of the few comprehensive studies performed to date, is that of the World Bank (1980). This identified a number of trends. Firstly, it indicated that molasses-based plants are likely to cost, on average, approximately 20% less than cane-based plants of similar ethanol production capacity. Secondly, it suggested that significant economies of scale were obtainable with larger plants. Furthermore, it predicted that considerable cost differences would occur between countries due to varying technical and infrastructural capabilities. Their analysis identified three broad classes of countries; those with low, medium and high capital costs. Relative costs of an ethanol plant of given capacity in these three classes were taken to be in the ratio of 1 to 1.25

to 1.9, respectively. Brazil is an example of a low cost country, due to its extensive experience in ethanol production. Kenya would at best come into the medium cost category. The estimated capital cost of ethanol plants of various capacities in a medium cost country are shown in Fig. 1, based on these World Bank figures.

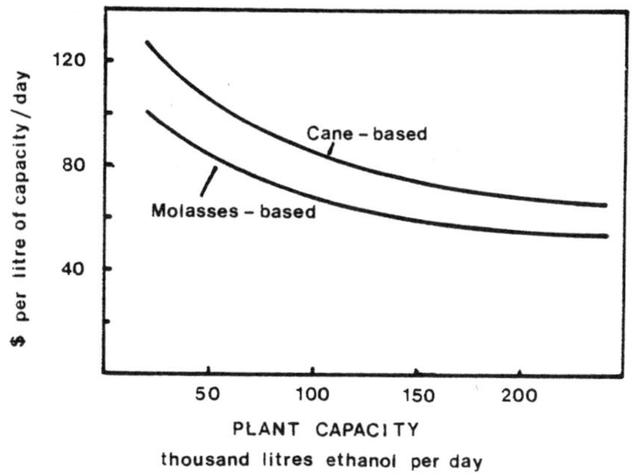

Fig. 1. Capital Cost Estimates of Ethanol Plants in a Medium Cost Country.
Source: World Bank (1980).

Looking briefly at the importance of transportation costs in choosing between the three system options mentioned, a number of simple calculations can be performed. Comparing the two molasses-based options, for example, it is interesting to calculate the additional costs that would be incurred under system option B due to the need to transport molasses from existing sugar mills to a centralised ethanol production plant. If one assumes that the ethanol plant was situated at the largest existing sugar mill, Mumias, the average transportation distance for molasses pooled from the four other nearby plants would be about 40 km. If transportation costs by road were 15¢ per tonne.km (Meta Systems, 1980), this would only add about $6 per tonne to the molasses cost, as compared with a current world price of approximately $100 per tonne. Alternatively, this can be regarded as adding 2 cents to the price of a litre of ethanol.

For a sugarcane based ethanol plant a similar calculation can be performed to provide a rough indication of the rate at which transportation costs increase as one goes to bigger conversion costs - which require larger areas of cane. Assuming a circular plantation zone in which half the land is used for cane, a crop yield of 50 tonnes.ha^{-1}.$year^{-1}$, transportation costs of 15 cents per tonne-km, and an ethanol plant that operates for 175 days per year producing 70 litres of ethanol per tonne of cane, a correlation between plant capacity and haulage costs can be derived. This is shown in Table 1.

TABLE 1 Sugarcane Haulage Costs for Ethanol Plants of Various Sizes

Plant Size litres.day^{-1}	Area of Plantation Zone Km^2	Average Haulage Distance Km	Haulage Cost cents.litre^{-1} ethanol
20,000	20	1.7	0.4
60,000	60	2.9	0.6
120,000	120	4.1	0.9
240,000	240	5.8	1.2

The conclusions from these two sets of calculations are similar and indicate that transportation costs are not likely to be a dominating factor in determining system choice. In neither case do differences in transportation costs amount to more than a few cents per litre. This is a relatively small proportion of the total production cost of the ethanol, which would probably exceed 30 cents per litre, even under optimistic conditions (Sheehan, 1978; Office of Technology Assessment, 1980).

FOREIGN EXCHANGE CONSIDERATIONS

For Kenya the prime motive for producing fuel ethanol is to reduce the foreign exchange burden of oil imports. For each litre of ethanol produced, roughly one litre of gasoline can be saved. At present Kenya has excess refinery capacity so this would enable a litre of gasoline to be exported - thereby earning foreign exchange. Assuming that Kenyan gasoline can be sold at say, 28 ¢ per litre[1], the net foreign exchange savings from ethanol production can be calculated.

It is also very important to remember, however, that if molasses or sugarcane are not used for ethanol production they themselves can be exported, thus earning foreign exchange directly. Sugar and molasses prices fluctuate considerably over time but at present are approximately $420/tonne for sugar and $100/tonne for molasses (these are both f.o.b. prices). It is, therefore, relevant to ask whether it would be more profitable in foreign exchange terms to export sugar and/or molasses directly, rather than producing ethanol.

To answer this question a simple calculation can be performed to find the foreign exchange earnings that can be derived from using a given quantity of sugarcane and molasses in different ways. This is shown in Table 2.

TABLE 2 Foreign Exchange Earned Through Various Uses of Sugarcane and Molasses

Feedstock	Alternatives	Foreign Exchange Earned
10 tonnes molasses	1. Export directly	$1,000
	2. Convert to 2780 litres of ethanol	$ 778
100 tonnes sugarcane	1. Export 10 tonnes sugar	$4,200
	2.7 tonnes molasses	$ 270
		$4,470
	2. Convert to 7,000ℓ ethanol	$1,750

[1] This is a typical ex-refinery value for gasoline produced from oil at £32 per barrel (World Bank, 1980).

From this analysis it is clear that at present prices, the foreign exchange arguments are against ethanol production from either feedstock. For sugarcane, in particular, exporting sugar directly yields more than double the foreign exchange that could be obtained from ethanol production. In the long run, the relative merits of ethanol production will be determined primarily by the relative prices of oil, sugar and molasses on the world market - ratios which change considerably over time.

To find when ethanol production would be preferable, based on foreign exchange criteria, an expression can be derived for the 'break-even point' at which earnings from ethanol production exactly equal those from sugar or molasses export. For the molasses case, this break-even point can be defined as follows, starting with 10 tonnes of molasses as before:

Earnings from ethanol production = Earnings from molasses export

or $$2780 \times C \times f = 10 \times M \qquad (1)$$

where C = Crude oil price ($ per barrel)
M = Molasses export price ($ per tonne)
f = Factor which relates ex-refinery gasoline price ($ per litre) to the prevailing crude oil price
2780 = Quantity of ethanol produced from 10 tonnes of molasses (litres)

Assuming that f has a value of approximately 0.28/32.0, in accordance with the current price ratio for gasoline and crude oil, break-even will occur when:

$$\frac{M}{C} = 2.4 \qquad (2)$$

In other words, producing ethanol from molasses will be preferable if the ratio M/C is less than 2.4.

For the sugarcane case, molasses and sugar are co-products. The calculation can be simplified, however, by ignoring the molasses and assuming that 100 tonnes of cane will yield either 10.5 tonnes of sugar or 7000 litres of ethanol. Thus, the break-even point will occur when:

Earnings from ethanol production = Earnings from sugar export

or $$7000 \times C \times f = 10.5 \times S \qquad (3)$$

where S = Sugar export price ($ per tonne)

Using the same value for f, the break-even point will occur when:

$$\frac{S}{C} = 5.8 \qquad (4)$$

Thus, producing ethanol from sugarcane is preferable when the ratio S/C is less than 5.8.

At present world prices $M/C \simeq 3.1$ and $S/C \simeq 13.1$. This confirms the previous results that ethanol production is not the most profitable way of earning foreign exchange from either sugarcane or molasses.

To indicate how these relative price ratios have varied over the last twenty years, a series of graphs of price trends are presented in Figures 2, 3 and 4. In Figure 2 the trends in actual world prices of oil, sugar and molasses are shown (in current dollars). The recent oil price explosion and the periodic sugar price peaks are both clearly illustrated. Molasses prices have shown a general increase in price during the 1970's, although not a smooth one.

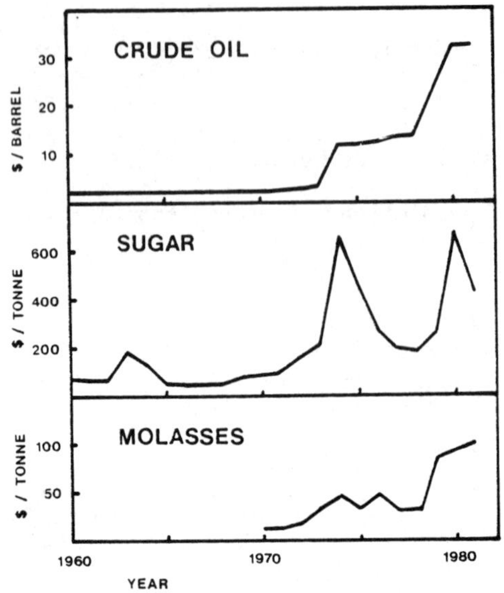

Fig. 2. Trends in the World Market Prices of Oil, Sugar and Molasses, 1960-1981 (f.o.b., current dollars)
Sources: World Bank (1978), Kanyi (1980).

In Figure 3, the sugar and oil price data has been converted to yield a timecourse for the sugar to oil price ratio. It indicates that sugar prices have never reached a sufficiently low level during the period in question, compared to oil, for it to be preferable in foreign exchange terms to convert sugarcane to ethanol. It also underlines the absence of any clear trends over time.

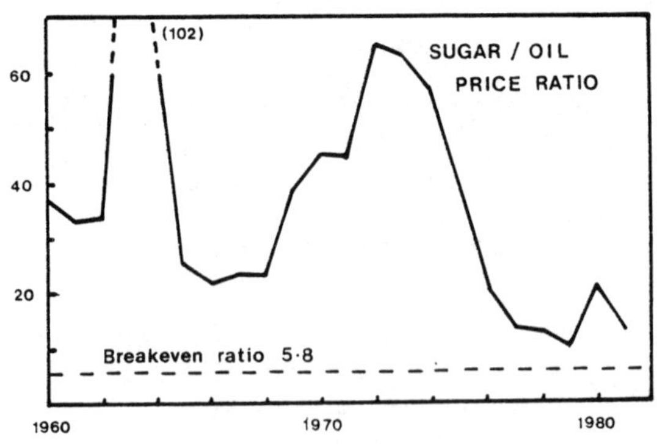

Fig. 3. Trends in the Sugar/Oil Price Ratio, 1960-1981.
(Sugar price in £ per tonne, oil price in $ per barrel).

A similar graph for the molasses to oil price ratio is presented in Fig. 4. In this case, molasses prices have been much closer to the break-even ratio than for sugar, and for a period in 1977/78 the ratio was actually below the break-even point. It was during this period that the Kenyan Government did in fact go ahead with its plans to build two molasses-based distilleries.[1]

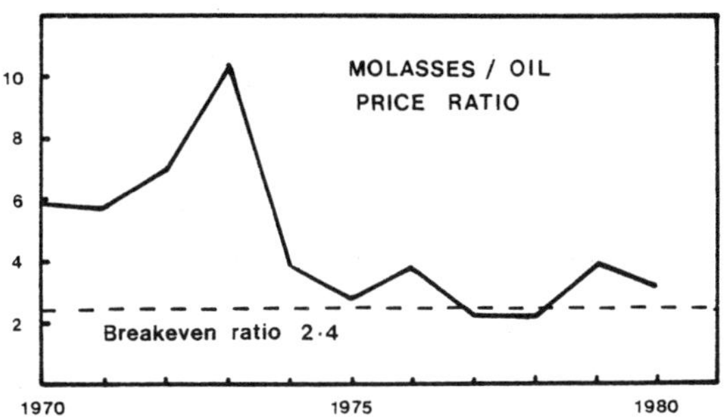

Fig. 4. Trends in the Molasses/Oil Price Ratio, 1970-1980. (Molasses price in $ per tonne, oil price in $ per barrel).

Looking to the future, predicting how sugar, oil and molasses prices will change is an exceedingly complex matter. Attempts are underway to stabilize the international sugar price but have so far met with only limited success. The official target range for sugar prices is presently $290-515 per tonne. Assuming that prices were to stabilize somewhere near the lower limit, oil prices would have to rise to $55 per barrel for the breakeven ratio to be reached. Whilst not impossible, the traditional volatility of the sugar market could well prevent such an occurrance from persisting for long.

The situation regarding molasses prices is somewhat more predictable. In the past, molasses prices have been dictated mainly by fluctuations in the world price of animal foodstuffs (wheat, corn soya beans, etc.), since the main use of molasses is as a feed additive. Because of the availability of molasses substitutes, therefore, the demand is more elastic than for sugar and price fluctuations are less dramatic.

In conclusion, therefore, when viewed from a foreign exchange point of view, the production of ethanol from molasses seems to be considerably more favourable than using sugarcane - at least in the near term future. It is crucial, however, that Kenya remains aware of world trends in molasses, sugar and oil prices, as well as fuel ethanol production. If a major shift occurs on a global scale, to convert sugar or molasses to ethanol, there would almost certainly be a significant market reaction, pushing up the price of these commodities on the world market and favouring exports. Conversely, as more countries try to cash in on potential earnings from sugar or molasses exports, the market would react in the other direction favouring ethanol production. Kenya is unlikely to cause any significant price changes by itself but could easily be caught in the trap of being lured by short-term profits into policies that could be unwise in the long-term.

---

[1] Two alcohol plants, at Muhoroni and Kisumu, are in the process of being built. Together, these are expected to produce 50 million litres of alcohol a year by 1982, using molasses as a feedstock. Escalating construction costs have seriously affected the profitability of the plants, however (Earthscan, 1981).

It is also important to note that one of the basic assumptions behind the previous calculations is that if a decision is made to export molasses, or expand the sugar industry, then it will actually be possible to get these products to the ports, and sell them at world prices. A number of factors could prevent this from occurring. In the case of sugar, for example, export quotas will prevent sugar exports beyond a certain level. Likewise, logistical difficulties or high transportation costs may prevent sugar and molasses getting to the ports, even though a potential market may exist. Considerable quantities of molasses have gone to waste in the past for this very reason.

In summary, therefore, although the theoretical foreign exchange earnings from sugar and molasses exports may be greater than those from ethanol production, these earnings may not be obtainable in practice. In such circumstances, a decision to go ahead with an ethanol programme could well be justified on the grounds that the more modest foreign exchange earnings that could be derived from ethanol production is better than nothing. It would also have the advantage of preventing the Kenyan economy from becoming dependent on the vagaries of the highly volatile international sugar market.

REFERENCES

Earthscan (1981). Earthscan Press Briefing Document No. 29. Gasohol: Energy and Environment in a Hungry World. 10 Percy Street, London WIp ODR, U.K.

Kanyi, D.M. (1980). Molasses Utilization in Kenya. Ministry of Industry, Kenya House, P.O. Box 30418, Nairobi, Kenya.

Meta Systems (1980). Potential for Fuelwood and Charcoal in the Energy Systems of Developing Countries. Report to U.S. Department of Agriculture, Forest Service, 10 Holworthy Street, Cambridge, Mass., U.S.A.

Sheehan, G.J., P.F. Greenfield and D.J. Nicklin (1978). Energy, Economics, Ethanol - A Literature Review. Alcohol Fuels, Sydney, No. 9.

Stansell, J. (1981). New Scientist 91, 1267, p.460.

World Bank (1978). Commodity Trade and Price Trends, Report No. EC-166/78. 1818 H Street, N.W., Washington, D.C. 20433, U.S.A.

World Bank (1980). Alcohol Production from Biomass in the Developing Countries. 1818 H Street, N.W., Washington, D.C. 20433, U.S.A.

TOPIC F

Solar Energy: Higher Latitudes

Experience with Domestic Solar Water-Heating

M. Hanlon

Electricity Supply Board, Ireland

ABSTRACT

The Electricity Supply Board monitored eight domestic solar water heating installations as part of its solar programme. The equipment and installation costs of these systems, problems encountered and performance analysis of the four best systems, are given in this paper. It is shown that, with a panel area $5m^2$ copper manifold, and a 50 gal store, up to 50% of an average householder's hot water requirements can be met. While economic assessment can be subjective , this would mean that, using a fuel inflation rate of 25% and a discount factor of 14%, it would take between 10 to 13 years for this system to pay for itself.

KEYWORDS

Installation; costs; problems; controls; direct; indirect; storage; performance; payback; auxiliary.

INTRODUCTION

In the mid-seventies, the Electricity Supply Board (ESB) began investigating solar domestic water heating systems. Early in 1979 it was decided to monitor the performance in actual houses as it had become obvious that both computer simulation and laboratory type work was not enough. Eight installations were monitored between August 1979 and August 1980. Three of these were existing and were suggested to use by either panel suppliers or the householders themselves while the remaining five were carried out under the ESB's direction.

OBJECTIVES

It was not intended to carry out a full and final assessment of solar water heating in the Irish climate but rather to concentrate on areas in the Irish context not covered in other research work. The following are the four main objectives which were limited to what was considered realistic -

: to determine what contribution solar energy could make to the Irish householder's domestic waterheating load using the panels and controls then available on the

Irish market.

A number of randomly selected panel types - all of which were supplied "off the shelf" - were monitored. Some of those chosen are no longer on the market but this could be more of a reflection on the financial status of the suppliers rather than the result of technical assessment.

: To become familiar with the problems encountered in both installing and maintaining solar systems.

If solar systems are to make any impact in Ireland they will have to be retro-fitted to existing houses. Not alone will the problem of fixing to an existing roof structure have to be examined but the random orientation of Irish houses will necessitate special consideration. The reliability long term performance and life span of systems can only be assessed after a number of years operation. Their combined effect on the yearly maintenance bill influences system economics and so will continue to be closely monitored.

: To investigate the different control equipment that was available.

Differential temperature controllers were used as standard in all of the pumped systems and they will be discussed later in 'Problems Encountered'.

A diverter valve, activated by a temperature sensor at the top of the solar cylinder, was installed in those systems with both solar and auxiliary stores. This enables the auxiliary cylinder to be by-passed when the water in the solar store is sufficiently hot for use in the taps without any boost. An economic assessment of this by-pass system is underway at present but will not be given in this paper.

: To compare the difference between direct and indirect systems and to consider the size of storage cylinder that should be used.

It was concluded by computer model at Trinity College Dublin (Ref. 2) that a direct system would have a significantly better performance than an indirect one and it was decided to see if this would be achieved in practice.

The size of store can affect performance and so a number of different store sizes were used.

TEST INSTALLATIONS

Four different panel types were involved in the eight installations -

Type W, A 'flat plate copper collector' of which there was three variations

- Type W1 consisting of manifold pipes
- Type W2 consisting of serpentine-shaped pipes and
- Type W3 a radiator type

Type X, A 'focusing collector' consisting of copper manifold pipes, each in its own parabolic - shaped groove.

Type Y, A 'heat pipe collector' consisting of a number of heat pipes leading into a top heat exchanger through which the system water flows.

Type Z, A 'storage collector' consisting of a number of tubes giving a total panel capacity of 44 gal (200 l) of water.

The panel area used in each installation varied from 2.5 m^2 to 5.2 m^2 depending on the panel type used. There were five direct and three indirect systems with panel Type Z accounting for two of the five direct systems.

All of the panels were installed with an orientation between 20° West of South and 20° East of South. It was not possible to depend on the random orientation of house roofs to achieve this and frames had to be constructed in some instances. Where the panels and/or the frame had to be fixed directly on the roof, the slate was drilled and the panel/frame was screwed to a cross-piece nailed across the rafters. The slate was then sealed. The installation and equipment costs for domestic solar water heating systems is given in Table I.

PROBLEMS ENCOUNTERED

Apart from the problems mentioned below all the systems have run smoothly. There were no problems with freeze-ups during the winter period (-5^oC max)

: One of the temperature differential controllers (supplied as part of a kit with one of the panel types) worked very erratically and replacements of the same type showed no improvement. The circuit design for a differential controller is quite simple, but examination of the faulty controllers showed that their circuit construction was of a very poor quality. Therefore, it was decided to install a different type of controller made by a well-known manufacturer of control equipment and this has worked consistently. Since the performance of a non-gravity feed solar system depends greatly on the efficient operation of its differential controller, care should be taken to ensure that its circuit construction is of a high standard.

: Minor leaks have occurred in two of the systems but these were easily repaired and have not recurred.

: At one installation, the 50 gal cylinder burst about 1 inch away from the side-seam but this was in no way associated with the presence of the solar system. There has been a history of corrosion fatigue failure in this area but as yet it is not known why it should be peculiar to here.

SYSTEM PERFORMANCE

Of the eight installations monitored only two could be considered to have performed satisfactory. This paper will give details for both of these along with two others - Houses A, B, C, D. The performance of none of the other four systems was better than the unsatisfactory ones covered. The system using panel Type X is one of those not covered but the manufacturers have stated that there was a misunderstanding with their agent about the area recommended - the original panel is being replaced by two of a new revised model, thus doubling the area.

At all installations, a heat meter was used to monitor the heat collected by the panels while a kWh meter recorded the immersion usage. In addition, both houses C and D had flowmeters which monitored both the total and kitchen hot water consumptions. All meters were read on a daily basis, usually in the evening. Solar radiation was not monitored at each installation, instead data from the meteorological office at Dublin Airport was used when analysiing the results. This is given in Table 2 below which shows that, for the test period, the global radiation on a horizontal plane was only 84% of the average yearly value while the hours sunshine had not such a marked difference being 94% of the average value. For the test period, the ratio of diffuse to total radiation was 52% while the average ratio is 60%.

For analysis purposes, the test year was divided into three parts -

: Summer - May, June, July, August (123 days)
: Spring/Autumn - March, April, September, October (122 days)
: Winter - November, December, January, February (120 days).

TABLE 1 Equipment and Associated Costs of Solar Domestic Water Heating Systems

EQUIPMENT	COSTS (IR£)		
	1979	1980	D.I.Y. 1980
Panels	500	500	500
Cylinder	55	135	135
Controller	45	50	50
Pump	25	25	25
Diverter Valve	75	75	75
Labour (including pipework)	300	650	
Pipework and Fittings	-	-	100
	1000	1435	985
*Frame (incl Labour)	250		
	1250		

NOTES : *A frame is necessitated where the panels have to be placed on a horizontal surface or on a roof slope but at a different angle to the latter.

TABLE 2 Radiation and Sunshine Data From Meteorological Office, Dublin Airport

	Global Radiation (MJ/m^2)		Sunshine Hours	
	Test Period	10 Year Average	Test Period	10 Year Average
Summer	1830	2190	641	722
Spring/Autumn	989	1139	489	503
Winter	305	383	262	262
TOTALS	3124	3712	1392	1487

Each of the four systems covered by this paper will now be discussed:

House A

This system, shown in Figure 1, was not installed by the ESB. Panel Type Z was used holding 44 gals (200 l) water. From the results, given in Table 3, it can be seen that -

: the average solar collection was very poor even in summer months;

: the household had a very high immersion usage indicating a greater than average daily hot water consumption - circumstances outside ESB's control did not permit the installation of flowmeters to confirm this;

: The solar system's contribution to the hot water requirements varied from zero in mid-Winter to 16% in Summer.

The three main limitations of this system were -

: it had to be shut down during most of the Winter and Spring for frost protection purposes;

: if the hot/warm water was not drawn off quickly from the panel at night then re-radiation to the atmosphere occurred;

: the system was operated manually, as recommended by the manufacturer, but it may have performed better if an automatic control system was used. However, the optimum control of a system having panel Type Z is not so easy. Many factors have to be taken into account, which include

- should water be drawn from the panel only when it is above a certain temperature?
- should the panel contents be dumped into a special store and if so, should the time of dumping be pre-set or dependent on weather conditions?

The poor performance of this system to-date does not warrant spending extra time and effort to decide which of these listed control alternatives would be best.

In conclusion, this system's contribution to the household's hot water requirements way very poor. However, it could probably be greater in a household with a lower daily hot water consumption and which would be prepared to change its lifestyle a little to use the solar hot water when it was available.

TABLE 3 House A : Solar Collection and Immersion Usage

	Solar Collection (kWh)	Immersion Usage (kWh)
Summer	181	908
Spring/Autumn	82	Central Heating + 849
Winter	22	Central Heating
TOTALS	285	1757

House B

This indirect system, shown in Fig 2, was installed by the ESB. It has a dual coil fitted in the 40 gal (182 l) store - both parts of which were used at all times. Planned tests on the use of the top and/or bottom coils were not carried out because of the poor panel performance.

This household uses only solar and electric immersion to heat its hot water and the results, given in Table 4, indicate that it was a lower than average hot water consumption. So this type of system may not even have given the overall contribution of 15%, varying from 2% in Winter to 24% in Summer, in a more 'average' household for the same period. ('Average' implying a total hot water usage of between 35 and 40 gals. per day).

TABLE 4 House B : Solar Collection and Immersion Usage

	SOLAR COLLECTION (kWh)	IMMERSION USAGE (kWh)	SOLAR CONTRIBUTION
Summer	114	369	24%
Spring/Autumn	99	448	18%
Winter	10	452	2%
TOTALS	223	1269	15%

Houses C and D

Both these systems, shown in Figs. 3 and 4, will be discussed together because they used the same panel type, copper manifold, and had the same panel area, 5.2 m^2. House C was an indirect system with a separate 50 gal. (227 l) solar store while House D was a direct one with both solar and auxiliary heating systems in the one 50 gal. store.

TABLE 5 Houses C and D : Solar Collection and Auxiliary Usage

	HOUSE C (INDIRECT)	(kWh)			HOUSE D (DIRECT)	(kWh)			
	Solar	Immer-sion	Other*	Solar Contri-bution	Solar	Immer-sion	Other*	Solar Contri-bution	Diffe-rence[1]
Summer	588	209	-	74%	886	899	-	49%	+50%
Spring/Autumn	578	118	542	46%	510	728	-	41%	-12%
Winter	353	-	722	33%	271	-	1702	14%	-30%
TOTALS	1519	327	1264	49%	1667	1627	1702	33%	+10%

Notes : * These values are estimated
1 Difference = % Difference of House D compared to House C.

The results, given in Table 5, will be discussed under the following headings -

: Percentage difference in solar heat collection and
: Percentage solar contribution to the domestic water heating needs in both cases
: Performance of direct versus indirect systems and storage size.

Percentage difference for the Winter period, the direct system in House D collected 3o% less heat than the indirect system in House C. This is contrary to what was predicted by the computer model devised by TCD (ref. 2) and seems to be due to the fact that all the heating systems are in the one store in House D. Because the solar system will only operate when the water temperature at the top of the panel is greater than the temperature at the bottom of the cyliner, this means that the system in House D will be less likely to collect on dull days, if the central heating is on, than that in House C. This indicates that it is probably best to have the solar and auxiliary heating systems in separate cylinders.

For the Spring/Autumn period, the 12% poorer collection by House D is due to the fact that this system was not functioning in October because of a faulty differential controller. If this month is ignored, then it collected 14% more heat than House C for this period.

The summer months showed a marked improvement of 50% in the collection of House D over House C. This was due to a number of factors -

: House D has a direct system whereas House C has an indirect. Again, reference is made to the TCD computer model (ref 2) which showed that an indirect system collected less per annum than a direct system and that the difference between the two became more marked as both storage and panel areas increased.

: House D has much higher average daily hot water consumption 55 gals (250 l), than House C, 42 gals (190 l), which means that the solar heated water is used more quickly in the former thus leaving potential for greater collection.

: In May, the solar store in House C was insulated to a far greater degree than it was formerly. Following this, a reduction of 25% in the hours run for the pump was noted. This would seem to imply that the heat loss from the cylinder was greatly reduced leading to lower solar collection values for this system.

Over the whole test period, House D collected 10% more heat than House C and if the 'faulty month' of October is ignored in both cases then this value rises to 20%. Obviously, this indicated that direct systems collect more than indirect - the percentage difference may be even greater if solar and auxiliary heating systems are in separate stores as explained earlier.

<u>Percentage solar contribution</u> For each period, the solar system contributed more to the hot water requirements of House C than to those of House D even though more heat was collected in the latter. This is due to the fact that House D uses on average 30% more hot water per day than House C. This highlights the fact that the heat collection and the resulting solar contribution are both very much dependent on the daily hot water requirements of each particular household. However, a contribution of between one third and a half of the daily hot water needs of an 'average' household could be reasonably expected using systems similar to those in Houses C and D.

<u>System type comparison and storage capacity</u> From the results it is obvious that direct systems perform better than indirect ones. However optimising the size of storage to use is a more complicated affair. In a project of this size it would be impossible to cover all combinations of panel and storage sizes and the poor performance of most of the systems tested made our information even more limited. Nonetheless, using data from Houses C and D, indications are that for a panel area of 5 m^2 a 50 gal (227 l) solar store should be used and solar and auxiliary stores should be separate.

ECONOMIC ANALYSIS

Two aspects must be taken into account when discussing the economics of domestic solar water heating installations -

: the initial cost of the system and

: the expected annual savings.

Table 1 gave details of both 1979 and 1980 costs. The large difference between these costs is slightly misleading in that the 1979 cost is based on a 'one-man-local-plumber-type' operation while the 1980 cost was the most reasonable quote obtained from a 'small-company-type' plumbing operation. Obviously, the latter costs are the more realistic and the D.I.Y cost given includes all materials required - only scaffolding and labour are excluded.

At the end of one years monitoring in August, 1980, the expected saving per annum for the two best systems was taken at IR £83 - this means a saving of 1650 kWh at 5 p/kWh which was then the current General Domestic Rate for electricity. The poorer installations were ignored when carrying out the economic analysis.

Pay-Back Period

How should the pay-back period on the initial investment be calculated? The following three methods were used:

Simple Pay-back method given in Table 6, where the initial investment was divided by the savings achieved in the first year of operation the number of years it takes for the system to pay for itself.

Pay-back periods allowing for fuel inflation but not for the time value of money using initial investments of £1000 and £1500 with fuel inflation rates of 15%, 25% and 35%, pay-back periods for this method are given in Table 7.

Pay-back periods allowing for both fuel inflation and the time value of money are given in Table 8. Fuel inflation rates of 15%, 25% and 35% were again used while the discount rate for money was considered fixed at 14%.

TABLE 6 Simple Pay-Back Periods

INITIAL INVESTMENT* 1980 (IR £)	PAYBACK (YEARS)
D.I.Y. 1000	12
Installed 1500	18

NOTE : *Refer to Table 1

TABLE 7 Pay-Back Periods with Fuel Cost Inflation Rates

g \ I	1000	1500
15%	8	10
25%	6	8
35%	5½	6½

NOTES : g = fuel cost inflation rate
I = initial investment, 1980 (IR £)

TABLE 8 Pay-back Periods with Fuel Cost Inflation Rates and A Discount Rate of 14%

g \ I	1000	1500
15%	14	20
25%	10	13
35%	8	10

NOTES : g = Fuel Cost inflation rate
I = initial investment, 1980 (IR £)

CONCLUSIONS

: There was a marked difference in the performance of the systems monitored which was due to a number of factors, such as, panel area and quality, storage size, quality of plumbing and hot water usage patterns. However, the results showed that it was possible, with a panel area of 5 m^2 and a 50 gal store, to meet up to 50% of the annual waterheating requirements of an 'average' family using the panels which were available on the Irish market in 1979. It is important to realise that this contribution would not have been achieved with some of the systems then available.

: To maximise the amount of energy collected in a direct system a store, separate to that for the auxiliary heating, should be used.

: Economic assessment can be subjective in so far as the rate of energy inflation, the life span of the system and the time value of money have to be assigned. Depending on the values choosen (or ignored) the payback period can vary from 5½ to 20 years. However, using a fuel inflation rate of 25% (1974 to 1979 the average rate was 23% p.a.) and a discount factor of 14% (the current bank rate in 1980) a solar domestic water heating system would take 10 to 13 years to pay for itself.

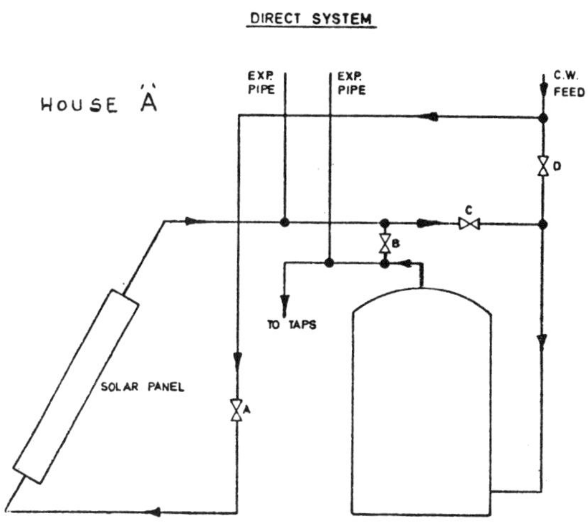

40 GAL. (182 L.)

PANEL : Type Z (Storage)

AREA : 2.3 m^2

TILT : 35° Approx.
ORIENTATION : 20° SE

NO. OF OCCUPANTS : 7

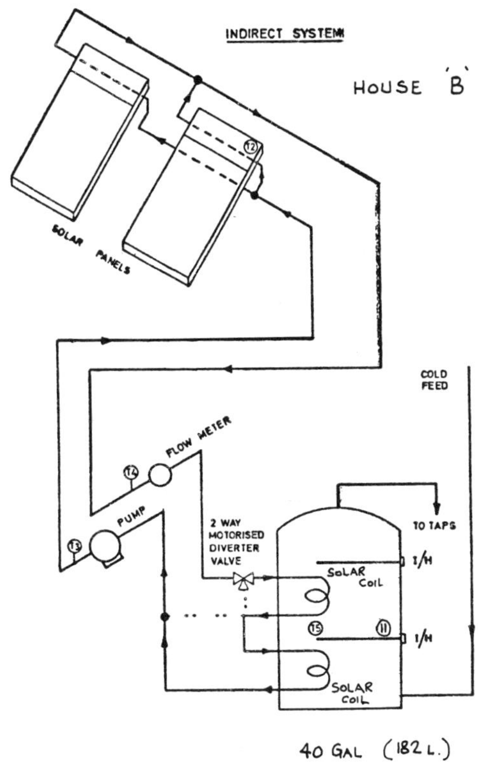

House B

Panel: Type Y (Heat Pipe)
Area: 3.5 m^2
Tilt: 32°
Orientation: 15° SE
No. of
Occupants: 4

House C

Panel: Type Wl, (Copper manifold)
Area: 5.2 m^2
Tilt: 33°
Orientation: 3° SW
No. of
Occupants: 5
Hot Water
Consumption: 42 gals./day

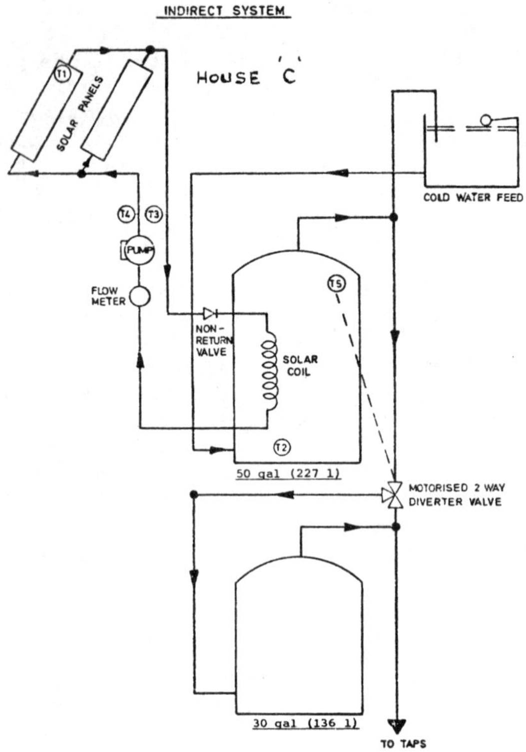

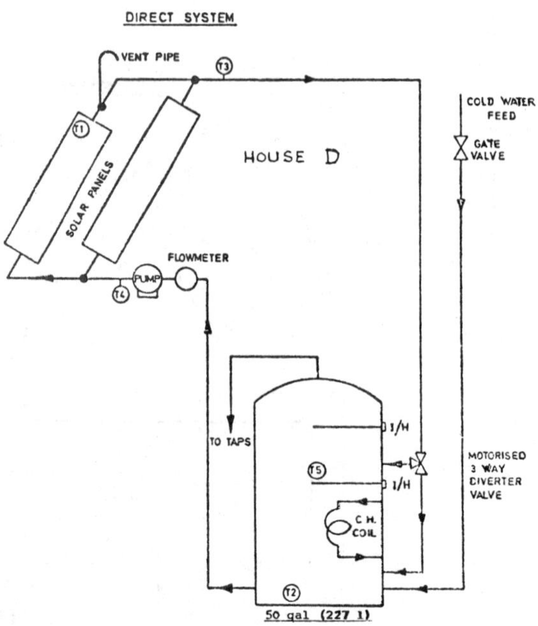

House D

Panel: Type Wl, (Copper Manifold)
Area: 5.2 m^2
Tilt: Winter (Mid. Oct.- Mid Mar.) 55°
Summer 35°
Orientation: 20° SE
No. of
Occupants: 7
Hot Water
Consumption: 55 gals./day.

REFERENCES:

1. Carroll, A.T. "The use of Flat Plate Solar Collectors for Domestic Water Heating" paper presented to Solar Energy Society of Ireland, March, 23rd, 1977.

2. Hand F, Asare B, Haslelt J. "Domestic Hot Water and Solar Energy in Ireland". Energy Research Vol. 1 249 - 257 April, 1977.

A Low Energy Consumption House with Solar Powered Ducted Hot Air Space Heating and Water Heating

F. W. Fleming

Department of Building, Heriot-Watt University, Edinburgh, UK

ABSTRACT

The paper outlines the development of the house design, the construction method used, the integration of solar air and water heating panels, the introduction of higher than normal standards of insulation, the control system to run the heating system, the sensors and data acquisition system leading to an economic analysis of the installation.

KEYWORDS

Solar-air heating, rock pile storage, solar water heating, insulation, low energy consumption.

INTRODUCTION

The house described was designed to fit the needs of a mixed adult and teenage family with privacy for individuals and areas for joint activities. The design had to incorporate high standards of insulation, good acoustic damping qualities and solar panels for space and water heating. The design brief imposed by the local planning authority limited windows to back and front walls, roof pitch to 30^0 and orientation had to be parallel to the road layout with the gable end facing the road. The inclusion of solar panels in the roof at latitude $+10^0$ met with considerable resistance and only threat of abandonment brought approval of the design and the required orientation shown in Fig. 1 Site plan.

The important constructional consideration was the author's wish to test the hypothesis that a reasonably successful system could be built on site utilising standard building materials, components and trade skills generally available.

THE HOUSE AND SYSTEMS DESIGN

The design started from the concept of various 'activity areas'. Figure 2. This progressed to the 'east-west wing' layout. Figure 3 shows this in elevation. Rationalisation of this gave the solution shown in Figure 4, again in elevation and this was simultaneously translated into the floor plans shown in Figure 5. The whole will be contained in what will appear to be a conventional envelope.

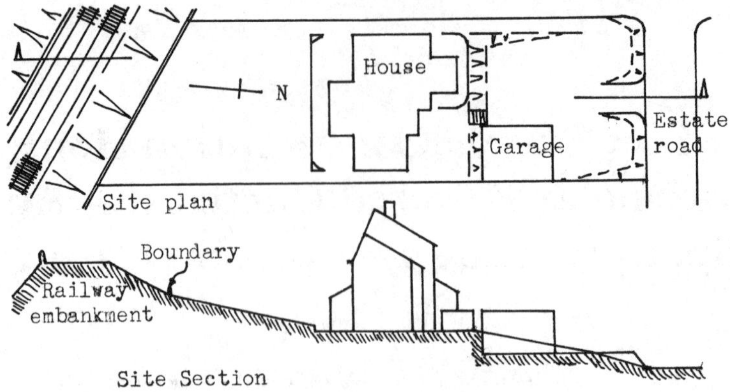

FIG. 1 SITE PLAN AND SECTION

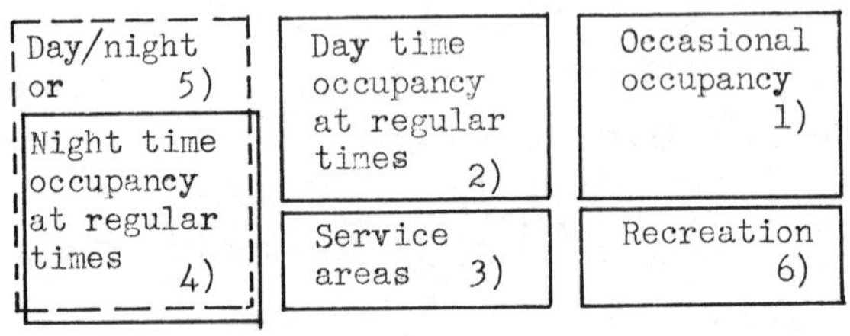

FIG. 2 BALLOON DIAGRAM OF ACTIVITY AREAS

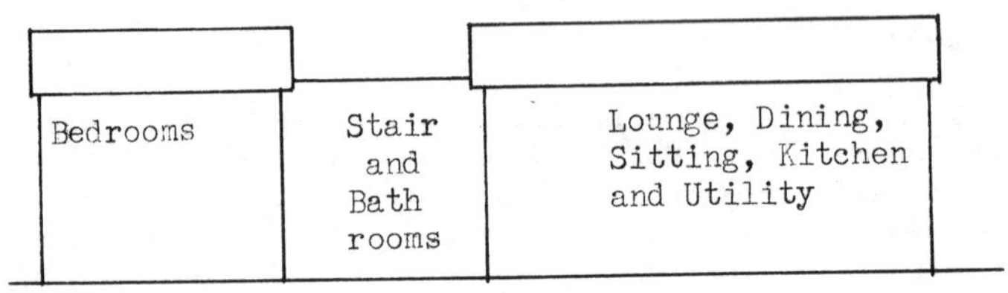

FIG. 3 EAST/WEST WING CONCEPT

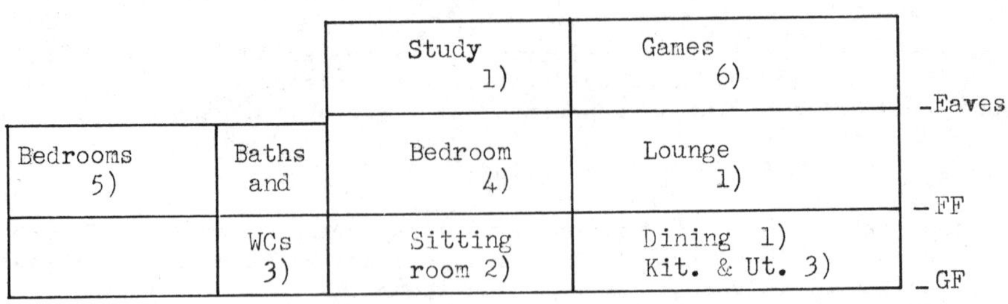

FIG. 4 RATIONALISATION OF FIG. 3 AND UTILISATION OF ROOF SPACE

The construction of the walls is in 225 mm thick autoclaved concrete block, dry lined inside and clad outside with 50 mm of fibreglass and a skin of 75 mm lightweight aggregate concrete block with a 3 coat dry dash render. The heavy inner wall gives mass for acoustic damping and thermal storage and the complete wall has a U value of 0.27 W/m^2.

The ground floor is of hung timber construction and incorporates 100 mm thick mineral wool batts. The underfloor space is ventilated to comply with the current Building Regulations but has a U value of 0.24 W/m^2.

The roof space has 150 mm thick fibreglass insulation in two layers with 12.5 mm thick plasterboard laid loose between. The plasterboard enhances acoustic damping and the construction gives a U value of 0.24 W/m^2.

Included in the design are double glazing to windows on the southern elevation, triple glazing to windows elsewhere and effective weather stripping to doors and windows.

Preliminary calculations suggest that a heating load of 0.225 $kW/^0C$ will be required. For a heating season from September to May inclusive, degree days averaged over the twenty years 1957-77 amount to 2474. This gives a total heating requirement for an average heating season of 13 360 kWh. Various sources put the available solar radiation over the same period as lying between 228 and 430 kWh/m^2 for surfaces from horizontal to vertical inclination and with a due south orientation. The roof can support 47 m^2 of solar-air panel but only 35 m^2 will be erected initially. Simple calculation might suggest to the uninitiated that anything from 60 to 100% of the heating requirement might be achieved by proper inclination of the panels. Nothing could be further from the truth. Of that total radiation, a portion will arrive at the wrong time. A portion will not be of sufficient intensity to be recoverable and of the portion 'collected' a further quantity will be lost due to panel and heat exchange inefficiency in the system. Detailed work, particularly a weekly simulation of a typical heating season, is being undertaken currently and should indicate whether the initial area of panel will be adequate and the matching store size to use.

The store will be a rock pile, a maximum of 16 tonnes of rock can be accommodated in a bin with a maximum height of 6 metres. Rock size will be 75 mm diameter and laboratory work has shown that a pressure drop of about 15 Pa/m can be expected. Larger rocks give a lower pressure drop but above 75 mm dia. heat transfer rates fall away. Efficiency of heat transfer and fan power must be balanced.

The air cooled collectors themselves have been the subject of comparative tests in the laboratory and Figure 6 shows six configurations the last of which, 6f, is being set up for test now and should prove to be best. All the configurations are based on a steel plate absorber, one Filon cover and a plywood casing insulated with glass fibre wool.

The delivery system will utilise standard ducting techniques but with slightly larger than normal ducts due to the lower delivery temperatures. An innovation on the domestic scale will be the inclusion of motorised dampers on all outlets. This will allow control of the heating to individual rooms on a temperature and a time basis. This should ensure more economical use of energy inputs and should save substantially on the preliminary projection of a 13 360 kWh heating load.

Water heating panels will be incorporated into the conservatory roof and being at a lower level than the storage tanks will allow a thermosyphoning and therefore self regulating system. The panels will be constructed using a copper pipe grid

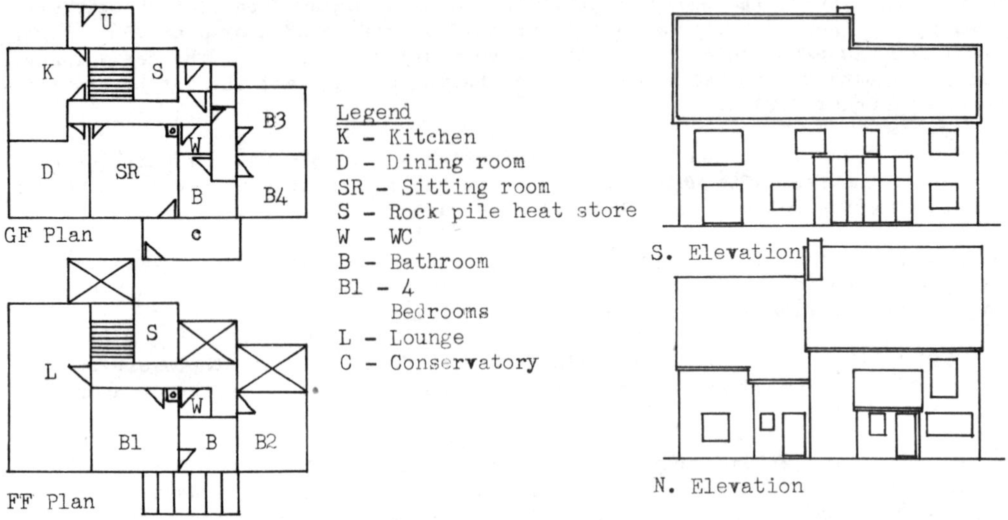

FIG. 5 GENERAL ARRANGEMENT PLANS AND ELEVATIONS

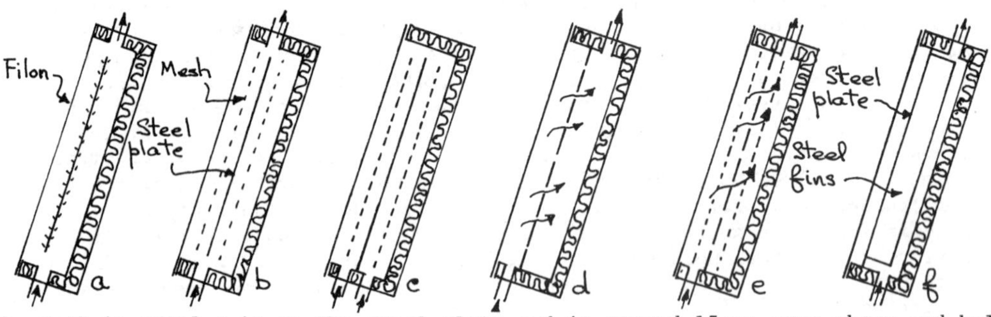

The mesh in panel a is on the steel plate and is spaced 15 mm away above and below in panels b, c & e. In panel d the plate was holed 25 mm dia. at 75 mm pitch.

FIG. 6 PANEL CONFIGURATIONS

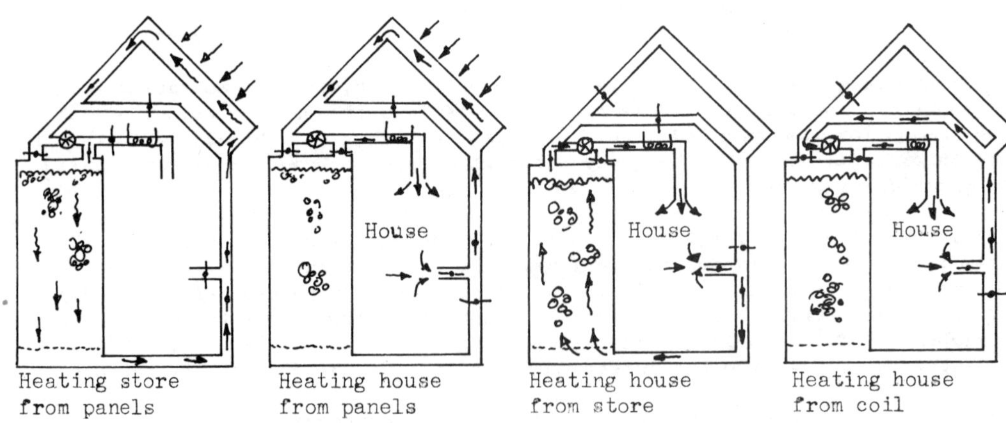

FIG. 7 SCHEMATIC OF HEATING SYSTEM

with clip on aluminium fins, one Filon cover plate and a plywood casing insulated with glass fibre wool. The transfer fluid will be a water/anti-freeze mixture and will operate through a coil in a solar pre-heat tank which will be directly coupled to the domestic hot water cylinder. A sealed header tank with a pressurised overflow bottle (similar to the automobile radiator system) will cut down on evaporative losses and will allow a slightly higher operating temperature.

A back-up system is provided in the form of a wood burning stove with a high output boiler. The boiler will feed the domestic hot water cylinder and/or a coil in the main air delivery duct as required. Figure 7 shows a schematic of the heating system and its modes of operation.

CONTROL AND DATA GATHERING

The control system for the space heating depends entirely on temperatures. These will be sensed using a 'silicon chip', more correctly described as a two terminal integrated circuit temperature transducer with current output proportional to absolute temperature. The manufacturer laser trims the i.c. to give an output of 298.2 micro-amps at 298.2^{o}K (+25°C). The i.c. is encapsuled in a TO-52 can and will be used to measure internal and external air temperatures; radiant temperatures in rooms (by placing at the centre of a 40 mm dia. black sphere - a globe thermometer); heat store temperatures; absorber plate temperatures; cover glass temperatures and transfer temperatures. The possibility of utilising the device for the measurement of cooling rate and wet bulb temperature is being investigated.

Heat mats will be used to measure heat flow through the structure. The mats comprise 40 pairs of opposed copper/constantan thermocouples either side of an acrylic strip and encapsuled in resin. A temperature difference across the strip creates a higher emf on the warmer side. The difference in voltage and the polarity can be noted and the heat gain or loss calculated.

Fuel usage will be monitored. Electricity meters will be adapted using phototransistors and l.e.ds on the alloy disc to measure consumption. Groups of circuits will be metered separately, ie lighting, small power, cooking, etc. Use of the back-up stove will be monitored using a temperature sensor. Gas consumption will be measured using a specially adapted gas meter on loan from the Research and Development Division of the British Gas Corporation and whose assistance is gratefully acknowledged.

Meteorological data will be recorded. Air temperature, relative humidity, wind speed and direction above the house, wind speed across the panels, precipitation and total solar radiation.

Other data being gathered include quantity and temperature of cold water intake, hot water temperatures in storage tanks, waste water temperature, moisture content of timbers and state of dampers controlling the heating system.

Data collection and control of the heating system will be carried out by a 16k microprocessor with twin floppy disc drive, cassette drive and printer. A control and data gathering program has been written but must undergo testing and debugging during the first heating season. Generally, all data from groups of sensors will be modified, stored if necessary and multiplexed through a number of units local to the sensors. The grouping will depend on the type of signal generated by the sensor. Each unit will be scanned every 5 minutes by the microprocessor. Air temperatures in the rooms will be compared with a 'table of required temperatures' and a heat input will be initiated if required. All the data will be stored on floppy disc and transferred at intervals to the University main frame computer for analysis. In particular we wish to know how efficiently and economically the system and its components are operating and how the high insulation standards effect the entrapment of moisture in the building.

ACKNOWLEDGEMENT

The author would like to express his gratitude to Professor V.B. Torrance and all colleagues in the Departments of Building, Physics and Electrical and Electronic Engineering for their invaluable assistance and encouragement during the development and construction of the house.

REFERENCES

Balcomb, J.D., J.C. Hedstom and B.T. Rogers (1975) Design considerations of air-cooled collector/rock bin storage solar heating systems. Los Alamos Scientific Laboratory of the University of California.

Close, D.J. (1963) Solar air heaters for low and moderate temperature application, Solar Energy 7(3), 117-124.

Close, D.J. (1965) Rock pile thermal storage for comfort air conditioning, Mech. & Chem. Eng. Trans. May 1965, 11-22. The Inst. of Eng. Australia.

Dunkle, R.V., (1975) Design considerations and performance predictions for an integrated solar air heater and gravel bed thermal store in a dwelling, Australian and New Zealand Section ISES, Melbourne.

Kreider, J.F., F. Kreith. (1977). Solar Heating and Cooling Engineering, Practical Design and Economics. McGraw-Hill Book Co., New York.

Lebens, R.M. (1980). Passive Solar Heating Design. Applied Science Publishers Ltd., England.

McVeigh, J.C. (1977). Sun Power - An Introduction to the Applications of Solar Energy. Pergammon Press, Oxford.

Wozniak, S.J. (1979). Solar heating systems for the UK: design, installation and economic aspects. Building Research Establishment Report, HMSO, London.

Institute of Plumbing. (1977). Plumbing services design guide. Hornchurch Essex.

Meteorolgical Office (1980). Solar radiation data for the United Kingdom, 1951-75 (Met. 0.18c) Bracknell Berks.

Active Solar Systems for Higher Latitudes

G. S. Saluja* and P. Robertson**

*Scott Sutherland School of Architecture, Robert Gordon's Institute of Technology, Aberdeen, Scotland, UK
**Energy Design Services, Aberdeen, Scotland, UK

ABSTRACT

Hot water supply and swimming pool heating are probably the two most important applications of active solar systems using flat plate water collectors. Higher latitude locations are not at a significant disadvantage when the collectors are oriented in the southerly direction and fixed at an optimum inclination.

This paper describes a solar hot water supply installation at a 15-person hostel in Newton Dee Village near Aberdeen at a latitude of 57.2°N and presents the results of extensive monitoring over a limited period in 1980. It has been shown that for a well designed system over two-thirds of energy required for hot water supply can be obtained utilizing under 1 m^2 collector area per person. The economics of such systems are discussed in a simplified manner.

KEYWORDS

Active solar systems; Solar water heaters; Solar energy for high latitudes; Economics of solar installations.

INTRODUCTION

Active solar systems have been in use for some decades now. Both water and air heaters are being used for a multitude of applications. Of these hot water supply and swimming pool heating are the two most widely used applications. The latter is already an economic proposition, while the economic viability of the former is questionable. Since the yearly total radiation on a horizontal surface decreases with increasing latitude, the application of such systems at higher latitudes is viewed with suspicion. This apprehension is partially misleading because an optimally inclined surface is a better receptor of solar radiation with increasing latitude. In this study two locations, namely Aberdeen (57.2°N) and Kew (London, 51.3°N) are chosen for comparison.

Absence of reliable data on active system performance at a high latitude location in UK prompted the authors to monitor a solar hot water heating installation in Aberdeen. This study presents the results of short term monitoring of the system and highlights the importance of carefully planned system design and installation.

The choice of solar system installers in Scotland, particularly in rural and northern locations, is limited. This study has attempted to set some guidelines for the economic viability of a solar hot water heating installation.

SOLAR RADIATION DATA

The measured total solar radiation data on the horizontal surface has been obtained, for Aberdeen and Kew, from the Meterological office in Bracknell and the Natural Philosophy Department of the University of Aberdeen. A computer program developed by Robertson (1981) was used to convert the data on south facing inclined surfaces. Table 1 shows the ratio of total radiation on inclined and horizontal surfaces for the two locations for 1975.

TABLE 1 Ratio of Total Solar Radiation on Inclined and Horizontal Surfaces

Location	Angle of incidence (o)									
	10	20	30	40	45	50	60	70	80	90
Aberdeen	1.13	1.24	1.32	1.36	1.37	1.36	1.33	1.27	1.18	1.05
Kew	1.07	1.19	1.24	1.26	1.26	1.25	1.20	1.13	1.03	0.90

Figs 1 and 2 show monthly totals on horizontal and optimally inclined surfaces for 1975 and 1976.

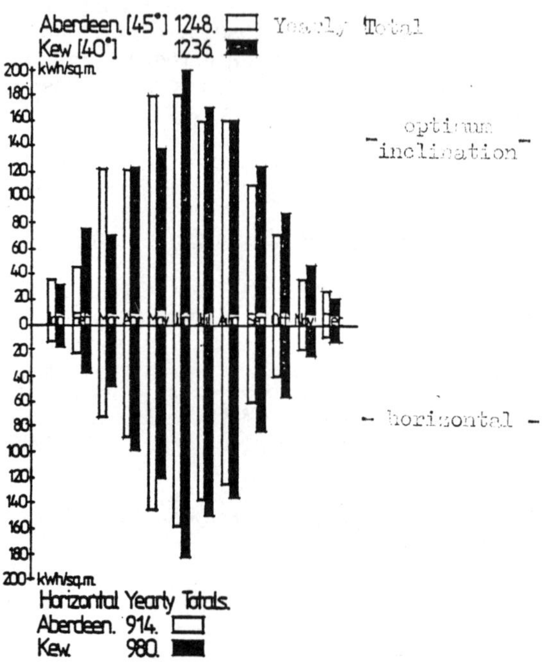

Fig. 1. Total monthly solar radiation - Aberdeen and Kew - 1975

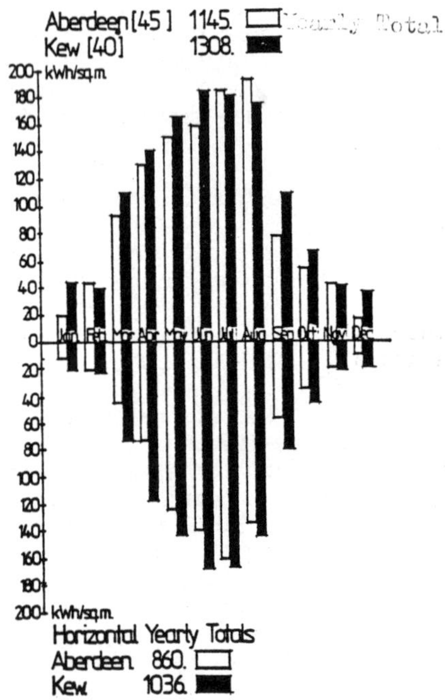

Fig. 2. Total monthly solar radiation - Aberdeen and Kew - 1976

Table 2 gives the ratio of total radiation at the two locations for horizontal and optimally inclined surfaces.

TABLE 2 Ratio of Yearly Total Radiation at Aberdeen and Kew

Collector Tilt	Period		
	1975	1976	Long Term Average
Horizontal	0.93	0.83	0.88
Optimum inclination	1.01	0.88	0.94

DESCRIPTION OF INSTALLATION AT NEWTON DEE VILLAGE

The building. Alcoyne house hostel unit for 15 adults, completed in 1980 and built for the Newton Dee Village community is located just outside Aberdeen at 57.2°N latitude and 2°E longitude. It is a detached building of 350 m^2 area on two floors with 10 bedrooms. Designed by Camphill Architects it incorporates economic insulation standards, strategically placed double glazed windows and a high thermal mass wall to maximise passive solar gains and a solar hot water heating installation.

Solar Installation. As illustrated in Fig. 3, the installation is a direct drain down system which preheats the water in a 450 litre solar storage tank before supplying to a 230 litre calorifier of an oil-fired boiler. The collectors are roof mounted at 45° and 70° inclinations and arranged in two parallel arrays of 6.4 m^2, each with independent circuits, circulating pumps and controls. Both arrays are connected to the same solar tank. A differential controller on each array controls the flow of water operating the respective pump when the controller temperature exceeds the storage temperature by a preset value of 3K. The collectors are drained into the tank automatically when the pump stops. The system is additionally protected by a frostat. A switch for each array enables the pump operation manually or automatically. One switch assembly, located on the kitchen wall, incorporates a digital read out thermometer indicating the temperature of the solar cylinder.

The monitoring equipment consists of solarimeters, electronic turbine flowmeters, a network of premium grade calibrated thermocouples and a data logging unit. The purpose of monitoring was to obtain the detailed information on collector and system performance and the solar contribution to the hot water supply energy requirements. Initial results showed some installation and monitoring problems (Saluja and Robertson, 1980). These problems have been satisfactorily resolved and it was found that the monitoring equipment was an essential part of successful commissioning of the system.

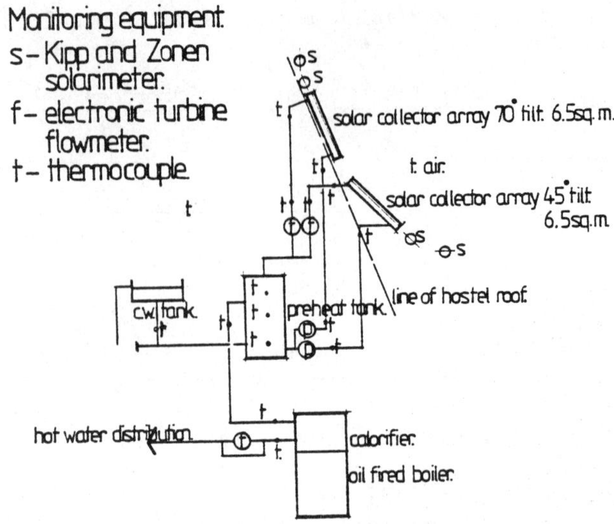

Fig. 3. Solar water heating installation

RESULTS AND DISCUSSION

Monitoring of the system consisted of logging solar radiation, flow rates in the collector-tank circuits and draw-offs from the calorifier, and temperatures at several locations at an interval of one minute. The system was monitored over the periods of 9-12 days each month during July-October 1980. The results of initial monitoring period (July) are included to emphasise the vital role of monitoring for proper commissioning of the system. However these results have not been included in subsequent analysis. Table 3 gives monthly average efficiencies for the two collector inclinations. Daily averages fluctuate considerably, however there is a definite trend to show a reduction in efficiency with decreasing average solar radiation for 45° inclination and clustering around 55% for 70° inclination at all radiation levels.

TABLE 3 Average Collector Efficiencies (%)

Collector Tilt	Period			
	July	August	September	October
45°	45	67	65	62
70°	44	57	55	57

Solar contributions to hot water supply energy requirements for the complete installation are shown in Fig. 4. The monthly figures have been obtained by averaging the monitoring results over the measurement periods. Owing to the considerable distance between the solar tank and the calorifier substantial losses incurred even with insulated pipework. Dotted lines on Fig. 4. represent the solar contributions if there losses were ignored. It is worth mentioning here that the architectural constraints of the building and nearby trees reduced the solar radiation on the location, due to shadowing, by around 20%.

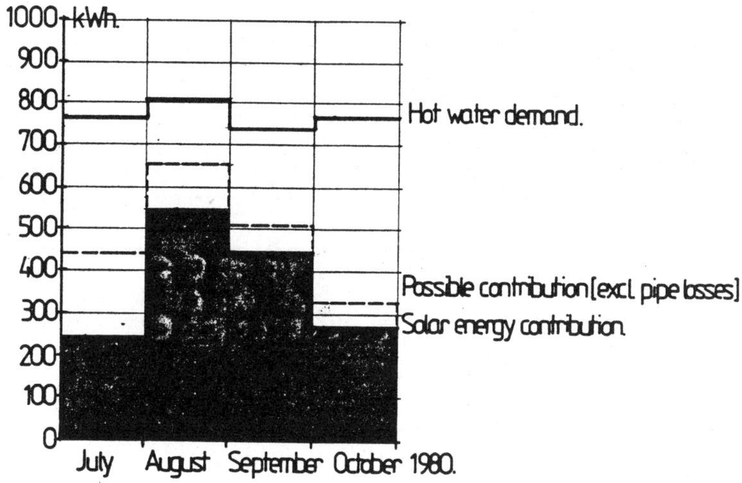

Fig. 4. Solar contribution (July-October 1980)

ECONOMICS OF SOLAR WATER HEATERS

The long term average incident radiation at the optimum collector inclination of 45° in Aberdeen has been calculated as 1134 kWh/year/m^2 surface area, based on average global radiation of 833 kWh/year/m^2 over a period of 1966-80. A well designed system with solar tank and calorifier of adequate sizes (combined storage capacity in the region of 70 litres/m^2 collector area) would be able to utilise approximately 55% of the incident radiation. Based on these figures a net utilisation of 600 kWh/year/m^2 is taken as the basis of economic evaluation. By the aid of present value concept (Kell and Martin, 1979) a simple analysis has been made for a range of installation costs and energy prices. The installation costs were taken as the present day value including minor maintenance costs over the postulated 20 year life of the collectors and other system components. The discount rate has been assumed fixed at 10% and the annual premium in energy costs as 0%, 5% and 10%. The results of the analysis have been presented in Table 4. Current costs indicate that a commercially installed domestic system (4 m^2 collector area) would cost around £400-£500/m^2, while the simplest swimming pool heating system utilising cheapest unglazed and uninsulated collectors, pool as the storage tank and filtering pump for circulation of water through the collector would cost just under £100/m^2. The d-i-y and larger scale systems would fall in the middle range, with £100-£200 for d-i-y system components and installations and £150-£300 for commercially manufactured systems and d-i-y installation. It should be pointed out here that for the swimming pool application the utilization factor is likely to increase from 55% to around 70% or higher (Spelman, 1978).

TABLE 4 No of years After a Solar Water Heating System is likely to be profitable

Installed cost/m² collector area	Energy cost per useful kWh								
	5p			3.5p			2p		
	% Annual premium			% Annual premium			% Annual premium		
£/m²	0	5	10	0	5	10	0	5	10
500	*	*	17	*	*	24	*	*	*
400	*	25	13	*	*	19	*	*	*
300	*	15	10	*	26	14	*	*	25
200	14	9	7	*	14	10	*	*	17
100	5	4	3	7	6	5	16	11	8
50	3	2	2	4	3	3	5	5	4

* over 30 years

ACKNOWLEDGEMENT

The authors wish to express their sincere thanks to Camphill Village Trust for providing unlimited access to the Alcoyne House to install and use the monitoring facilities. Thanks are due to the Science Research Council and Robert Gordon's Institute of Technology, including Scott Sutherland School of Architecture, for providing financial support and for the use of their numerous facilities and equipment. Acknowledgements are also due to Washington Engineering Limited (now Sundwel Solar Limited) for providing a flow meter.

REFERENCES

Kell, J. R., and P. L. Martin (1979). Heating and Air-conditioning of Buildings (Chapter 24, Energy Conservation). The Architectural Press, London, 585-90.

Robertson, P. (1981). Influence of environmental and architectural form on the design and thermal performance of the flat-plate solar collector system. Ph.D. Thesis (Section 3), Robert Gordon's Institute of Technology, Aberdeen.

Saluja, G. S., and P. Robertson (1980). Monitoring of solar water heating installation in a 15 person hostel. Proceedings, U.K. - ISES, Midland Branch Conference on Solar Energy in the 1980s, Birmingham.

Spelman, N. A. C. (1978). The estimation of size and performance of solar collectors for heating swimming pools. Proceedings, U.K. - ISES Conference, London.

Simulations to Determine the Effect of Irradiance and Demand Sequence Structure on the Solar Contribution of a Domestic Hot Water Heating System

K. P. Donnelly

Statistics Department, Glasgow University, Glasgow G12 8QW, Scotland, UK

ABSTRACT

The development of a fast, numerically accurate simulation program for a basic solar domestic hot water heating system is described. The program is suitable for gaining a theoretical understanding of the effect of the stochastic nature of irradiance, ambient temperature and hot water consumption sequences. Particular emphasis is placed on the adoption of a standard data file format, on the development of a suite of programs to construct and modify data in this format, and on methods adopted to ensure that the simulation program is flexible and convenient to use in an exploratory manner. Some results are presented, showing for example that the large day to day variation in irradiation and hot water consumption has less effect than might have been expected on the annual average solar contribution, but that the regular diurnal ambient temperature variation in summer has more effect.

KEYWORDS

Solar heating; simulations; store-mapping; stochastic process; stratified annual storage.

THE PROGRAM

This paper describes a solar simulation program developed for use with a project to model mathematically solar heating systems, and presents some of the results obtained from it. Solar simulation programs have already been developed in the U.K. (Courtney, 1977; Rosenfeld, 1979?), as well as in the U.S., for example the notable TRNSYS program (Klein, Beckman and Duffie, 1976), but the requirements for the present program differed in many respects from the usual:

1. Only a single basic system need be modelled because the aim was to understand and model this mathematically first. For this reason only a direct solar domestic water heating system with an unstratified tank was modelled. The collector transmittance-absorptance product and heat loss coefficient were assumed to be constant. Tank and pipe losses, collector heat capacitance, the effect of wind speed, pumping power requirements and the required temperature differential for operation were all assumed to be zero. Volume demand was assumed independent of temperature (an effectively infinite supply temperature). Flow rate was assumed to be infinite. However the cold feed temperature was modelled as a cosine

function varying between 5°C on hour 1000 of the year and 15°C six months later. Real hourly average radiation and ambient temperature data for Kew 1968 were used.

2. The program had to be numerically very accurate. Usually inaccuracies of a percent or two are of no importance in solar simulations, but when attempting to understand and model the effect of factors such as hot water demand patterns it is essential to be sure that changes in results are due entirely to changes in the factors rather than to numerical inaccuracies.

3. The program had to be fast because it was to be used interactively, in an exploratory manner, and because it would have to be run over long time periods to increase precision when simulated random data was being used.

4. The program had to be flexible and easy to use. This was achieved by using a single standard format data file for each variable, and by developing a suite of utility programs to modify data in this format, as well as by getting the program to read a specifications file, which could be screen edited beforehand.

5. Storage space in the program was of no consequence because the computer had virtual memory. Therefore the program read in (or store mapped) complete data files before beginning the calculations.

Integration methods

The speed and accuracy of the simulation program depends on the method used to integrate the system differential equation. Defining the following quantities (constant over any hour):

$$A_r = (F_R(\tau\alpha)G + F_RU_LT_a)/C,$$

$$B_r = F_RU_L/C,$$

$$A_d = T_cD/V,$$

$$B_d = D/V,$$

where $F_R(\tau\alpha)$ and F_RU_L are collector constants, G is the incident irradiance, T_a and T_c are ambient and cold feed temperatures, C is the tank heat capicity (per unit of collector area), D is the demand flowrate and V is the tank volume, the differential equation for the tank temperature T as a function of time t is

$$\frac{dT}{dt} = A - BT$$

where $A = A_r + A_d$, $B = B_r + B_d$ if $T > A_r/B_r$ (pump on),
and $A = A_d$, $B = B_d$ otherwise.

This is integrated numerically by calculating T over a regular series of time steps. The simplest method of doing this, by assuming the average value of dT/dt over the time step to be equal to the value at the beginning of the step is called the Euler method. Alternatively a naive estimate of T half way through the time step can be made and the average value of dT/dt assumed to be the value calculated at this point. This is called the Crank-Nicolson or modified Euler method (Duffie and Beckman, 1980). A third method is to integrate the differential equation analytically over each hour, giving

$$T = \frac{A}{B} + (T_o - \frac{A}{B})e^{-Bt}.$$

The solar contribution when the pump is running, $\int(A_r - B_r T)dt$ is then obtained from

$$\int T dt = (At+T_o - T)/B.$$

The value of T at the end of the hour must be checked to see whether the pump switches during the hour, and if so the calculation is performed again in two stages with a break at the switch time. The analytic method gives an exact solution.

Table 1 compares the accuracy and speed of the different methods. The average solar contribution per unit collector area (Wm^{-2}) is calculated over a year for three demand patterns: (a) uniform demand over every hour of the year (b) some real hourly hot water consumption data, and (c) demand concentrated over a single hour at the middle of each day, each pattern having the same annual average. The first two methods are repeated with various numbers of time steps per hour, and the CPU time in miliseconds taken on the calculations is shown in square brackets. In each case editing and reading the specifications file, reading the data and printing the results took about an additional 500 ms. The programs were run with an optimising compiler on an ICL 2980 running under EMAS. In each case an effort was made to write an efficient algorithm.

TABLE 1 Comparison of integration methods

	Euler				Crank-Nicolson				Analytic		
Steps	(a)	(b)	(c)	CPU	(a)	(b)	(c)	CPU	(a)	(b)	(c)
1	30.657	31.017	35.455	[239]	30.109	30.070	29.270	[324]			
2	30.386	30.544	32.453	[376]	30.124	30.112	30.192	[545]			
3	30.299	30.399	31.698	[514]	30.127	30.119	30.323	[766]			
4	30.256	30.329	31.352	[651]	30.127	30.122	30.364	[987]	30.129	30.125	30.413
5	30.230	30.287	31.153	[788]	30.128	30.123	30.382	[1208]	[506]	[366]	[262]

Because the error with the Euler method decreases only linearly in the length of the time step, it is not possible to acheive adequate accuracy with this method in a reasonable time. With extreme demand patterns such as (c) the error can be quite serious. The Crank-Nicolson method is a big improvement, but can only give adequate accuracy for our purpose in all cases if five or more time steps per hour are used. Although the analytic method can be difficult to extend to cases such as non-linear collector curves, it is clearly the superior method for our purpose and was adopted for all future calculations.

Store mapping and a standard data file format

Initially it was found that the program was spending nearly all its time reading in data. It had to read in a year's data for four variables and took about 4000ms CPU time for each with a non-optimising FORTRAN 7.7 compiler. This was reduced to about 2000ms when the data was stored in binary (unformatted) form. However, with the EMAS operating system a procedure called store mapping is available whereby instead of reading the data, the program is told where it is to be found on disc and works with it in situ. Although store mapping is available only in the language IMP, which meant that an IMP module had to be written to copy the data to FORTRAN arrays, the time per variable was reduced to 110ms, or 50ms with an optimising compiler.

The original version of the program required the data to be stored in columns in a single file. However, this was found to be inflexible and so a new standard data format was adopted. Each file contains hourly data for a single variable for one year, and is unstructured (that is, not divided into records or lines). First there is a 256 byte explanatory heading. Then there is 264 hours (11 days) data from the end of the previous year, for the purpose of allowing the simulation to reach equilibrium. Numbers are stored end to end in 8 byte format, double precision being required for adequate accuracy. Finally data for the year follows, either 8760 or 8784 hours according to whether it is a leap year or not. This format is also efficient in terms of storage space

256 byte heading	264 hour lead in from prev. year	8760 hrs. data	or 8784 for leap year

A suite of programs

Once a standard data format is adopted it is possible to develop a suit of programs to work with such data and modify it for exploratory purposes. The following programs have been developed:

PACK converts a filed formatted data in free format to the standard data file format.
UNPACK does the opposite.
UNPACKN unpacks a small sample, including the heading, for a quick look.
YEARPLOT graphs a year's data, month by month, with a heading read from the file.
ANGLES generates standard format files containing the solar altitude, azimuth and declination.
INCIDENT takes beam irradiance, diffuse irradiance and solar angle data and generates a file contining the irradiance on a surface at specified tilt angle with specified ground reflectance. It assumes isotropic diffuse radiation and non specular ground reflectance.
DAILYMEAN replaces each day's data by the average for that day.
MONTHLYMEAN does the same for months.
MEANPROFILE replaces each day of each month by the mean 24 hour profile for that month.
MEANYEARPROFILE replaces each day of the year by the mean 24 hour profile for the year.
JUMBLE randomly reorders the hourly data for the year (to destroy any autocorrelation while preserving the average and variability)
JUMBLEDAYS randomly reorders the days of the year.
SOLARSIM is the simulation program using analytic integration.

All the programs except PACK, UNPACK and YEARPLOT are very fast. Programs which alter or generate data files register the fact in the file heading.

Specifications file

In order to make the simulation program as flexible as possible, the system parameters and names of data files, etc. are not fixed in the program but are read in from a specifications file. The standard specifications file is given below.

```
Collector area        :4
Collector FRta        :0.75
Collector FRUL        :6.
Tank volume           :200
Radiation file        :SOLARDATA_K68S45R20P
Ambient temp. file    :SOLARDATA_K68TP
Demand file           :CALL UNIFORM5
```

Cold feed temp. file :SOLARDATA_COLDTEMPP
Results file :RESULTS-MOD
Print months :No
Results title :Read

The program asks for the name of the specifications file, adopting the standard file by default if return is pressed. It then asks if the file is to be edited. By default the screen editor is called and the file is edited, by moving the curser around, to a temporary file which is read in to the program. Alternatively "No" can be typed if editing is to be skipped, or the name of a permanent file to hold the edited specifications can be typed. The default radiation data was for Kew in 1968, at a 45° tilt to the south, with 20% ground reflectance, as indicated by the file name above. The specifications file can have in place of the name of a data file the name of a subroutine to generate the data, as for the demand data above, which by default is a uniform 5 litres for every hour of the year (120ℓ/day). By default the results are appended to a file called RESULTS. One of the biggest problems with exploratory analysis is in keeping track of the results. For this reason the specifications file contains a title which is printed out with the results. If this is given as "Read" then the program asks for the title from the terminal.

RESULTS

Some results obtained using the program are illustrated in Fig. 1. The quantity represented is the average solar contribution in Watts per metre squared of collector area.

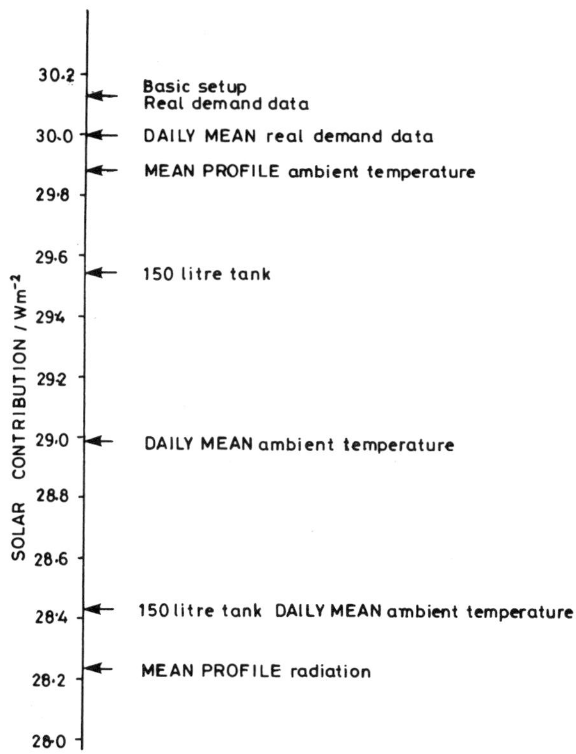

FIGURE 1 VARIOUS RESULTS

Demand pattern

Replacing the uniform demand of the basic setup by real measured demand data (actually only 100 days data repeated enough times to make up a year and rescaled to average 120ℓ/day) gave almost exactly the same result. Although too much attention should not be paid to the precise figure, which is partly due to the chance coincidence of irradiance and demand, this result is surprising and fortunate in view of the large hour to hour and day to day variation in daily totals. (The data used - from the first house in Table 2 - had a coefficient of variation of daily totals of 0.32.) The higher than expected solar contribution was not due to autocorrelation in demand between days - jumbling the days in fact produced a marginally higher result. Nor was it due to having a tank size equal to average demand for 1.7 days - the real demand data produced marginally better results, relatively, with a 100 litre tank. Nor again was the result due to the 24 hour demand pattern - the "DAILYMEAN real demand data" result in Fig. 1, obtained using the real data averaged over each 24 hours, is barely lower. Hourly demand data from some of the other 4 houses available (Table 2) did produce slightly lower average solar contributions, but this was due to the 24 hour pattern, and when DAILYMEAN data was used, all gave similar results.

Figure 2 shows the effect of concentrating the same demand over a single hour of each day. The best time for demand to occur is about 13.00 or 14.00 hours (local apparent time), after the water has been heated but while the sun is still shining so that the tank temperature is lowered and further heat can be collected. The benefit of this compared to demand at night is 3.6 percent, or slightly more than this if demand is spread over more than a single hour. The graph shown is level at night because no tank losses are assumed. Although this is not completely realistic, tank losses should be low with economic levels of insulation.

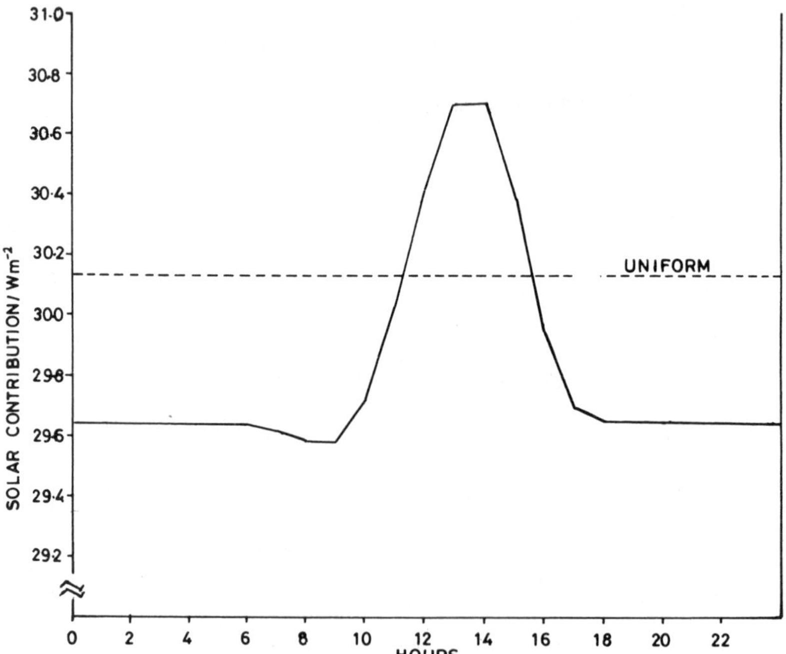

FIGURE 2 SOLAR CONTRIBUTION WHEN DEMAND IS CONCENTRATED ON A SINGLE HOUR, CENTRE ON THE TIMES INDICATED

Table 2 summarises hot water daily consumption data from 5 houses collected by the Building Services Research Unit of Glasgow University (Gibson, 1978a,1978b). Although the effect of average consumption on solar contribution is not considered here, it is much larger than other effects. In view of the large range of average consumption in the five similar houses of Table 2 - a factor of 3.7 - it would seem that one of the most important things for a general economic assessment of solar water heating is an investigation of the extent and reasons for variations in average consumption between households, and a prediction of future levels.

TABLE 2. Hot water consumption in 5 Scottish local authority houses, metered in 1970 and 1971

Occupancy	no. days	average (litres/day)	st. dev.	average/ person	autocorr. lag 1	autocorr. lag 7
2 adults, 3 children	100	166.3	52.6	33.3	-0.17	0.17
2 adults, 1 child	79	89.6	38.9	29.9	-0.23	-0.14
3 adults	111	84.7	38.3	28.2	-0.05	0.09
2 adults	111	45.3	19.6	22.7	-0.01	0.26
2 adults, 2 children	84	154.9	54.4	38.7	-0.18	0.11

Irradiance and ambient temperature patterns

One of the most interesting results seen in Fig. 1 is the quite large beneficial effect of the regular diurnal variation in ambient temperature which occurs in summer. Replacing ambient temperature by the daily mean reduced the solar contribution by 3.8 percent, and this was little affected by tank size. The MEANPROFILE ambient temperature results, by contrast, are little different from the basic setup, showing that any correlation between irradiance and ambient temperature on a monthly basis has little effect. Replacing irradiance data by the monthly mean 24 profile reduces solar contribution by 6.3 percent. From a fuel saving point of view a typical solar system is better suited by a mixture of high and low irradiance because in the bad times it can switch off.

Seasonal tariffs

Assessment on the basis of average fuel saving can be criticised on the grounds that the true cost of providing backup heating, especially by electricity, is greater in winter than in summer. Countries such as Sweden are beginning to introduce seasonal tariffs. Figure 3 illustrates an exploratory analysis weighting the solar contribution by the factor $1 + aC(t)$, where C is an annual cosine function with maximum on hour 1000 of the year (11 February). Thus $a = 0.5$ implies a maximum to minimum tariff ratio of 3. Naturally the optimal collector tilt is increased, although perhaps less than expected. This may be because the season tariff curve chosen does not have a sharp enough peak in winter or a broad enough trough in summer, even if the choice of curve must be speculative.

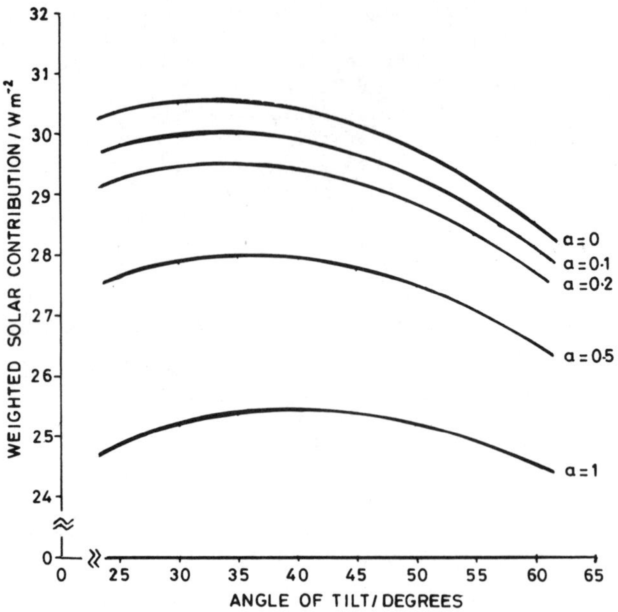

FIGURE 3 SOLAR CONTRIBUTION WEIGHTED BY SEASONAL COST OF BACKUP FUEL.

REFERENCES

Courtney, R.G. (1977) A computer study of solar water heating. Building Research Establishment. Current paper CP30/77; BRE, Garston, Watford WD2 7JR, U.K. Reprinted from Building and Environment 12, 73-80, 1977.

Duffie, J.A. and Beckman, W.A. 1980. Solar Engineering of Thermal Processes, Wiley, New York.

Gibson, E. 1978a. Building Services Research Unit, Glasgow University, Report no. 202.

Gibson, E. 1978b. Ph.D. thesis, Glasgow University.

Klein, S.A., Beckman, W.A. and Duffie, J.A. 1976. Solar Energy 18, 113-127. "A design procedure for solar heating systems".

Rosenfeld, J.L.J. 1979? Performance assessment of solar heating systems. Shell Research Ltd., Thornton Research Centre, P.O. Box 1, Chester CH1 3SH, U.K.

TOPIC G

Solar Energy: Tropical

Field Test of a Solar-powered Micro-irrigation Pump* in the Village of Basaisa, Egypt

C. Nelson, S. Arafa and R. Pearson

The American University in Cairo, Cairo, Egypt

ABSTRACT

A solar powered pump for irrigation was made available by the Catholic Relief Services in Cairo, for a field test in the village of Basaisa which is a locale for the project on Utilization of Solar Energy and Development of an Egyptian Village: an Integrated Field Project. The Basaisa Project is supported by the US National Science Foundation and the American University in Cairo.

The theme of the project at Basaisa Village is that villagers must actively participate in all aspects of the project and that the acceptance of the new technology and their incorporation of it into their living pattern must come from them.

The SEI-15 Sun pump has a 2-square-meter silicon solar cells array which convert sunlight directly into electricity. The metal framed array is positioned on a tralley and combined with a small verticial turbine pump with a permanent magnet d-c motor.

The sun pump was transported to Basaisa on the evening of Thursday, May 3rd, 1979, explained to the farmers the same evening, demonstrated by villagers to visitors and villagers from nearby villages the following day, May 5th, and the Basaisa villagers started its actual operation in the field on the morning of Saturday May 6th, 1979.

This paper reports on the operation of the sun pump on site and the farmers response, evaluation and recommendations.

KEYWORDS

Solar powered; micro-irrigation; village; Egypt; Rural community development.

INTRODUCTION

Through the cooperation of the Catholic Relief Services; the Basaisa Solar Energy Project has been given the opportunity to test a new, solar powered water pump in an Egyptian Delta agricultural setting. The pump, consisting of a photovoltaic solar panel, submergable pump, wire and hoses had previously only been demonstrated for scientific and government groups such as at the Egyptian National Research Center. It was during that demonstration that; Dr. Salah Arafa, Co-principal investigator of the project suggested the equipment to be used and field tested in the Basaisa Village, Sharkia Governorate. The purpose of this demonstration, in a typical Egyptian agricultural environment was to see how third world farmers might use the equipment; measure their responses and evaluate whatever problems might arise. As this proposal fulfilled both the interests of the Catholic Relief Services, the manufacturers of the sun pump and the Basaisa village; a three week trial period was agreed on and begun May 3rd, 1979.

THE SEI-15 SUN PUMP

The Problem

Over 50 million of the world's smallest farms currently exist on less than 50 million hectares of the fertile alluvial valleys; deltas and coastal plains of the Third World. Most of these areas experience extended dry seasons during which the weather is warm enough but cropping impossible without irrigation. Though water is frequently available from shallow wells, the costs of lifting water by conventional means are becoming prohibitive for these very small farm units.

One Solution

A possible solution is the use of solar-powered micro-pumps. The SEI-15 Sun Pump was developed specifically to irrigate about 1.0 ha of land. Its four principal components; carefully matched to optimize overall system performance, are:

The solar cell array. This is a 2-square-meter array of silicon solar cells (semiconductor devices) which convert sunlight directly into electricity. (This is not a thermal process and no heat transfer is involved).

The trolley. This is a ruggedly built tubular steel frame which supports the array; allows proper orientation to the sun; and facilitates transport from house to field, and field to field, as required.

The pumpset. The SEI-15 combines a small vertical turbine pump with a permanent magnet d-c motor. It will fit into a 150 mm well casing, or can be fitted with a floatation collar for use in an open well or canal.

The maximum power tracker. This is an electronic device added to improve output by adjusting the operation of the pump during periods when either sunlight or lift diverge from the base values for which the system was originally optimized. It also ensures correct pump start-up as the sun rises.

Performance

The pump will produce between 50 and 150 m^3/day depending on local insolation and lift conditions. The attached diagram shows projected output under a typical monsoon climate weather pattern, against lifts which vary between 2 and 4 m. over the year. (The designation SEI-15 relates to the unit's capacity to produce a total of about 15,000 m^3 during the periods in the year when irrigation is needed).

Operation and Maintenance

The SEI-15 Sun Pump is quiet, safe and simple to operate. The farmer wheels it to the field in the morning and positions it alongside the well or canal, with the array pointed in a southerly direction. He then tilts the array 35° to the east (against a rubber stop), lowers the pump into the water, and moves a switch to the "On" positions. At noon the array is turned, by hand, about its central axis so that it is tilted 35° to the west during the afternoon. Several times during the year the rear biped legs are adjusted, to predetermined positions, to follow the sun's change in elevation as the seasons progress.

Routine maintenance involves dousing the array occasionally with a bucket of water to wash off any accumulated dust or dirt. The array consists of six modular panels with a life expectancy of 14-16 years. These can, however, be individually replaced with simple hand tools in the event of damage. The brushes in the motor of the pumpset (which will last 6-7 years) may need to be replaced every couple of years.

Economics

Solar electric pumpsets can now be manufactured and sold in Third World countries at prices which are more than competitive economically with conventional diesel and electric units (where substantial rural electrification costs are involved), for farms below 1.5 ha. and lifts less than 5 m. Steps to establish the first manufacturing facility in India, with a capacity of 12.500 units/year, are already in progress. The critical point is that providing affordable solar micro-pumps to the farmers of the Third World is now a manufacturing or industrial development problem--no longer a research or experimental task.

Availability

Prototype SEI-15 Sun Pumps can be made available from within 90-120 days of order, for testing, evaluation and demonstration at established agricultural engineering research institutions.

Background Information

Basaisa village was an ideal setting for this demonstration in Egypt. Not only does it represent a very typical agricultural community in the Delta using tradition agricultural practices, but the ideas of technological change have been well planted in the minds of the villagers since the beginning of Dr. Arafa's visits to the village (some 4 years). In accordance with the principles of the project, all proposed technological innovations have been proceeded by demonstrations, training programs in appropriate technologies plus classes developing the concepts of the use of solar energy. These preparatory efforts have resulted in a general village understanding of the possibilities of solar power, its uses and even the maintenance of equipment already installed in the village. The solar water pump, for example, was a well understood concept even before this demonstration period was proposed. A small working model of a solar panel operating a D.C. motor pump was first introduced last winter. The pump, running off a solar storage battery, was 100% locally made and could lift 15 liters of water three meters high every 6 minutes. The villagers in a demonstration of solar equipment to another village (Shimbarra; 25 km from Basaisa), set up and explained the pump completely themselves. A natural flow of comprehension has grown from this; the villagers understand not only the mechanics of new solar technology; but also its advantages, uses and possibilities for their future.

Thus, on the evening of May 3rd, (about 6.30 p.m.), the arrival of a larger solar pump to Basaisa came with no surprise to the farmers. Dr. Salah Arafa took advantage of the gathering of a large group of farmers for an engagement party that evening, to explain the use and handling of the new equipment. Responsibility for the pump was placed in the hands of certain villagers with Seoudi as a responsible person. Dr. Arafa then left the village about 9:00 p.m. Upon arriving back to the village the next morning with a group of faculty members and students of AUC and other visitors, the pump was completely set up on the edge of the road to the village and was ready to be turned on.

Operation, Problems, and Response

It was not difficult to stir enthusiasm for this silent machine that pumps water with solar energy. Farmers came from many nearby villages to see the new pump and villagers were more than eager to try it on their own land. Demonstrations continued all day on May 4th, so real use didn't begin until the 5th. The anxious farmers agreed that a certain isolated patch of clover needed immediate irrigation and would benefit most from the first use of the new pump. It had been impossible to water this patch for fear of watering the drying fields of wheat surrounding it. Although water in a nearby canal was easily accessible with the new pump, it had been impossible to reach this clover patch for 60 days. The immediate practicality of the pump is evident from its first use. Throughout this trial period, the best use of the pump was always in small, otherwise inaccessible patches, that had been recently watered. Even the first day, however, clarified the fundamental weakness of the system; the pump lacks the strength needed for general irrigation. This section of clover, being so dry, required 2 1/2 hours to be watered. The same job, however, normally requires 3 hours, if irrigated by Sakia (Egyptian buffalo-powered water wheel). Do not assume from this data that the pump irrigates more quickly. Rather, the time normally required is purely a function of the field's distance from the Sakia. Generally, irrigation of land near the Sa-ia takes an hour/feddan which increases up to 3 hours, at a distance of 1 km. This plot of clover sits at about 1 km from the nearest Sakia (which is the greatest possible distance from any one Sakia in this village). The practicality of the pump is thus shown not in absolute terms of liters pumped, but through relative terms of effort and time needed for individual plots compared to the traditional sakia. General operations for the next 2 weeks proved relatively successful or discouraging dependent upon the size of the field being watered and the length of period since it was last irrigated.

The farmers themselves decided when and where to use the pump. For the first week they were highly supervised in operations but later left on their own. During this time certain mechanical/operational difficulties arose that required special assistance. Other problems were made clear by the farmers and many suggestions came forward. Special mention needs to be made about the exlamations of amazement and joy of most observers. Praises to God were often given for this new hope of the future. This pump has in truth, excited the farmers more than any other equipment placed in the village. Irrigation remains one of the most key issues of their lives and the possibilities for solar irrigation are existing for all. So in the interest of improving this system, an outline of the major problems, as a result of our experience, will follow. These can be divided into mechanical problems of the present system and conceptual problems for greater use of solar irrigation in the future.

The primary mechanical problems occurred with the cart for the solar panel and the connection from the maximum power tracker to the cord running to the pump. The design of the trolley for the solar panel and the tilt mechanism are very efficient with the exception of the wheel. The plastic ball wheel was very easy to manoever over all village paths, but did not endure a week. We were then without a wheel for 6 days, three of which were too cloudy to get much benefit from the pump. The replacement wheel was still too small and fragile. What is required is a large, firm wheel (like that of a motorcycle) that has a wide base (3-5 inches) and fairly large diameter 1 1/2-2 ft.).

The second problem, with the electrical connections, needs careful consideration.

Any connections that will be handled often by unskilled hands need to be firm and well-designed. The two plug connections should be replaced by sturdy well sealed plugs and sockets.

A third operational improvement involves the hose. The original light plastic hose was neither long enough nor sturdy enough. We replaced it with a firm rubber hose of 9 meters which greatly improved pumping and ease of reaching the fields. Unfortunately the diameter of the new hose was slightly smaller than the original and may have affected pumping quantitites. A 10 meter, firm rubber hose of appropriate diameter would solve this problem. With regard to conceptional problems, several points deserve mentioning. Principal among these is the complaint that the pump is not strong enough. Farmers opinions varied greatly on this point. Some considered a pump twice as powerful sufficient, while others suggested up to 40 horse-power. What is clear is that this present size pump can only be used in specific circumstances when distance from the Sakia is great or water is only available through on the spot pumping. Several examples of the latter occurred. For a period of three days no water was being supplied to the canals and sakias could not be used. This is a periodic occurance of the water distribution system in each region and can cause some hardship. The pump's greatest triumph was the recovery of low lying water in various trenches which normally remains unusuable. Often, as well, some one has finished irrigating their land and the entire length of the irrigation ditches remained full of inaccessible water. Here also the solar pump was a great advantage. Yet the problems remains, for general irrigation, the pump is not large enough. Certain crops such as cotton and wheat require not only large quantities of water but also a strong flow of water to create the spot flooding preferred in their traditional agriculture. A weaker flow of water can produce waterlogging or simply too much penetration of water into the soil, especially with the new under ground tiling systems for leaching. Theoretically, efficiency of any water pump, with regard to land surface irrigated could be more than proportionally increased through a strong pump. The water that now soaks too deeply into soil through slow pumping could be conserved through more rapid irrigation.

A consensus has been reached on an appropriate sized pump and panel. Ideally the quantity of water pumped by a 6 horse-power gas pump satisfies all needs. This would be twice the size; in pumping power, of the present day sakia. As the villagers made most suggestions with regard to their own village, they suggested that a stationary pump of about this size be installed in the same location that 3 sakias now occupy. This, if operational would fulfill almost 70% of the village's irrigation needs. Realistically, a back-up system of sakias will have to remain to use in periods of extended cloud cover or with maintenance problems.

There is little doubt that this simple solar pump is an exciting demonstration of future possibilities. "By the Great God," said one farmer, "its a beautiful thing." And just as we hoped, our understanding of this technology was enhanced by this demonstration period. A larger machine is needed with sturdy parts to handle both the desired quantity of water and the rough treatment it may receive. Time is needed as well to bring solar energy costs down to a practical level.

Table (1)

Date	Land Area	Crop	Land Owner	Time	Hours	Comments
5/5	3 kirats	clover	Rifaat Mustafa	12:15-3:50	3 1/2	land previously not water for 60 days.
6/5	to fill ditch for clothes washing			9:30-5:15	7.3/4	the water never completely filled the 1 km long ditch.
7/5	1 kirat	new rice pitch	M. Hammad	8:15-4:30	8 1/4	very dry earth (no water for 40 days).
8/5	3 kirats	clover	Abdalla Foda	8:30-4:00	2 1/2	lack of sufficient water in the trench, ran pump 1/3 of time.
	1 kirat	vegetable	A.M. Mansour	4:30-6:00	1 1/2	
9/5	4 kirats	cover + vegetable	A.M.Mansour	8:30-2:30	4	the pump was run 2/3 of the time and clouds slowed pumping. The cart wheel out today.
15/5	1 kirat	new rice pitch	M. Shafie	10:00-6:15	6	pumped water of a ditch otherwide not useable used new 9 meters hose & one connection broke (elec)
16/5	1 kirat	new rice pitch	M. Shafie	9:30-11:00	1 1/2	The reverting of a rice patch takes little time.
	3/4kirats	rice	A. Fouda	2:30-5:30	3	Insufficient water in the trenches stopped pumping.
17/5	1 kirat	rice	M. Aly	7:30-10:30	3	Farmers take complete control of setting up pump.
	1 kirat	rice	Abdel Wahed	11:00-12:30	1 1/2	operating and moving it to new locations.
18/5	1 kirat	rice	M. Aly	7:30-11:00	3 1/2	Both patches were water with-n the last 3 days.
	2 kirats	rice	A.M. Ismail	11:30-4:00	4 1/2	

RECOMMENDATIONS

Following is a summary of recommendations expressed by the villagers for a better use of the pump.

1. The wheel must be stronger with a wide circular shape.
2. The length of cord and hose must be increased, to 10-15 meters.
3. Wooden box to be installed (underneath) the panel (for pump, hose, and tools).
4. The control box must be put underneath the panel and the electrical connections must be stronger.
5. The pump must be adjusted for the 1-2 meter depth to increase the discharge.
6. Provision must be made to use the panel in charging batteries for other uses in the village.

ACKNOWLEDGEMENTS

The authors and the rest of the AUC-Basaisa project team acknowledge the assistance of the CRS staff in Cairo and Dr. R. W. Matlin, the designer of the pump to lend us the solar pump to carry out this study.

The enthusiasm of the Basaisa villagers, their support and their interest in experimenting with us is highly recognized.

On the Use of Solar Energy in Irrigation

W. A. Kamal

Faculty of Engineering, Alexandria University, Egypt

ABSTRACT

The two basic thermodynamics conversion techniques of solar water pumping are analysed. The study revealed that an engine working on the organic Rankine cycle and of ouput within few kilowatts (maximum 20 KW) would be most suitable for pumping purposes. It is also concluded that a pumping rate of 1.5-2 cubic meters per hour is obtainable for every square meter of the flat plate collector area in an engine employing Freon-11 as the working fluid.

INTRODUCTION

Developing countries are mostly situated in sunny regions, and are in general not self sufficient for their conventional energy requirements. Therefore, it is very vital that they should develop their utilisation of solar energy.

The intermittance and variability of solar radiation reaching the surface of earth inherently limit its useful conversion by requiring either intermediate storage of energy or application to a task where intermittency is acceptable. The pumping of irrigation water is such an operation.

A review of past efforts to develop solar powered irrigation pumps (Pytlinski, 1978) revealed that successful attempts were made as early as 1850. One of the largest (37KW) and most important solar irrigation plants was constructed and installed by Shuman & Boys in Meadi, Egypt in 1913 (Ackermann, 1914).

In this paper, the basic mechanical techniques for solar water pumping are critically reviewed and assessed. The Rankine cycle-powered solar pump is then studied in detail and the different design alternatives outlined. Thermal calculations of the cycle are made to find the most suitable working fluid.

SURVEY OF THE POSSIBLE METHODS OF SOLAR PUMPING

In principle, a solar pump is a conversion system with different single converters, which transforms the energy in the solar radiation to mechanical energy used to drive the water pump. As can be seen from figure 1, solar energy can be converted to mechanical power in a conventional or a specially designed thermodynamic system.

In the solar-thermal scheme, a solar collector or a concentrator is employed to produce a fluid at high internal energy. This fluid may be utilised directly in any of the conventional cycles such as Rankine, Stirling or Brayton to produce mechanical energy which is used to operate a conventional pump. Alternatively, the high internal energy fluid may be utilised in specially designed systems for water pumping. In the following sections the thermodynamic conversion processes are summarized and assessed.

A) Solar Pumps Working On Conventional Cycles

i) The solar Rankine cycle. The cycle consists basically of an evaporator, an expander, a condenser and a liquid feed pump. Vapour is obtained when a liquid working fluid is heated by solar radiation either directly in the collector or indirectly by exchanging heat with the primary fluid as shown in figure 2. The vapour expands in a turbine, a reciprocating engine or a rotary displacement engine, doing work. From the engine it flows to the condenser which is either air cooled or is cooled by the pumped water. The condensate is re-injected by the feed pump (usually operated by the solar engine itself) to the evaporator. The solar Rankine cycle will be analysed in the second part of this paper.

ii) The solar Stirling cycle. Two of the hot-air engines designed to operate on the Stirling cycle were adapted to use solar energy as their input heat source. These are the Farber (1965) engine and the Khana (1970) engine. The Farber engine is shown schematically in figure 3. The engine has an efficiency of 9% (not including collector efficiency) and develops maximum power at crank speed of 150 rpm. When the solar radiation was concentrated directly within the engine through a quartz window by using a parabolic concentrator a brake efficiency of over 30% was accomplished (Trayser and Eibling, 1966).

iii) The solar Brayton cycle. The cycle is shown in figure 4

The main disadvantage of hot-air engines, either Stirling or Brayton is that they require high temperatures for their operation. Therefore, flat plate collectors can not be used and the engines have to be used in conjunction with two and preferably three-dimensional concentrators. Since there is a practical limit to the size of dish-type concentrators, due to the wind loading and their sun tracking, one may conclude that such systems are for small capacities, lower than 5KW.

B) Special Designs of Solar Pumps

In this section, special designs of expander and pump that have been developed specially for solar energy as their input source are presented. Basically, they all make use of the fact that the change of phase of any substance (from solid to liquid or from liquid to vapour) upon heating, results in a change in its volume. The volume increase at a given pressure may be utilised to displace water to a higher elevation and the volume reduction at a lower pressure corresponds with a suction of water from a depth.

Three different designs of solar pumps are presented in this section, namely those of Boldt (1978), Rao (1976) and Michaelis (1978). The first two use the change of phase from liquid to vapour while the third uses the change of phase from solid to liquid.

i) The solar powered thermopump. Figure 5 shows the basic idea of a thermopump using water as its working medium. If a working medium other than water is used, a collapsible rubber lung is used to separate it from the pumped water, see figure 5b. Normally a liquid with a low boiling point such as Freons, propane or ethyl-alkohol can be used. Figure 6 shows a schematic diagram of a solar powered thermopump designed by Boldt (1978) and tested in Tanzania. The pump is cheap, reliable and can be produced locally from readily available materials and requiring no high standard of accuracy. However, it has a very small capacity and is limited to shallow wells (less than 15m deep).

ii) The Birla pump. In 1976, Rao and Rao presented the first prototype of their air-cooled and water cooled solar water pumps having no moving parts except for check valves. The pumps are specifically designed for rural lift irrigation and use an inexpensive mixture of petroleum liquids with a boiling temperature range of 35-40°C. In the case of the water-cooled pump, figure 7, water can be pumped continuously, while the air cooled pump pumps only once in a day, thus its pumping capacity is limited by the water tank size.

Generally, the control of the system of valves is the major disadvantage of Birla pumps since the proposed electrical control does not meet the requirements of a rural solar pump which should be independant of any external source of power.

In 1980, Sudhaker et al (1980) modified Rao's original design to solve the problems of contamination of the pumped water with the working fluid.

iii) A pump employing a thermal linear motor. This pump, shown schematically in figure 8 gives a discharge of $1m^3/hr$ with a flat plate collector of area $2m^2$, (Michaelis, 1978). The solar thermal linear motor depends on the increase of volume of wax when changing from solid to liquid. The two linear motors (5) & (6) each consists of a cylinder inside which wax is put and the cylinder is encased in a jacket through which the alternately hot (80°-90°C) and cold (20-30°C) water flows and brings the expansion wax to the point of melting or setting. Therefore, the sliding sealed pistons (7) & (8) are forced out or in again, operating the water pump pistons (9) and (10). The pump is relatively cheap and durable but the pumping rate and efficiency are low.

From the previous survey, it is clear that the efficiency of the conventional cycles (Rankine, Stirling and Brayton) as well as their daily ouput are higher than those of unconventional designs. The special designs, on the other hand, have the advantages of being cheap, easy to maintain and can be produced locally, which makes them suitable for developing countries. However, since in solar power systems, the cost ratio of the collector to the engine is usually high, the engine must have a high efficiency. Thus, the concept of a simple, cheap engine is not economically viable.

Among the pumps operating on conventional cycles, that of the Rankine cycle is recommended for the following reasons:-

i) It has a high efficiency even when operating with the moderate temperatures (80-200°C) attainable by a flat plate collector.

ii) It is highly reliable and durable because of its sealed construction.

iii) It is adaptable for use over a wide power range (1KW to hundred of KW^s).

DETAILED STUDY OF THE SOLAR RANKINE CYCLE

A) Controlling Design Parameters Of The Cycle

In selecting a Rankine-cycle solar water pumping system one has the following choices for the system components and operating conditions:-

i) The type of solar collector. The maximum temperature attainable from a flat plate collector is less than 100°C, however, if pressurised it might reach 150°C. Normally, temperatures of 100-250°C for the parabolocylindrical concentrator and 150°-450°C or more for the paraboloid concentrator are attained. The flat plate collector has the advantages that it traps direct and diffuse radiation and it needs no sun tracking and of low production cost. However, it has an efficiency of 10-60% depending on its operating temperature, while a concentrator may have an efficiency of 40-80% depending on its concentration ratio.

In a recent paper, Wood and Platt (1979) concluded that shallow solar ponds make possibly the lowest cost solar thermal power system for irrigation pumping.

ii) The type of expander. For the power range required in solar pumps (less than 50 KW) the expander may be: a turbine, a reciprocating engine or a positive displacement rotary engine. A turbine is favoured for its easy regulation, small size, high efficiency and the possibility of large expansion to condenser press. (subatmospheric). However, small power turbines have low efficiences, high relative cost and the requirement of a gear mechanism to match the pump speed. A reciprocating engine is much cheaper, needs no speed reduction and has high efficiency in small sizes and good part load operation. But, it contains many moving parts thus needs internal lubrication and is very bulky when the vapour pressures are low. A positive displacement rotary engine (the spiral expander) which has only a few moving parts and no valves and needs no gear box is suggested for use in solar thermal power units by Bhide (1978).

iii) The type of working fluid. The primary fluid going through the collector and the secondary (or working) fluid going through the power plant may be the same or different from each other. The primary fluid is in most cases water, while the working fluid is usually an organic fluid, either halocarbons (refrigerants) or hydrocarbons. The choice of the working fluid depends on the operating temperatures in the evaporator and the condenser. For small capacity turbines a fluid of high molecular weight is desirable.

iv) The type of evaporation. The working fluid may be evaporated either directly in the solar collector or indirectly through a heat exchanger by means of water circulating in the collector. In the latter case, a temperature difference of about 5°C between the water outlet from collector and fluid outlet of the boiler is needed.

v) The operating temperatures. The condenser temperature depends on whether the condenser is water cooled, evaporative cooled or air cooled, i.e. it is dictated by the availability of cooling water. Condensing temperatures of 28°, 38° and 49°C are typical for the three cases respectively. The selection of the boiling temperature (maximum temp.) is an optimisation process between the collector efficiency and cost and Rankine cycle engine efficiency and cost since

$$\eta_{s\,(\text{system})} = \eta_{c\,(\text{collector})} \times \eta_{R\,(\text{Rankine})}$$

As can be seen from figure 9 η_c decreases as the collector temp. increases, while the Rankine cycle efficiency improves. The resultant is that there is an optimum operating boiler temperature (around 100°C for R-11) resulting in the highest overall system efficiency, (Ishikawa and Watanabe, 1975).

B) The Thermal Analysis of the Rankine Cycle

The calculations were performed to estimate the discharge of the pump operated by a Rankine cycle at the time of peak solar radiation intensity per square meter of the collector area, when different types of the working fluid are used. The most suitable fluid is that giving the highest rate of water pumping (for a fixed lift of 3 meters), highest cycle thermal efficiency and highest overall plant efficiency.

It is assumed that solar collector leaving termperature and efficiency are 80°C and 43% for a flat plate collector, 200°C and 60% for a parabolo cylindrical concentrator and 450°C and 60% for a paraboloid concentrator. The condensing temperature is fixed at 30°C while the incident solar radiation per m^2 is 800 watts. Besides the following efficiency values are assumed for the different components of the plant:- reciprocating engine 75% turbine 80% (when expanding steam) and 70% (when expanding refrigerant), water pump 70%, mechanical transmission 95% and heat exchanger 95%.

The results for the case of indirect evaporation of the working fluid using water heated in a flat plate collector are given in table 1. From the table, it is clear that Freon-11 is the most suitable organic fluid for the Rankine cycle pumping plant.

The results for a flat plate collector evaporating Freon-11 directly, are compared in table 2 with those for concentrators of the trough and dish types evaporating water. The clear increase in pumping capacity and efficiency is accompanied by a big increase in cost which makes use of concentrators viable only for large plants. Due to the scattered nature of deep-well irrigation, several small irrigation pumps, rather than a large unit, are generally needed.

CONCLUSIONS

Solar energy can be readily used in pumping water for irrigation at a relatively high initial cost and nearly no running cost. Due to the scattered nature of deep wall irrigation several small pumps, rather than a large unit, are generally needed. This makes it possible to mass produce small solar pumps and therefore reduce the production cost making them competitive with the alternative sources.

REFERENCES

Ackerman, A.S. (1914). The utilization of solar energy. The Society of Engineers. London, Trans. 103: 81.

Bahadori, M.N. (1978). Solar water pumping. Solar Energy. 21: 307-316.

Bhide, V.G. (1978). Conversion of solar into mechanical or electrical energy, Indian experience. In (Solar water pumps, technical potentialities), published by GATE, S 1/4: 35-41.

Boldt, J. (1978). Solar powered water pump for the rural third world. Laboratory for Energetics report RE 78-3, Techical University of Denmark.

Farber, E. and Prescott, F. (1965). Closed-cycle solar hot air engines - part 1, a 1/4 HP engine. Solar Energy. 9: 170-176.

Ishikawa, S. and Watanabe, M. (1957). Organic Rankine cycle engine development and its application to solar energy utilisation. Presented at COMPLES Conference, Dahran, Saudi Arabia.

Khana, M.L. (1970). Studies on hot air engines run with solar energy. Paper No. 7/14 presented at ISES Conf., Melbourne, Australia.

Michaelis, F.H. (1978). Water pump operated by a solar thermal linear motor. In (Solar water pumps, technical potentialities) publisehd by the German Appropriate Technology Exchange (GATE). S 1/4: 62-67.

Pytlinski, J.T. (1978). Solar energy installations for pumping irrigation water. Solar Energy. 21: 255-262.

Rao, D.P. and Rao, K.S. (1976). Solar water pump for lift irrigation. Solar Energy. 18: 405-411.

Sudhakar, K., Krishna, M, Rao, D.P. and Soin, R. (1980). Analysis and simulation of a solar water pump for lift irrigation. Solar Energy. 24: 71-82.

Trayser, D. and Eibling, J. (1966). A 50-Watt portable generator for remote area use employing a solar powered Stirling engine. Presented at Int. Soc. Energy Conversion Conference.
Wood, R. and Platt, E. (1979). Shallow solar pond powered irrigation pumping - A feasibility study - Lawrence Livermore Laboratory report, University of California.

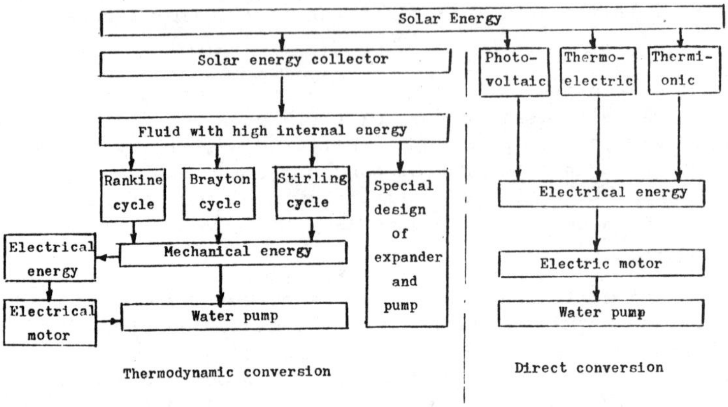

Fig. 1. The solar water pumping chart

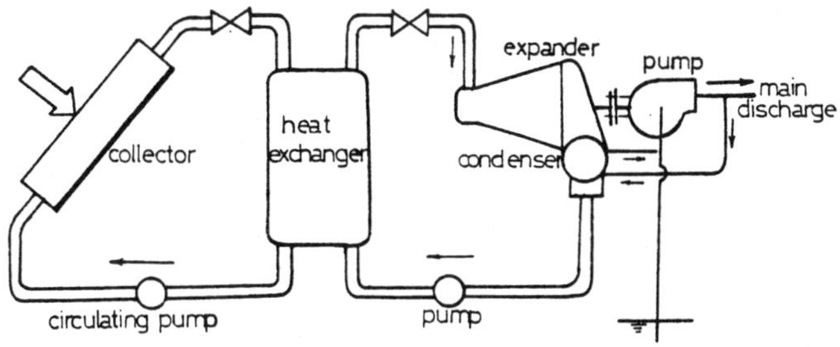

Fig. 2 The solar Rankine cycle.

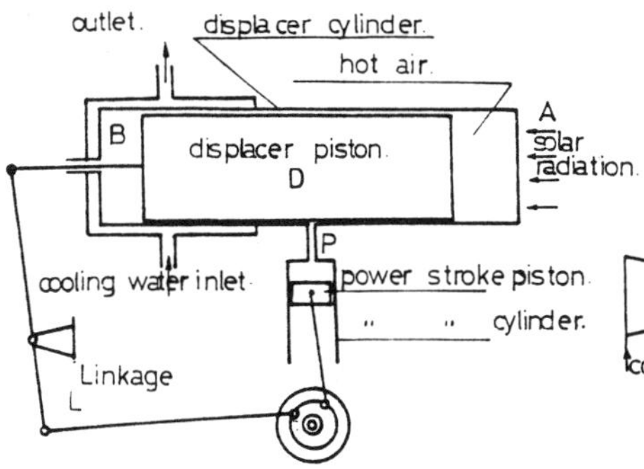

Fig. 3 The solar stirling engine.

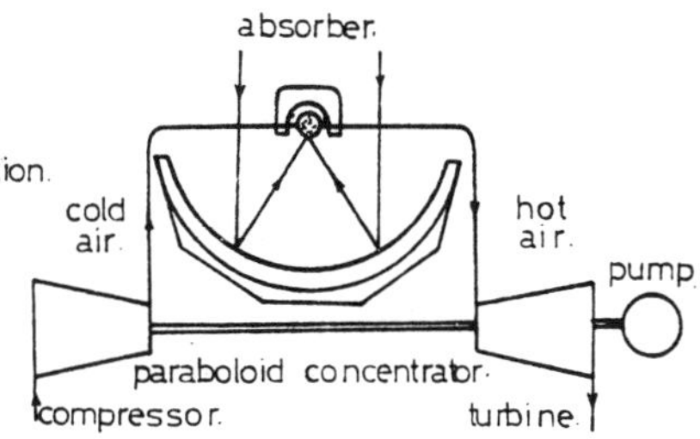

Fig. 4 The solar Brayton cycle.

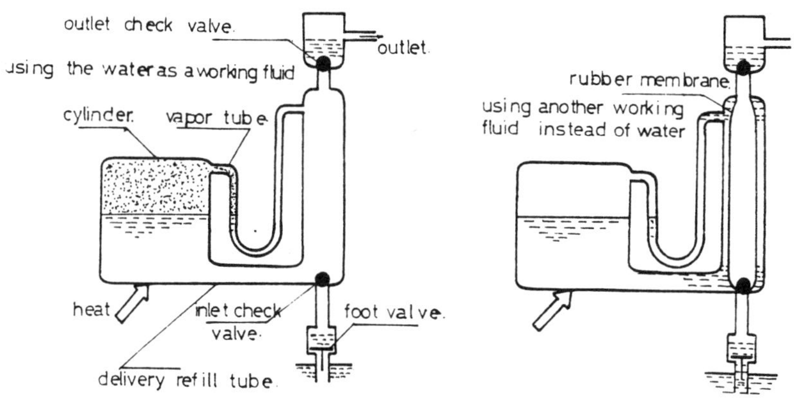

Fig. 5 The basic idea of a thermo-pump.

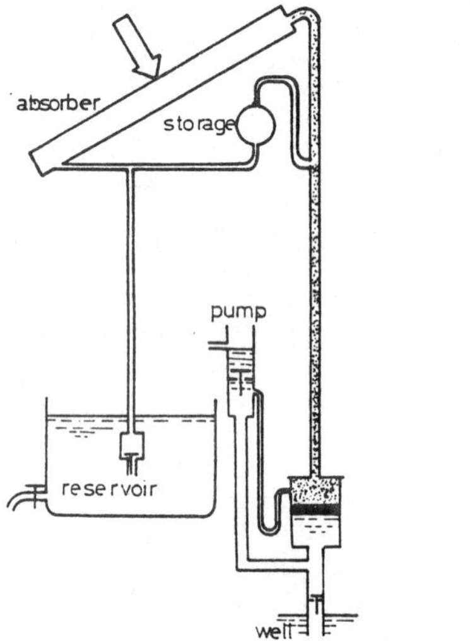

Fig. 6 Principle diagram of the solar powered thermo-pump.

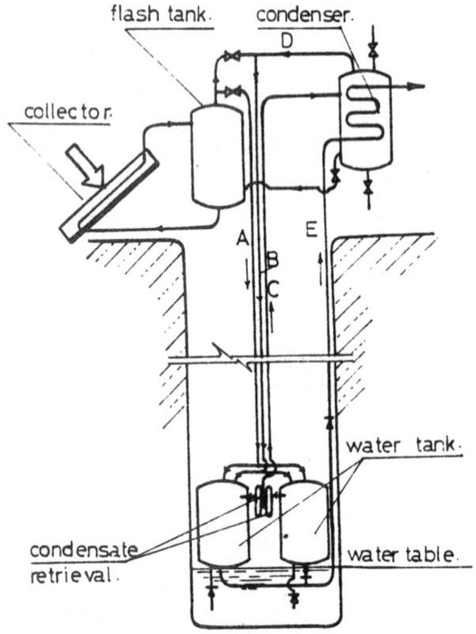

A=high pressure vapor. B= spent vapor. C=condensate return. D=equalisation E=pumped water.

Fig. 7 Water-cooled Birla pump.

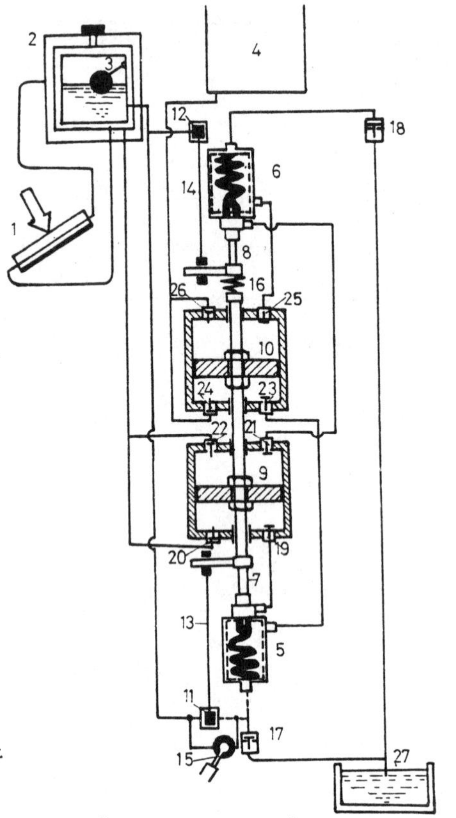

Fig. 8 Solar thermal linear motor pump.

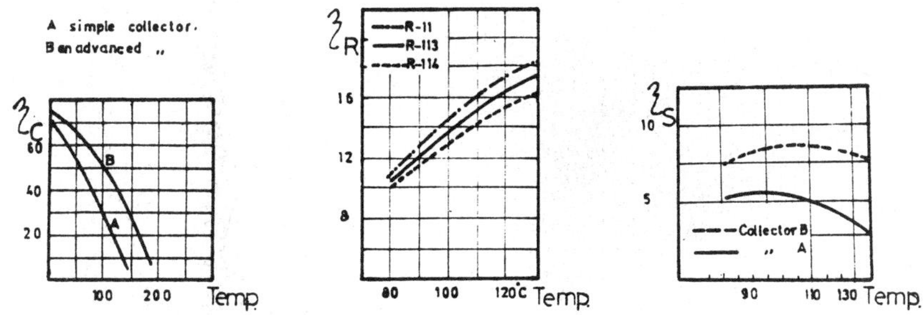

Fig.9 Variation of collector, Rankine cycle engine and system efficiency with collector outlet temperature

Table 1. Effect of type of working fluid on the performance of a Rankine cycle solar pump (indirect evaporation)

Working fluid	Expander	Pump Discharge* m^3/ hour	Cycle thermal efficiency %	Overall system efficiency %
Freon-11	Turbine	2.262	8.49	2.31
	Rec. Eng.	2.424	9.09	2.47
Freon-12	Turbine	1.970	7.46	2.00
	Rec. Eng.	2.000	8.00	2.17
Freon-22	Turbine	1.860	7.08	1.92
	Rec. Eng.	1.990	7.60	2.06
Freon-113	Turbine	2.178	8.20	2.22
	Rec. Eng.	2.333	8.80	2.38
Freon-114	Turbine	2.149	8.08	2.19
	Rec. Eng.	2.303	8.65	2.35
Butane	Turbine	2.130	8.00	2.17
	Rec. Eng.	2.280	8.56	2.30
Methyl Chloride	Turbine	1.676	6.07	1.74
	Rec. Eng.	1.796	6.50	1.

* Pump discharge is for 1 m^2 of collector area. Lift = 3 meters
Water is the primary fluid.

Table 2. Comparison between the results for cycles using a F.P. Collector, a trough concentrator and a dish concentrator.

Collector/ Concentrator	Working fluid	Pump Discharge m^3/ hour	Rankine cycle efficiency %	Overall system efficiency %
Flat plate	Freon-11	2.99	10.7	3.0
Paraboloo-cylindrical	Water	10.53	27.0	10.8
Paraboloid	Water	11.90	30.4	12.1

Application of Small Scale Solar Crop Driers to Maize Drying in Kenya

H. Othieno, W. Grainger and J. W. Twidell

Department of Applied Physics, University of Strathclyde, Glasgow G4 0NG, UK

ABSTRACT

Maize drying and marketing patterns in Kenya have been investigated to determine the point at which drying could most effectively be improved. Local methods of determining moisture content are compared with instrumented methods. Two small-scale, natural convection, closed bin solar crop driers were designed and constructed using locally available materials. One system employs a single flat plate collector, the other has a multilayer collector. Test results are compared with results from traditional open air sun drying. Measurement methods and operating routines are described.

KEYWORDS

Natural convection; solar collector; crop drier; maize; moisture content.

INTRODUCTION

The practice of preserving foods against times of shortages is universal. It has now become of major importance as a result of increased population, urban living, trade and transportation. Food must be treated to prevent decomposition and growth of fungi. Open air sun drying is a common method of drying agricultural products and is the dominant practice in the Third World. In developed countries high cost of labour and relatively low cost of fossil fuels and electricity encouraged the development of large heated air driers with forced convection. The price of fossil fuels is now so high that there is revived world wide interest in the use of solar energy for drying. This is an encouragement to Third World countries which due to poverty and lack of advanced technology continue with simple methods of grain drying. (See general reference and bibliography.)

This paper considers the application of small scale solar crop driers in Kenya. The project was funded by the U.K. Overseas Development Agency to design and test closed bin small scale maize driers. Theoretical and laboratory experimental work was done at the University of Strathclyde (Grainger, 1981) where two types of natural convection air heaters were built for the driers. One type used a flat plate collector and the other a multilayer collector. The sun's heating effect was simulated by heating the collectors electrically. The following laboratory investigations were made:

(i) Effect of angle of elevation of the collector on output air temperature and velocity, and hence the determination of the best angle of elevation for both flat plate and multilayer collectors for equatorial regions.

(ii) Effect of gap between absorbing surface and transparent cover on output air temperature.

(iii) Effect of collector length on output air temperature.

In Kenya, the demand for maize drying and methods of marketing were first investigated. Once the details of the needs had been established, an appropriate scale and design of drier could be considered. Two types of collector were designed and built for the Kenyan situation using the experience gained at Strathclyde. The collectors were each connected to similar grain bins, Figs. 1 and 2. The systems were tested and their performance compared with the traditional sun drying method. Other methods of drying, e.g. open air tray drying, were not investigated.

Kenya, being on the equator is excellent for solar energy devices. The capital, Nairobi, is situated in the cold region of the country and receives about 32 MJ m^{-2} on a horizontal surface on a clear sky day. The country now spends about 25% of her export earnings on oil import (UNEP, 1980). This bill can be reduced by the use of solar energy devices.

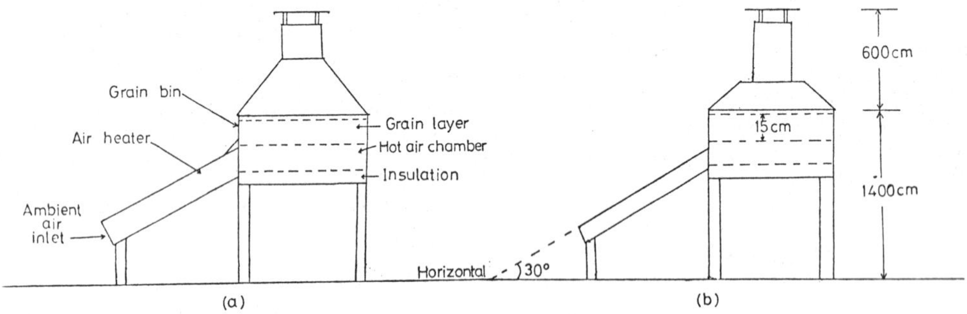

Fig. 1 Fixed orientation natural convection solar driers for equatorial regions. The collector slope of 30° was established experimentally at the University of Strathclyde as the optimum angle of performance where the sun passes directly overhead. (a)multigrid, (b) flatplate.

MAIZE DRYING AND MARKETING IN KENYA

The farmers are always eager to leave their maize drying in the fields as long as possible before harvesting. However if the rainy season comes earlier than expected or if the farmers wish to plant and harvest twice in a year, maize must be harvested as soon as it matures and then dried. Usually the newly harvested maize is sun dried by spreading it out on mats or bare ground in a layer 1-2cm thick. This method is still practised but the grain losses incurred due to rodents, birds, spillage and contamination can be as high as 10-15%. Moreover, rewetting and overdrying in variable weather can cause cracking of maize kernels which reduces the quality of the grain. Grading details and specification for dry shelled maize (Kenya Bureau of Standards, 1977) must be adhered to by the organisations handling maize, and some drying in large oil fired driers is necessary by grain merchants, Fig. 3. Solar driers should therefore be beneficial to the community. However, as with any other technology, a solar drying system must be properly chosen and

Fig. 2. Two natural convection solar crop driers constructed at Kenyatta University College, Nairobi. On the left is the multilayer system on wheel castors, on the right is the fixed orientation drier with a flat plate collector.

designed to meet the particular requirements and give satisfactory performance. We therefore had to understand the operation and requirements of the various organisations involved in handling maize in the country.

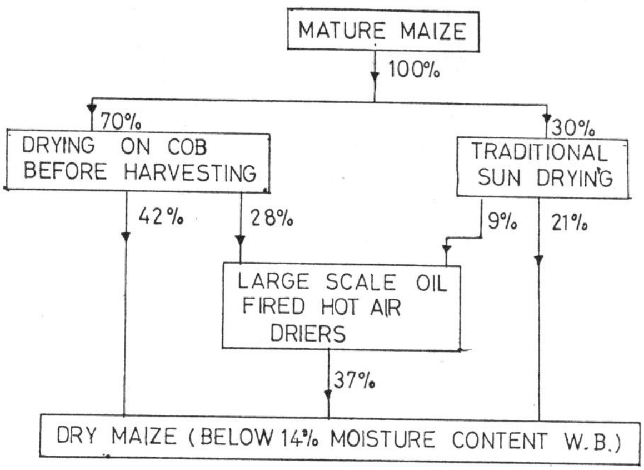

Fig. 3. Estimation of present drying pattern of maize in Kenya (excluding losses).

Figure 4 shows the organisation involved in maize marketing. The width of the arrows indicates the relative probability of choice. The chart is a simplified example of the infrastructure. Maize growers are also consumers and only sell the surplus of their harvest. So the maize farmers' choice of customer depends on the quality of surplus and the availability of transport. The National Cereals and Produce Board (N.C. & P.B.) offers the highest price for a standard bag of maize (90kg), while the local trader offers the lowest. Obviously the farmers' choice would also depend on these and other factors such as transport. Due to lack of finance and space for storage, the local traders cannot handle large quantities of maize and may turn farmers away because of this. On the other hand N.C. & P.B. and F.C.S. are strict about the condition of the grain and base their grading techniques on Kenya Bureau of Standards specification for dry shelled maize. Consequently N.C. & P.B. centres which do not have driers normally do not accept wet maize (with moisture content above 13.5% wet basis). The few centres which have large oil fired driers accept both wet and dry maize but charge the owners for every 1% moisture content above 13.5% per 90kg of maize. The wet grain is then dried in the imported oil-fired driers. In 1978 the centre at Nakuru received about 90,500 bags (8,150 tonne) of maize with moisture content of 13.5% wet basis. Because of overproduction of maize in that year, the centre anticipated a longer storage period than usual and decided to dry all the 8150 tonne of maize down to 12.5% wet basis. About £4,500 (British) was spent on fuel which means that it cost 1% of maize price to reduce moisture content of the maize by 1%. With oil price having doubled in the last 3 years while maize price remained constant, the fuel cost for reducing moisture content by 1% has also doubled. During the recent 1980/1 shortage of maize, the drying centres accepted maize with moisture contents ranging from 14-20% and had to dry it down to 13.5%. This meant spending even more on fuel.

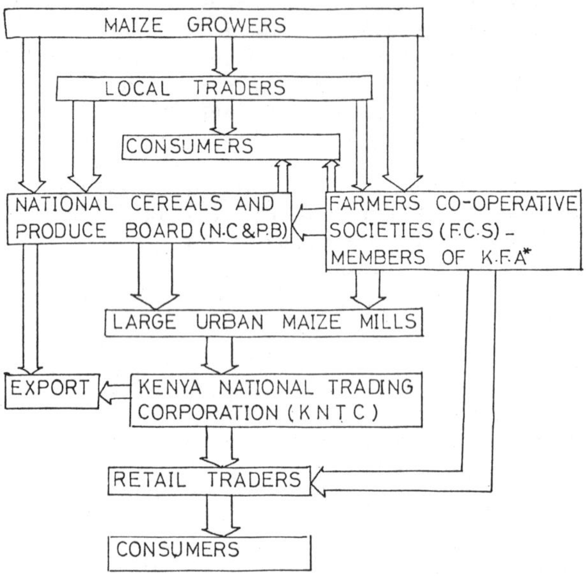

Fig. 4. Maize marketing in Kenya. *K.F.A. = Kenya Farmers Association.

Most of N.C. & P.B. depots scattered around the country are located at the provincial or district administrative headquarters - far from the grain fields. When maize arrives at these centres, random samples are taken from a few bags and moisture content measured with a meter, e.g. Burrows Moisture Computer. Farmers Cooperative Societies have no moisture meters and have to rely on the traditional methods which include (i) driving a finger nail through grain kernels to sense the force applied, (ii) dropping a handful of maize and listening to the sound, (iii) hand weighing a sample of maize and (iv) looking through the grain kernel against the sun so wet areas appear darker than the drier areas. We tried these traditional methods on various samples of maize and checked a local expert's estimations against our Marconi moisture meter. We were satisfied that these methods give a surprisingly good estimate of the condition of the grains, representing an accuracy of moisture content estimation to ± 2%. Thus the difference between 15% and 20% wet basis moisture content is determined reliably, but that between 13% and 14%

(the important region for storage) is uncertain. Moreover, the "local" method is not associated with a method of registration or calibration.

SMALL SCALE SOLAR CROP DRIERS: DESIGN AND CONSTRUCTION

Having studied the maize marketing pattern and the demands at each stage, we recommended that:

(i) maize should be dried after shelling,

(ii) driers should be designed to dry about 90kg of shelled maize per day from about 18% to 13% moisture content wet basis,

(iii) driers should be used at the F.C.S. centres where training programs and constant interaction with farmers have been established,

(iv) the temperature of the air reaching the grain should not exceed 60°C if milling and baking qualities of the grain are to be preserved. Collector air temperature may be considerably higher however, see Grainger, 1981 .

With the experience gained at Strathclyde University in investigating the performance of natural convection air heaters, we designed a single flat plate solar air heater to meet the requirements mentioned above (Grainger, 1981). The heater was 1m x 1m extendable to 1.8m long. The air gap between collector plate and transparent cover was 5cm, Fig. 1. A second air heater based on the design of the first was built. It had a wider air gap (15cm) in which a multilayer wire mesh collector was fitted. A total of six layers of wire mesh was used. This is the second type of air heater investigated at Strathclyde and found to be more efficient than a flat plate air heater. The complete driers were made by connecting the air heaters to drying bins containing the grain. The whole assembly was designed from empirical analysis as described elsewhere (Grainger, 1981). The single flat plate system was permanently fixed on the ground while the multilayer system stood on castor wheels to allow for rotation should it be necessary. The grain bins were the same in size and grain capacity, but had different cover designs, Fig. 2.

Locally available materials were used in the construction of these driers. The cost of construction excluding labour is summarized in Fig. 5. Note that these driers were for experimental purposes and are more expensive than the driers built for normal use.

Items	Qty	Drier No. 1 Single flat plate Collector	Drier No. 2 multigrid Collector
		*£	*£
3m x 1.5m sheet of hardboard	8	20.00	20.00
1m x 1m window glass (4mm thick)	1	10.00	10.00
Packets (25kg each) of diatomite insulation	6	15.00	15.00
wood, screws, paint	-	50.00	50.00
1m x 1m Coffee tray wire grids	9	- -	25.00
Rubber hooks	4	- -	10.00
Wheel castors	6	- -	15.00
Sundries	-	15.00	15.00
Total		110.00	160.00

Fig. 5 Construction materials and cost. Figures are given in British pounds. One British pound was equivalent to 18 Kenyan shillings. Kenya's average yearly income per capita is estimated to be around £50. (Burrows, 1975).

With the cost of labour included, these experimental models would cost as much as £200 each. As Fig. 5 shows, the most expensive items apart from unavoidable wood were hardboard, coffee tray wire, rubber hooks and wheel castors. Some of these, as the table shows, were expensive because large quantities were used. Hardboard is readily available in Kenya and is cheaper than other commercial materials available. The quantity of hardboard for commercial driers could be reduced by using cheaper material for the construction of the grain bin top cover and chimney assembly. If rubber hooks and coffee tray wire are replaced by cheap sisal ropes and wood slats respectively, and wheel castors omitted, the multilayer prototype solar crop drier would cost about £110 without labour.

MEASUREMENT AND DATA RECORDING

The flat plate drier system was repeatedly used to dry maize of various moisture contents down to 13% wet basis in different weather conditions (i.e. varying wind and insolation - there was no rain during the drying runs). We analysed the air temperature profiles across different sections of the collector. The multilayer system was also used for drying wet maize down to 13% wet basis.

Measurements recorded included:

(i) Wind speed: recorded with Cassella sensitive anemometer type T16110/1 connected to a calibrated counter.

(ii) Insolation: instantaneous insolation was measured by a Kipp Solarimeter Type CM5 facing south at the same angle of 30° as the collector. The solarimeter output was recorded by a Solartron Compact data logger type 3430B. A Delta-T integrator type T1-1750 connected in parallel with the logger recorded integrated insolation over intervals of 30 minutes.

(iii) Output air velocity: measured with (a) Prosser air velocity meter type AVM 502 with the probe fixed in the chimney above the grain bed, (b) Cassella air meter type T16354 attached near the top of the chimney.

(iv) Collector output air temperature: measured at the top of the collector and bottom of the grain bed using swg 40 copper/constantan thermocouple junctions connected in parallel.

(v) Wet and dry bulb air temperature: measured at the entrance to the air heater, below and above the grain bed.

(vi) Grain temperature: measured at the bottom, middle and top of the grain bed. The temperature of directly sun-dried grain was also measured.

(vii) Air temperature inside grain bed: measured at the bottom, middle and top layers of the grain bed.

(viii) Grain moisture content: measured by sampling with a Marconi moisture meter type TF 933C and Macfarlane Robson moisture determination unit type IR-400 using a 300W infrared bulb as a heater.

All temperature measurements were monitored by the data logger at intervals of 30 minutes. Wind speed, insolation and air velocity were also recorded at the end of each interval. For all temperature measurements in both flat plate and multilayer systems, swg 40 copper/constantan thermocouples were used. Wet bulb temperatures were measured by thermocouple junctions wrapped in a layer of cotton thread kept wet by a wick immersed in a reservoir of distilled water, Fig. 6. The justification that air velocities are sufficiently high to give full depression for this method is given by Monteith, (1954).

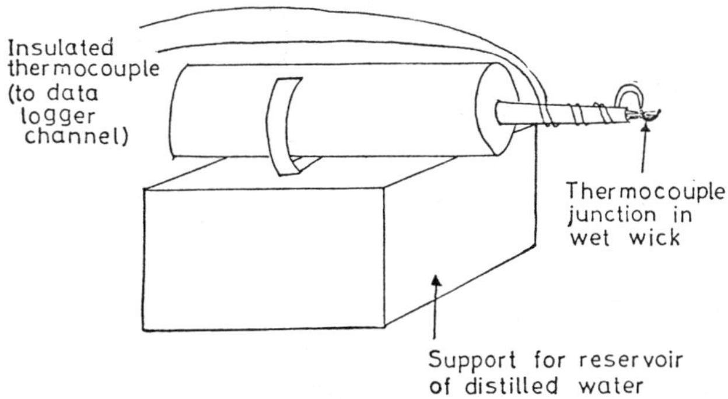

Fig. 6. Wet bulb temperature measurement technique.

Grain kernel temperature was measured by fixing thermocouple junction into a small tightly fitting hole drilled in the grain, Fig. 7.

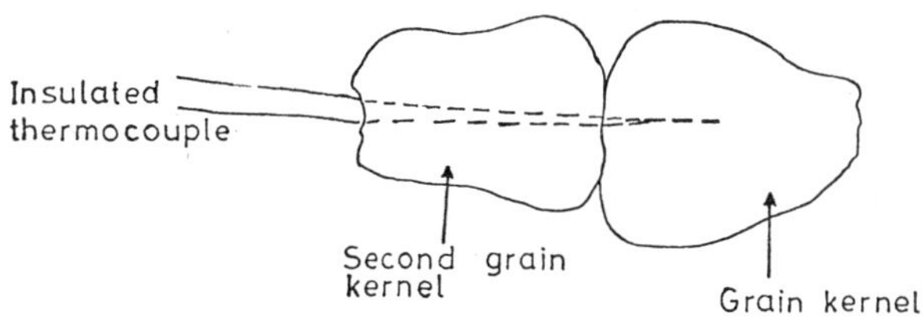

Fig. 7. Grain kernel temperature measurement technique.

RESULTS

Figure 8 shows the drying process of 80kg of maize (9cm deep) in a bin of cross section 1m x 1m, using warm air from a multilayer solar collector elevated at an angle of 30° to the horizontal. The collector was 1m wide and 1.5m long.

The drying rate of in-bin maize was compared with the drying rate of 10kg of maize traditionally sun-dried on the ground during the same period of time. Maize spread on the ground dried at the same rate as the middle layer of grain in the bin. The bottom layer of maize in the bin dried faster than the other layers and the maize on the ground. The "open air" maize was spread thinly (1-2cm thick), and so dried uniformly. However the layers of maize in the bin dried at different rates. The drying rate of the grains at the top surface of the bin was slow and quite unpredictable.

The other drier using a single flat plate collector produced similar results. Both driers were able to reduce moisture content from 18% wet basis to 13% wet basis in two days. However, it was observed that moisture content of the grain increased at night if the ambient air inlet was left uncovered. This effect delays drying time. No condensation was observed inside the systems at any time, but there was considerable condensation occurring on the top of the glass cover overnight which slowed down the temperature build-up at the beginning of the day.

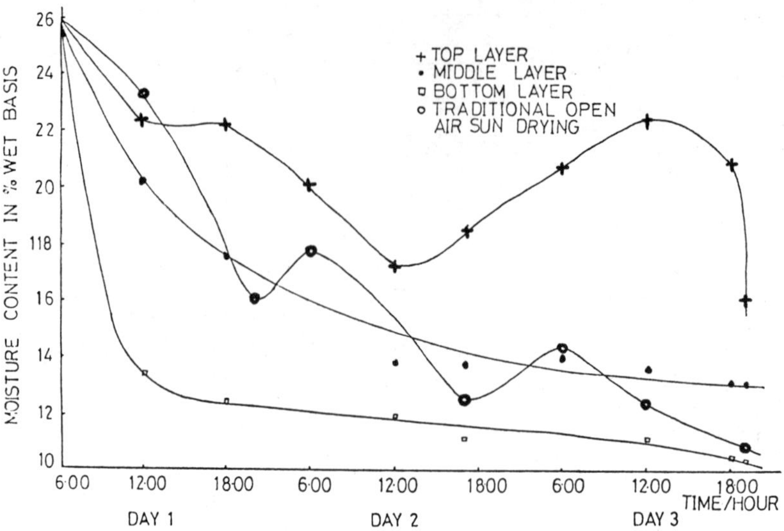

Fig. 8

CONCLUSION

The experience gained in this work convinced us that a specific design criteria and operating routine must be followed for successful drying. The following recommendations are made for operation of the driers located at Nairobi on the equator.

Criteria for Design and Construction

1. The grain load should be about 90kg of shelled maize.
2. An important design constraint is the maximum temperature of the air around the grain. It should not exceed 60°C to preserve the baking and milling qualities of the grain (Brooker, 1974).
3. Double glazing may be considered on sections of the collector beyond one meter of length (Grainger, 1981) but care must be taken not to exceed the maximum air temperature for the grain.
4. The gap between the cover and the collector should not be less than 5cm for a single flat plate collector. It should also be noted that the larger the gap beyond this the lower the air temperature. For a flat plate collector of length 1.5m, a gap of 5-6cm is recommended for maximum air temperature. This is based on a maximum instantaneous insolation of 1 kW m^{-2} on a horizontal surface.
5. Layer separations for a multilayer collector may be as small as 1cm. The number of layers is determined by the size of layer perforations and impedance to air flow. A mathematical method of determining the optimum number of layers is being developed (Othieno, thesis in preparation)
6. The sides and rear of the air heaters and the grain bin should be insulated to reduce heat losses. If a material of thermal conductivity k and thickness t is used as an insulator then $k/t \leqslant 2$ Wm^{-2} K^{-1}. This gives conduction losses of less than 10% of input power.
7. The whole drier system should be airtight so there is no leakage of warm air to the environment.
8. Convective heat loss from the transparent cover is significant (Grainger, 1981). The collector should be shielded from the wind.

Criteria for Operation

A carefully designed and constructed drier may not satisfy the user if incorrectly operated. We strongly recommend the following routine for the two driers.

1. The glass cover, collector surface(s) and the bin, including warm air chamber, should be regularly and carefully cleaned.
2. The solar collector of a fixed orientation drier should be elevated at an angle of 30° to the horizontal and face south for drying during the months of October to March. From April to September it should face north.
3. A moveable drier should be orientated to the east from sunrise to midday and to the west from midday until evening. The angle of slope of the air collector may be greater than 30° (the optimum angle for this mode of operation is being determined).
4. All access windows (except chimney and collector entrance) should remain tightly closed during the operation of the system.
5. Maize should be stirred once a day if a uniform drying is to be achieved. The time of stirring should be about 6 hours after filling the bin or the start of drying.
6. Moisture content should be determined at the end of each day.
7. Ambient air inlets (collector entrance and chimney) should be closed at night to prevent grain from absorbing moisture from the cold air.
8. It is most important to avoid unnecessary heat loss from the grain especially during the night.

Future Investigation

The construction of our multilayer solar air heater has been based on the design factors for a single flat plate collector. Work is currently going on to model mathematically and fully analyse multilayer collectors (Othieno, in preparation for thesis). The designs can then be fully compared with other designs, e.g. open air tray drying.

ACKNOWLEDGMENTS

We are very grateful to the U.K. Overseas Development Agency (O.D.A.) who funded the project in Kenya and to Kenyatta University College who provided the working space and accommodation. The S.R.C. and Association of Commonwealth Universities provided the scholarships given to W. Grainger and H. Othieno respectively.
Dr. J. MacLellan is kindly continuing experimental tests on the driers at Kenyatta University College.

REFERENCES AND BIBLIOGRAPHY

A.S.H.R.A.E. (1977). Fundamentals Handbook, American Society of Heating, Refrigeration and Air-conditioning Engineers, New York.

Auer, W.W. (1980). Solar energy systems for Agricultural and Industrial process drying. In A.S. Mujumdar (Ed.), Drying '80 Vol. 1, Hemisphere Publishing Company, Washington, pp. 280-292.

Brooker, D.B., Bakker-Arkema, F.A., and Hall, C.W. (1974). Drying Cereal Grains, A.V.I. Publishing Co. Ins. Westport, Connecticut.

Burrows, J. (1975). Kenya into the second decade, World Bank report. The John Hopkins University Press.

Duffie, J.A. and Beckman, W.A. (1974). Solar Energy Thermal Processes. Wiley, New York.

Exell, R.H.B. and Kornsakoo, S. (1978). A Low-Cost Solar Rice Dryer. Appropriate Technology, 5, 23-24.

Exell, R.H.B. (1980). A Simple Solar Rice Dryer: basic design theory. Sunworld, 4, 186-191.

Grainger, W., Othieno, H., and Twidell, J.W. (1981, to be published). Small Scale Solar Crop Driers for Tropical Village Use: theory and practical experience. In Proceedings of Solar World Forum 1981, Pergamon Press, Oxford.

Gupta, C.L., and Garg, H.P. (1967). Performance Studies on Solar Air Heaters. *Solar Energy*, *11*, pp. 25-31.

Harris, K.L. and Lindblad, C.J. (1978). *Postharvest grain loss assessment methods*. American Association of Cereal Chemists, St. Pauls, Minnesota.

Johnston, J.C. (1979). Solar Roofs Dry African Crops. *Sunworld*, *3*, 161-163.

Kenya Bureau of Standards (1977). *Specification of dry shelled maize, KS01-42*. Kenya Bureau of Standards, Nairobi.

Monteith, J.L. (1954). Error and accuracy in thermocouple psychrometry. *Proceedings of the Physical Society*, *B*, *67*, 217-226.

Moody, T. (1980). Drying Maize for storage in the Humid Tropics. *Appropriate Technology*, *7*, 4-6.

Roa, G. and Macedo, I.C. (1976). Grain Drying in Stationary Bins with Solar Heated Air. *Solar Energy*, *18*, 445-449.

Saif-Ul-Rehman, M. (1967). Solar Energy Utilization in Developing Countries. *Solar Energy*, *11*, 98-107.

U.N.E.P. (1980). *Energy and Environment in East Africa* (*Proceedings of an international workshop*) UNEP, Nairobi.

TOPIC H

Photovoltaic

Residential Photovoltaic Applications in the United States

S. J. Strong and R. J. Osten, Jr.

Solar Design Associates, Lincoln, Massachusetts, USA

ABSTRACT

The U.S. Department of Energy predicts that residential photovoltaics will become cost-effective during the 1980s, perhaps as early as 1986. This paper will present the results and conclusions of our work in the design and engineering of residential photovoltaic systems, with special emphasis on the integration of PV systems with passive solar residential architecture, using the performance results of our first lived-in residence, in Carlisle, Massachusetts, as a case study.

KEYWORDS

Residential photovoltaics; PV module configuration; PV/passive interface; utility-grid interaction.

INTRODUCTION

Even in the colder regions of the United States, in well insulated residences with modest but effective passive systems, general diversified electrical demand can easily exceed net auxiliary space heating demand on a kWh-for-kWh basis. When the extremely low overall efficiency of central electrical generation and distribution is taken into account, a residence's diversified electrical demand may consume three to four times more gross resource energy than its auxiliary space heating. Thus on-site generation of electricity can make an important contribution both to the energy independence of the individual homeowner and to the overall reduction of world fuel consumption.

Solar Design Associates has completed the architectural design and engineering of six full-sized single-family passive solar residences with photovoltaic arrays for construction in various areas of the continental United States.

These residences have been designed with state-of-the-art passive solar and energy conservation features. Five of the PV arrays are roof-integrated, and all use high-density, high-efficiency flat plate PV modules. The PV systems are sized to provide all or most of the residences' electrical requirements, depending on geographic location and the energy-use patterns of the occupants. All systems employ utility-interactive inverters with no on-site electrical storage.

Our first lived-in residence was completed in January 1981 in Carlisle, Massachusetts. Our second lived-in residence is now nearing completion in Santa Fe, New Mexico, a third is under construction, and we have completed and tested additional prototype PV systems in the Northeastern and Southwestern regions of the United States to verify residential PV systems performance.

The Carlisle residence features a 7.3 kWp PV array using 126 63.5 cm x 122 cm glass superstrate high-density PV modules with 100-mm^2 semicrystalline square cells from Solarex Corporation. The residence features 270 m^2 of net living area, super-insulation and direct passive solar gain.

Appliances are all electric, including the high-efficiency dual compressor heat pump that provides supplemental heating for space and domestic water. Primary domestic water heating is supplied by a 10-square-meter solar thermal array integrated with the south roof.

The Carlisle PV system has been in operation for more than eight months, and the initial results of field monitoring show that its output is a good match for the electrical demands of this large, energy-conserving residence, and that the house will likely generate a net surplus of energy annually.

THE PASSIVE/PV INTERFACE

As noted in the introduction, the PV and passive systems complement one another well in energy terms, and in combination may produce a residence which can be truly energy-independent. In architectural terms, the combination requires careful design. One issue often mentioned is the competition for aperture area, which in our experience has proven to be a valid problem but a surmountable one. In general, in most U.S. and European latitudes, the passive aperture will tend to occupy the vertical south walls and the PV system the roof. The pitch required by the PV array is usually compatible with sunspace roof glazing and domestic hot water collectors, so the principal issues become the division of roof area and the avoidance of self-shading. High-density and high-efficiency arrays are favored here, as they can produce a given output with less gross area.

PV modules must be designed, physically and electrically, to maximize flexibility. Key issues here are shading tolerance and string length as well as module dimensions. An array which tolerates some occasional localized shading without back-biasing, overheating and disproportionate power loss will allow a departure from the monolithic plane and a much greater freedom in building form. String lengths influence the available choice of array proportions and shapes. A system which requires 13 modules in series to achieve operating voltage can be easily configured in only one way--rows or columns of 13 modules. A 14-module string can be arranged in two ways, 1x14 or 2x7. A 12-module string (heartily recommended) can be configured in many ways--1x12, 2x6, 3x4, doglegs--to allow the array to be designed to fit the building and not the other way around. When a designer must deal with the simultaneous requirements of passive space heating and the PV array, this kind of flexibility becomes critical to the generation of aesthetically pleasing buildings which people will enjoy.

THE CARLISLE HOUSE

Located in Carlisle, Massachusetts, about 30 kilometers northwest of Boston, the Carlisle House faces south with a 98.4 m^2 PV array on the 45-degree sloping roof; 32.5 m^2 of floor-to-ceiling, double-glazed, sliding glass doors allow sunshine to warm the living room, dining area and family room. A concrete slab finished with dark quarry tile and containing 11.5 m^3 of mass, stores heat absorbed from the sun. A large fireplace in the center of the living room offers woodburning heat with additional thermal mass in a massive central chimney. A woodburning stove in the family room also contributes to heating the house. The kitchen, a laundry, study and bath are also on the ground floor, with the master bedroom suite with dressing room, whirlpool tub room and bathroom, three more bedrooms and another bathroom on the second floor.

Double stud outer walls have 20.5cm of fibreglass batts plus 2.5cm of rigid styrofoam under the siding (R-30); 30.5cm of fibreglass insulation in the ceiling (R-40) and 20.5cm under the floors (R-26). Triple-glazed windows on the east, west and north walls complete the low-heat-loss structure. Earth-berming and styrofoam insulation against the foundation outer walls also keep heat loss to a minimum. The predicted net heat loss for the house is 3.5 Btu/DD/ft^2, about two and one-half times better than conventional houses.

Architect's rendering of the Carlisle House

Much of the space heating comes from passive solar gain. Auxiliary space heating is provided by a 38,500 Btu/hr, 11.3 kW, air-to-air, dual-compressor heat pump with a low-pressure air circulation system for the entire house. Supply registers are in the floor near the outer walls, and return air enters the chimney at the top of the cathedral ceiling and travels down a chase in the masonry core to the air handler, picking up chimney heat on the way. The low-speed fan may operate continuously to circulate the passively heated air throughout the house. South-facing clerestory windows along the ridge line provide additional solar gain. In the summer, these windows are shaded by an overhang, and they may be opened for natural-convection ventilation and cooling. The heat pump can also be used to provide air conditioning if desired.

The south roof is divided into two sections for the PV array, one with 42 modules (6 rows of 7) and one with 84 modules (12 rows of 7). The 18 rows are connected 2-rows-in-series to provide 9 strings of 14 modules each to match properly the input voltage of the inverter. Each module is comprised of 72 semicrystalline cells (2 paralleled sets of 36 cells) and produces about 58 watts of power under full sunshine at nominal temperatures.

The 9 strings are paralleled and fed to the DC-to-AC inverter manufactured by Windworks Company. This utility-dependent inverter converts the DC current from the array to AC power to match the 60-cycle house electric service. Table 1 gives characteristics of the PV power system.

Table 1 PV Power System Characteristics

ARRAY	
Power	7.3 kWp (58 Wp/module)
Energy	10 MWh/year
Voltage	190 VDC, 38-A nominal
Configuration	126 modules 18 rows of 7 modules 9 strings of 14 modules
INVERTER	
Power	8 kW DC input (200 V,40 A)
Input voltage	100-200 VDC
Output voltage	240-VAC single phase
Efficiency	95%

Under full sunshine conditions at normal operating temperature, the PV array is rated at a nominal 7.3 kW peak; and produces somewhat more power in the winter cold and somewhat less in the summer heat. There is no electric energy storage available at the house, so when more power is available than is needed, it is "banked" with the utility--that is, fed to the utility grid for use elsewhere and credited to the house account. When the power needs of the house exceed the output of the PV array, such as when heavy loads coincide with cloudy weather or at night, the utility then provides the electricity required to make up the difference.

Architect's rendering of the Santa Fe PV residence

A computer simulation of energy flow was performed for the house using a conservative load senario and a typical meteorological year. Table 2 gives the predicted energy production and percent of load the array furnishes. On an annual basis, the house is expected to be virtually self-sufficient; with winter requiring considerable utility dependence, and summer providing a large surplus of electricity.

TABLE 2 Predicted annual energy production of the Carlisle House.

	Array kWh/mo.	Array % of load
January	521	40
February	619	58
March	862	77
April	885	95
May	1011	123
June	1071	145
July	1007	140
August	927	131
September	933	140
October	732	85
November	507	54
December	459	37
Total	9534 kWh	94%+

The house was built during the fall and winter of 1980-81, and the PV power system was first tested in January. The power output from the array was measured on a near-perfect, cold, clear day in March when the solar insolation was 1.03 kW/m^2 and it was found to have a peak output of 7.87 kW.

Operational test data was taken for a continuous 13 day period in April and a continuous 38 day period in May and June. The average power output to the inverter was 27.2 kWh/day for the April period and 26.3 kWh/day for the May-June period. The cloudiest day produced 2.6 kWh and the sunniest day produced 39.9 kWh.

CONCLUSION

Residences are currently being financed with mortgages that stretch out over twenty-five or thirty years. During the financial lifetime of these homes, we will see major changes in the way the world produces, distributes and consumes energy. While no one is sure of the exact nature of these changes, one thing is certain; the era of abundant and inexpensive fossil fuels and electricity is over.

Perceptive clients and designers have begun to understand this; and the amount and type of energy a residence consumes and the prospect of energy self-sufficiency are now receiving considerable attention from all sectors.

The Carlisle House demonstrates that energy self-sufficiency on an annual basis is possible without compromising lifestyle in a well designed passive solar residence using a utility interactive PV system with no on-site electrical storage.

Future designs will integrate PV and wind systems with hydrogen production and on-site electrical storage to create energy producing organisms--homes which provide the comfort and amenities consistent with our currant standard of living without need for purchased energy.

ACKNOWLEDGEMENT

The authors would like to acknowledge the U.S. Department of Energy and the Massachusetts Institute of Technology's Lincoln Laboratory for their sponsorship and support of the Carlisle House project as part of the U.S. Department of Energy's residential photovoltaic program. Special thanks are due to Mr. Paul Maycock, former director of the U.S. DOE PV program and Mr. Burt Nichols program manager at MIT.

The World's Largest Photovoltaic Concentrator System — a Case Study

H. V. Smith

Mississippi County Community College, Blytheville, Arkansas, USA

ABSTRACT

The world's largest photovoltaic concentrator system has been developed and is now in place at Mississippi County Community College in Blytheville, Arkansas. The system consists of 270 single-axis tracking parabolic collectors providing a concentration ratio of 30 suns optical. The design output of 320 kilowatts (DOE standard) peak is to be provided by approximately 65,000 single crystal silicon cells. The active cooling system allows 22 million Btu's per day thermal.

The ongoing debate within the photovoltaic industry concerning the advantages of one sun photovoltaic panels versus large, tracking concentrating systems is a matter of major interest. This large-scale DOE demonstration project is intended to investigate the various aspects of a large-scale concentrator system. Various interface relationships, hardware reliability, solar cell matching, inverter operation, and many other aspects of the complex system are topics for discussion.

Interface parameters essential to integrating an energy-conscious architectural design with a total solar system are additional topics for consideration.

KEYWORDS

Active cooling; concentrator photovoltaic system; total energy system; solar electricity; energy-conscious architectural design.

Mississippi County Community College is located in the rich farmland of Northeast Arkansas. Only a few miles from the Missouri Bootheel and within a stone's throw of the mighty Mississippi River, this is the youngest of the six community colleges of the state. Founded in 1974, funded for operation by the state legislature and for construction by a local property tax, the institution has enjoyed a steady and healthy growth to a recent fall enrollment of some 1200 credit students. Although not a comprehensive community college by the broad definition (in Arkansas, vocational programs are handled through a separate institution), it does embrace the community college philosophy in offering occupational programs, community services, and a broad-based learning skills program, as well as the usual freshman and sophomore years' study. The institution has full membership in the North Central Association.

In 1976, when the institution began its in-depth planning for the first construction phase, the possibility for federal funding was a prime consideration. Working with our architects and a group of private consultants, a proposal was made through the office of the late Senator John McClellan to make this institution a large-scale energy experiment featuring the emerging high technology of solar electricity.

In 1977, the Department of Energy issued a grant for some $6.8 million, and a local bond issue brings the total to approximately $10 million. The project includes a building designed for solar electricity, a power plant consisting of a total energy photovoltaic system, and featured two experimental developments.

The first of the experiments concerned a decision to feature a tracking concentrating system, and in 1976 no such concentrators for photovoltaics existed. The Research Branch of Honeywell was employed and developed a unique flat plate collector using 3M reflecting Fresnel lenses.

Unfortunately, when we reached the prototype stage, construction estimates far exceeded our budget; therefore, we proceeded with a national competitive procurement and the successful bidder was Solar Kinetics of Dallas, Texas. They have provided and installed 270 single-axis tracking concentrating collectors. These are parabolic troughs 20 feet by 7 feet and provide a concentration ratio of about 28 suns optical and about 42 suns geometric. The reflective material is a thin film (aluminized milar) of FEK 244. The collectors are ganged in groups of six per row. Each row tracks east/west, driven by a hydraulic system. The shadow band tracking mechanism is mounted on each row and is activated by photo-diodes over a time interval of approximately 30 to 45 seconds.

Concentrating collectors in stow position.

A second major research effort was the development of a unique electrical storage system. General Engineering Laboratory of Durham, North Carolina, owned and operated by Dr. Ralph Zito, had filed patents for an iron-redox battery some 10 to 15 years earlier. The concept embodied two carbon plates separated by a porous membrane. Ferrous chloride from two separate sources was then pumped through the two chambers thus created. When a d.c. current is passed across the chambers, it causes the transfer of electrons through the membrane, a deposit of a molecule of iron on the wall of the plate, a change from ferrous to ferric chloride in one chamber, and thus the chemical storage of electrical energy. Dr. Zito had only built several single-cell prototypes; therefore, the problem was to bring this concept from design to a production battery that would provide two to three megawatt-hours of electrical storage.

We succeeded in building a four-kilowatt prototype, but a multitude of engineering problems complicated the very tight budget situation, and finally time ran out. Estimates for completion far exceeded our budget, and we were forced to abandon this aspect.

One of the most unique features of this proposal was the integration of the college building program with a total solar energy source. The two were to function as one unique system. Based on the premises that energy consciousness was to be the major architectural parameter, the firm of Cromwell, Neyland, Truemper, Levy & Gatchell, Inc., of Little Rock, Arkansas, provided a most provocative design concept. In fact, it was so provocative that DOE did not believe its value. As a result, DOE required that we simulate the design in order to ascertain the truth of the argument. This simulation was done by Ayres Associates of Los Angeles on the major computer at Berkeley, California. Concurrently, DOE did a similar study using a computer at the University of Chicago. Both studies showed that the energy budget for the proposed structure was 35 to 40 percent better than if we had built by usual standards. We won the argument, but it cost us a serious delay.

Front view of energy-conscious building.

In late 1977, we conducted a national competitive procurement which resulted in the largest purchase of solar cells in the history of the technology. Some 65,000 silicon concentrator cells were contracted with Solarex Corporation of Rockville, Maryland. The final design calls for a single crystal cell with minimum efficiency

Vaulted glass concourse of energy-conscious building.

of 12 percent at 55° C over the range of 1 W/cm^2 (10X) to 5 W/cm^2 (50X). Each cell is 2.5 x 5 cm.

The cells are laid down on a hollow V-shaped aluminum extrusion. Necessary cooling is provided by a solution of ethylene glycol flowing through the hollow extrusion, and maximum temperature of 55° C is maintained by the flow rate.

Electrical interface between the solar system and the building is provided by a d.c.-a.c. inverter built by Delta Electronic Control Corporation of Irvine, California, and provides peak power tracking as well as controlling the mixture of power from the local utility and the photovoltaic system.

When the iron-redox battery system was rejected because of cost and time, the college negotiated a contract with the local electric utility - Arkansas-Missouri Power Company (now Arkansas Power and Light Corporation), an operating company of Middle South Utilities, Inc. - for a one-for-one electron exchange. In other words, if we generate more power than we need over the weekend, the utility will accept that on its power grid, and in essence, run the meter backward or return the power to us during the week when needed. I believe that this is an exceptionally significant accomplishment and will have an impact on other alternate energy source relationships in the future.

The byproduct of cooling the cells, thermal energy, is interfaced with the building's heating system through a standard heat exchanger and provides the primary heat source for winter heating. Included within the building heating loop are two 40,000-gallon, well-insulated and buried hot water storage tanks. This provides storage for some three to four days of heating capability.

Control of all aspects of the solar system is provided by a Hewlett-Packard desk top computer, Model 9845-B. This provides wake-up capabilities, emergency stow, and a multitude of redundant safety measures. A second Hewlett-Packard desk top computer gathers data for the National Data Bank located in Seattle, Washington. A complete weather station, as well as some 135 data gathering points, provides constant input to this system.

The design of the system calls for a peak electrical output of 320 kw (DOE standard) with a peak thermal output of 22 MBtu/day. Total cost of installed energy system (all components) is $9.70/watt electrical peak (1976 dollars). This does not take into account the value of the thermal energy, which could amount to about $25,000 per year. If our calculations are correct and if the system performs as designed, we should realize about 85 percent of our total yearly energy budget from the solar system. Let's hope we are correct!

Over the past several years, there have been many discussions concerning the advantages and disadvantages of flat panel arrays versus concentrating arrays. Most experts agree that flat plate arrays are the simplest, most reliable, and lowest cost systems for residential and remote applications. Excellent results are obtained when flat plate arrays are fixed at some selected angle and are passively cooled.

However, actively cooled concentrating systems may have greater long-term cost effectiveness for some applications, particularly those requiring both electrical and thermal outputs.

Because the MCCC project is a large-scale actively cooled concentrating system and the thermal output is essential for winter heating of the integrated buildings, it is an excellent source to study the effectiveness and long-term payback of such a system. The United States Department of Energy recognizes this potential and has agreed to support a maintenance, operation, and study effort over the next five years. We believe data gathered in this effort will have great influence on the future choice of such systems.

If you are a potential user of a photovoltaic system and your choice is for an actively cooled concentrator system, the MCCC project should be carefully studied. There are many pitfalls to be avoided, particularly in the installation and troubleshooting stages. Feel free to visit our site and examine this large-scale experiment.

Solar Electric Systems for Domestic Housing — the Australian Experience

P. Golding

Department of Physics, Monash University, Clayton 3168, Victoria, Australia

ABSTRACT

The use of direct conversion devices for meeting the electrical power needs of domestic housing in rural and island communities is demonstrated to be practical. The solar electric system installed at the Bos Solar House is described, as an example of the pioneering introduction of such systems in Australia. The widespread use of photovoltaics for generating domestic electrical power is foreseen.

INTRODUCTION

Electricity generating systems for rural and island communities need to satisfy definite criteria. The generating capacity should be carefully sized to match the loads and the storage capability needs to be tuned to the pattern of demand. Together they should combine with interfacing into a total system giving predictable and reliable performance. Further, the climate of location dictates that all components require low maintenance and be exceptionally durable.

If one adds to this list an additional criterion that power generation should draw on renewable and environmentally sound energy resources, a number of systems may be suggested. The alternatives include the use of wind, solar thermal (including the production of hydrogen) and solar electric generators. All have been used or are currently under development in Australia.

Wind generators were common sights in Australian rural communities, until replaced by diesel generators and A.C. main-grid electricity supply in the late 1950's and the 1960's. Only one Australian wind generator manufacturer survived the consequent market depression and is today an exporter of wind generating systems. The establishment recently of the Australian Wind Energy Association [1] is evidence of the renewed interest in wind energy use in Australia. Tasmania, the most southerly Australian state, is particularly well-suited in terms of prevailing winds, for the siting of wind electric generators.

A solar thermal power station, incorporating tracking parabolic concentrator arrays, is currently under construction in White Cliffs, New South Wales. It will supply the electricity needs of a rural community of 2,500 people. At the

Australian National University in Canberra, Australian Capital Territory, a research team is investigating commercial hydrogen production. The dissociation of ammonia at high temperatures, generated by solar concentrators is being considered. As well as these alternative methods of useful energy production, the use of direct solar-to-electricity production using photovoltaic technology, will have an enormous impact in the decade ahead.

DIRECT PRODUCTION OF ELECTRICITY.

The utilization of devices for direct conversion of sunlight into electricity has some very distinct and practical advantages. Direct conversion, solar electric systems require a minimum of plant. Once installed they require no significant maintenance and are extremely durable, even in harsh environments. Though initially very expensive, they provide a practical solution to meeting the needs of remote power consumers. The largest installation of direct conversion devices in Australia is located in the Northern Teritory, where Australian Telecom have installed photovoltaic arrays to power a remote telecommunications network between Alice Springs and Catherine. Photovoltaics and 'container cargo' size battery stores were chosen for these installations because the line of microwave repeater stations are truly remote from state electricity grids. Similarly, photovoltaic arrays may be considered for providing electricity generation to meet many needs, including domestic consumption of electricity.

SOLAR ELECTRICITY FOR DOMESTIC HOUSING.

In the past the economics of these installations was unquestionable - they were not realistic alternatives. However, this situation is rapidly changing. A practical example of the incorporation of photovoltaics for domestic housing is seen in a unique house located on the Mornington Peninsula in Victoria.

The Bos Solar House, located at Robinsons Road, Pearcedale, was built by Judy and Michael Bos. A solar powered electric system provides all the electricity requirements of the house. The system is seen to consist of three main parts : electricity generation, storage and distribution.

Eight commercial 'Solarex' photovoltaic cell panels, each panel containing 36 cells, provide direct conversion of sunlight into electricity. These generating panels are mounted on a hinged frame and located on the aluminium house roof. A simple reflector is placed behind the cells in order to increase available solar radiation input, particularly during the short winter days. In addition to the solar cell panels, a small wind generator provides some charging capacity.

The electricity generated by the photovoltaic panels is stored in 18 pure lead-acid 'Dunlop' batteries. These batteries have an estimated 25 year life and are the type used by Telecom Australia in their remote photovoltaic-power transmitting stations. The batteries are located in a ventilated battery store against the south wall (southern hemisphere) of the house. Space is provided for including additional batteries if the generating capacity is ever increased.

Electricity supply is distributed about the house via two separate circuits, enabling 12V D.C. or 240V A.C. appliances to be used. The use of D.C. appliances is favoured since they obviate the need for the use of an inverter. Some 80 fluorescent lights, refrigerators, stereo and water pumping equipment all run off the 12V circuit. An automatic washing machine will soon be included, the heavy starting load having caused some initial problems as a standard A.C. appliance. It should also be mentioned that the Bos family are very concious of their power consumption and do not waste valuable power unnecessarily. They are not, however, thrifty to the extent of going without electric mixers, stereo equipment and the like, which are used as the occasion arises.

THE INTEGRATION OF SOLAR ELECTRIC SYSTEMS.

The Bos House demonstrates the integration of many solar energy systems into an efficient and interesting low-energy house complex. The only direct connection with external community services is the telephone. The house is self-sufficient for water, electricity, heating and cooling and waste disposal. The house has a wide variety of energy collection and distribution systems.

Passive design has greatly enhanced temperature control. An active solar heating system complements the passive design features. The Bos family have never been short of electricity in their three years of living on-site. A monitoring programme is proceeding to assess the performance efficiency of the various systems. The house was privately funded, with the solar electric system complete costing £5,000 sterling.

THE POSSIBILITIES - CONCLUSIONS.

The Bos Solar House has been a pioneering example in Australia. The possibilities for the introduction of solar electric systems are enormous. At present photovoltaic installations are sensible for domestic houses in remote locations. At the recent Solar World Forum, held in Brighton, England, Dr. P.D. Maylock of the U.S. Department of Energy [2] made the prediction that photovoltaics would be comparable to oil, in terms of cost per unit of power production, in just three years. Regardless of the specific dates, the next decade will bring new scales of efficiency and unit cost. Such changes are certain to benefit the people of rural and island communities requiring electricity at work and at home.

REFERENCES.

1. Mark Diesendorf, Secretary, The Australian Wind Energy Association, C.S.I.R.O. Division of Math. Stat., Canberra (1981).

2. P.D. Maylock, U.S. Department of Energy, 'Results of the U.S. Photovoltaics Program', I.S.E.S. Solar World Forum, Brighton, England 23-27 Aug. (1981), to be published by Pergamon Press, Oxford.

FIG. 1. North face of house, facing Sun.
Showing: Greenhouse and glazed ceiling in living room.

FIG. 2. Roof installations - photovoltaic panels and water heating panels and storage.

List of Participants

AGNEW, Mr P.A.	Department of Mechanical Engineering University of Glasgow GLASGOW G12 8QQ
AL AZZAWI, Mr A.R.H.	Jad P O Box 255 Baghdad IRAQ
AL SARRAJ, Mrs Karawan	Department of Applied Microbiology University of Strathclyde GLASGOW
AMADO, Prof. Pierre	Ferolles 77330 Ozoir La Ferriere FRANCE
ARAFA, Prof Salah	Department of Materials Science The American University of Cairo 113 Kasr El-Aini St Cairo EGYPT
AUDU, Mr	Department of Electrical Engineering, Kaduna Polytechnic, P.M.B. 2021, Kaduna, NIGERIA
BANNISTER, Mr Scott	Napier College Colinton Road EDINBURGH

BARNARD, Mr G	Department of Chemical Engineering Imperial College Prince Consort Road LONDON SW7 2BY
BARP, Mr Bruno	Escher Wyss A6 Ch 8023 Zurich SWITZERLAND
BENTZEN, Mr Erik J.	Danish Trade Office 50 East Fettes Avenue EDINBURGH
BOA, Mr Thomas W.	Northern Lighthouse Board 84 George Street EDINBURGH
BOAK, Dr David	British Council 3 Bruntsfield Crescent EDINBURGH EH10 4HD
BREIDENBACH, Mr H.	Schulstrasse 38/43 D-8356 Speilgelau W. GERMANY
BROADBENT, Mr S.	5265 W Quarels Drive Littleton Colorado 80123 U.S.A.
BURNESS, Mr G M	9 Churchill Drive Dingwall Ross-shire
BURNETT, Mr Norman	Messrs Crouch & Hogg 18 Woodside Crescent GLASGOW G3 7VU
BURSTOW, Mr Frank	c/o Hurd & Partners Old Sickbay Kyle of Lochalsh Ross-shire
BUTERA, Prof F	Istuto di Fisica Tecnica/Viale Delle Scienze Palermo ITALY
CAMERON, Mr David A.	Countryside Commission for Scotland Battleby Redgorton PERTH
CAVE, Dr Peter R.	Department of Mechanical Engineering University of Dundee DUNDEE

CHANDLER, Mr Charles	14 Hart Road Byfleet Weybridge KT14 7N ENGLAND
CHOUDHURY, Prof S H	Department of Crop Botany Bangladesh Agricultural University Mymensingh BANGLADESH
COUTTS, Dr Tim J	School of Physics Newcastle Upon Tyne Polytechnic Ellison Place Newcastle ENGLAND
CRAWFORD, Mr Mark	Department of National Parks and Wildlife Conservation Katmandu Nepal INDIA
EGHOLM, Mr. E.	Orkuradid J C Suaboes 635 F.R. 3800 Torshaun FAROE ISLANDS
DEAN, Mr W F	Harcraig House 26 Dundee Road Broughty Ferry DUNDEE DD5 1LX
DONNELLY, Mr Kevin	University of Glasgow GLASGOW
EL GAMMEL	Mechanical Engineering Department Paisley College PAISLEY
ELLIOT, Mr George	Marine Operations National Engineering Laboratory East Kilbride GLASGOW G75 0QU
EMERY, Mr Michael	Physics Department University of Tasmania GPO Box 252C Hobart, Tasmania AUSTRALIA
EVANS, Mr F C	Physics Department University of St Andrews North Haugh St Andrews KY16 9SS SCOTLAND

EZEH, Mr Chuks	Wye College Wye Ashford KENT TN25 5AY
FARAGALLA, Mr M	Department of Electrical Engineering University of Dundee DUNDEE
FLEMING,Mr F W	Heriot Watt University Department of Building Chambers Street EDINBURGH EH1 1HX
FRANCIS, Dr E	Department of Energy Thames House South Millbank LONDON SW1P 4QJ
FRASER, Dr A I	International Forest Science Consultancy 21 Biggar road Silverburn PENICUIK, EH26 9LQ
GADDY, Prof J.L.	Department of Chemical Engineering 227 Engineering Building University of Arkansas College of Engineering Faetteville, Arkansas 72701 U.S.A.
GAIR, Mr S	4 Broompark Gargunnock STIRLINGSHIRE
GALT, Ms Jean	Wind and Water Power Ltd 1 Huntly Street EDINBURGH EH3 5HB
GANAPINI, Dr Walter	Comitato Nazionale per l'Energia Nucleare Piazzale R Morandi 2 Milano ITALY
GARSIDE, Mr A J	School of Mechanical Engineering Cranfield Institute of Technology Cranfield BEDFORD MK41 7YX
GINN, Mr G W C	'Red Deer' 48 Strath Gairloch, Ross-shire

GIUFFRIDA, Prof M
AMN
Via G D'Annunzio 113
Genoa
ITALY

GOLDING, Dr Peter
Department of Physics,
Manash University,
Clayton, 3168,
AUSTRALIA

GRAINGER, Mr W
Department of Applied Physics
University of Strathclyde
GLASGOW.

GRANT, Mr S R
Architecture Services
Highland Regional Council
Regional Buildings
Glenurquhart Road
INVERNESS

GUNN, Mr R
Newton House
Lybster
CAITHNESS

HASSOON, Mr. H.
34/19 Dour Al-Muaalimeen Al-Thanyah,
Al-Zaafaranyan,
Baghdad,
IRAQ

HAAHR, Mr Jan
Volund Glasfiber & Windtechnik A/S
Marsk Stigsvej 4
DK 8800 Viborg
DENMARK

HALLY, Mr Martin
Project Department
Electricity Supply Board
St Stephens Court
DUBLIN 2

HAMILTON, Mr A R
Planning Department
Shetland Islands Council
Planning Department
Victoria Buildings
The Esplanade
Lerwick
SHETLANDS

HANLON, Miss Marion	Design & Development Section Project Department Electricity Supply Board St Stephens Court DUBLIN 2
HANNA, Mr Robert	NBST Shelbourne Road DUBLIN 4
HARRISON, Mr J D L	Energy Technology Support Unit AERE, Harwell Oxfordshire OX11 ORA
HARDY, Mr	International Forest Science Consultancy 21 Biggar Road Silverburn PENICUIK EH26 1LQ
HAUGHEY, Dr Douglas	Scottish Institute of Agricultural Engineering Bush Estate Penicuik MIDLOTHIAN EH26 OPH
HENDERSON, Mr David M	Highlands & Islands Development Board Bridge House Bank Street INVERNESS IV1 1QR
HOPE, Mr D R	Kirkwood House Biggar, Lanarkshire ML12 6PD
IRVINE, Mr James C	Shetland Council of Social Service 5 Harbour Street LERWICK
JAMIESON, Mr P	James Howden & Co Ltd 195 Scotland Street GLASGOW G4 8PJ
JANUARY Mr M	Blaenau Ucha Farm Treuddyn, Nr Mold Clwyd, N. WALES
JANUZZI, Mr G de Martino	Cavendish Laboratory University of Cambridge CAMBRIDGE CB3 OHE

KAMAL, Dr W A
Engineering Faculty
Alexandria University
Alexandria
EGYPT

KANYI, Mr D M
P.O. 51056,
Nairobi,
KENYA

KELLEDY, Dr Elaine
Department of Statistics
University of Dublin
Trinity College
DUBLIN 2

KERR, Mr Andrew
The Dower House
Melbourne,
Nr DERBY

KIRKLAND, Mr David
Ottersgill
Stenness
ORKNEY

KUROKAWA, Mr K
New Energy Development Organisation
3-1-1 Higashi-Ikebukuro
Toshima-Ku,
Tokyo
JAPAN

KWANT, Mr Kees
Waalstraat 115
7523 RC Enchede
The Netherlands

KYALAANI, Mr F
68 Beresford Avenue
HULL, HU6 7LS

LEMMA, Miss T
College of Teacher Education,
P.O. Box 81248,
Addis Abbaba, Ethiopia.

LEWIS, Mr A W
Department of Civil Engineering
University College
Cork
EIRE

LIDDELL, Mr H
Achloa Centre
ABERFELDY
Perthshire

LIGT, Mr J J de
Advicebureau V E K, B.V.
Veiling CCWS 48
Burgemeester Elsenweg
2675 ZV Honselersdijk
Porthus 567
NETHERLANDS

LINDBERG, Dr R	LBES/UCLA 900 Veteran Ave Los Angeles CA 90025 U.S.A.
LIPMAN, Prof N	Rutherford Laboratory & Reading University Department of Engineering Chilton Didcot Oxfordshire
LOCKERBIE, Dr N	Department of Applied Physics University of Strathclyde Glasgow
LUNDSAGER, Dr Per	Riso National Laboratory Roskilde DENMARK
MacAULAY, Mr D	Western Isles Island Council Landwick Road STORNOWAY PA87 2BW
MacDONALD, Mr John	Western Isles Health Board South Beach St STORNOWAY
MacGREGOR, Mr Kerr	Mechanical Engineering Department Napier College Colinton Road EDINBURGH EH10 5DT
MacKAY, Mr Ian	Assistant Director, Department of Physical Planning Grampion Regional Council Woodhill House ABERDEEN AB9 2LU
McKELLAR, Mr D R	Y-Ard Ltd Charing Cross Tower GLASGOW G2 4PP
McLEAN, Mr Darrell	NBRI, CSIR P O Box 395 Pretoria 0001 Republic of SOUTH AFRICA
MacMILLAN, Dr D	Highlands & Islands Development Board 9 Castleacre Campbeltown ARGYLL
McPHEE, W.R.	19 Leys Drive INVERNESS

McPHERSON, Mr D J	Cromarty Mains Cromarty ROSS-SHIRE
McWHIRTER, Mr W	Architecture Services Highland Regional Council Regional Buildings Glenurquhart Road INVERNESS
MAGUIRE, Mr T	Tom Maguire & Co Ltd Sunderland House The Dockyard Pembroke DYFED SA72 6TD
MAGORIAN, Mr D	dhr. incorporated, 1055 Thomas Jefferson Street, N.W., Washington, DC 20007, U.S.A.
MARDON Mr R	Cuillich Lodge Ardross Alness ROSS-SHIRE
MARKS, Mr B	17 Banchory Avenue Inchinnan RENFREW PA4 9PX
MARTIN, Dr D J	Pentland Scotch Whisky Research Ltd 84 Slateford Road EDINBURGH EH11 1QU
MAYS, Dr I D	Sir Robert McAlpine & Sons Ltd 40 Bernard Street LONDON WC1N 1LG
MEYNELL, Mr P J	White House Little Bedwyn Marlborough WILTSHIRE SM8 3JP
MILLAR, Mr M	Department of Mechanical Engineering Napier College Colinton Road EDINBURGH EH10 5DT
MITCHELL, Dr C P	Department of Forestry St Machar Drive OLD ABERDEEN AB1 2UU
MOWLE, Mr Alan	Nature Conservatory Council 12 Hope Terrace EDINBURGH

MUNRO, Mr J S	Kilmore Wester Balblair Beauly INVERNESS
NAUNTOFTE, Mr	Volund Glasfiber and Windteknik A/S Marsk Stigsvej 4 DK-8800 Viborg DENMARK
NGOKO, Dr Nelson	Department of Environmental Engineering University of Ife NIGERIA
OBLADEN, Prof N L	Director, Laboratorio de Hidraulica Saneamento e Mico Ambiente University Catolia de Parana Rua Imaculada Conceicao 1155 80 000 Curitiba Parana BRAZIL
OL'OTABA, Mr S O	Makerere University, cc Education, P.O. Box 7062, Kampala, UGANDA
OLSEN, Mr H	Energy Council of Faroe Islands P O Box 72 Torshava via DENMARK
OTHIENO, Mr H	Physics Dept Kenyatta University College University of Nairobi KENYA
OTI, Mr	Department of Mechanical Engineering, Kaduna Polytechnic, P.M.B. 2021, Kaduna, NIGERIA
PACETTI, Prof Marco	University of Ancona Via della Montagnola 30 I 60 100 Ancona ITALY
PEATFIELD Dr T	Lancaster Polytechnic COVENTRY

POPE, Mr I T	Hydro Electric Board 16 rothesay Terrace EDINBURGH
POUND, Mr B	3 Rivey Way Linton CAMBRIDGESHIRE
RADWAN, Dr M S	Mechanical Power Dept Faculty of Engineering & Technology University of Helwan El-Edfairi 81 CAIRO Egypt
RATTO, Dr C	University of Genova Instituto di Scienze & Fisiche Viale Benedelto X45 Cap 16132 ITALY
RIDDELL, Mr John C	Ardyl House Hollybush Green Collingham Wetherby YORKS LS22 5BE
REID, Mr A U	Kyleakin Norrels Drive East Horsley SURREY
ROBERTSON, Mr P	The Scott Sutherland School of Architecture Garthdee ABERDEEN AB9 2QB
SALUJA, Dr G	The Scott Sutherland School of Architecture Garthdee ABERDEEN AB9 2QB
SARGENT, Mr M	Orkney Island Council Council Offices Kirkwall ORKNEY KW15 1NY
SCOTT, Mr R	The Institute of Terrestrial Ecology Merlewood Research Station Grange-over-Sànds CUMBRIA LA11 6JU
SCHNADT, Mr L	Weatherwise Homes The Park FORRES IV36 OTZ
SEEBY, Mr N	Community Forestry Development Project Katmandu Nepal INDIA

SHANKS, Ms DD	I.I.E.A. 1708 13th St Boulder Co 80302 U.S.A.
SHRESTHA, Mr S	MCR, Wye College University of London Ashford KENT TN25 5AH
SMITH, Mr H V	Mississippi County Community College Blytheville Arkansas 72315 U.S.A.
SMITH, Mr J	Whatlings Energy Conservation Services 14 North Claremont St GLASGOW
SINTUNOWA, Mr C	Energy Studies Unit University of Strathclyde 100 Montrose Street GLASGOW
SINGH, P N	ORP Windmill Project Kusunih Kalan Post Box 62 Chazipur INDIA 233001
STEVENSON, Mr W	Hydro Electric Board 16 rothesay Terrace EDINBURGH
STRONG, Mr S J	Solar Design Association Conant Road Lincoln Mass. 01773 U.S.A.
SWET, Mr C J	704o Woodville Road Mt Airy Maryland 21771 U.S.A.
THOMSON, Mr J S	Assistant Director of Development Highland Regional Council Regional Buildings INVERNESS

TONKINSON, Mr T	Agricultural Consultant Control Data Corporation 8100 34th Ave S Minneapolis Minnesota 55440 U.S.A.
TOWRIE, Mr J	Orkney Island Council Council Offices Kirkwall ORKNEY KW15 1NY
URASSA, Dominic	68 Beresford Ave Beverley Road HULL
URQUHART, Mr D C M	The Scott Sutherland School of Architecture Garthdee ABERDEEN AB9 2QB
VARDY, Mr A	Civil Engineering Department University of Dundee DUNDEE
VAN LIEROP, Mr W E	University of Technology P O Box 217 7500 Enschede NETHERLANDS
VAN de VUSSE, Mr T	Shell International Petrol Co Shell Centre LONDON SE1 7NA
VERHAART, Mr P	WES 00 12 The P O Box 513 5600 MB Eindhoven NETHERLANDS
VIS, Mr T	Advicebureau V E K B V Veiling CCWS 48 Burgenmeester Elsenweg 2675 ZV Hanselersdijk Postbus 567 NETHERLANDS
VOLK, Mr M	Sulzer Brothers Ltd Dept 15/1531 8401 Winterhur SWITZERLAND

WATES, Mr N	Commonwork Bore Place Bough Beech Edenbridge KENT TN8 7AR
WATSON, Mr G R	Northumbrian Energy Workshop Tanners Yard Gilesgate Hexham NORTHUMBERLAND
WALDEMARIAM, Mr K	26 Kirk St Leith Walk EDINBURGH EH6 4EZ
WASSERTHAL, Dr W	P O Box 764 Arusha TANZANIA
WEINGART, Mr J	Lawrence Berkeley Lab University of California Berkeley California 94720 U.S.A.
WELSH, Mr K	15 Robson St DUNDEE DD4 7JQ
WHITE, Mr R	Llwyngwern Forge Machynlleth Powys WALES
WHITEHURST, Mr B	1983 Hoods Drive New Bern North Carolina 28560 U.S.A.
WILSON, Mr T	Watlings Energy Conservation Services 14 North Claremont St GLASGOW
WILSON-GOUGH, A K	Dartington Institute Dartington Trust Totnes DEVON
WOODBRIDGE, Dr K	New Manse North Ronaldsay ORKNEY
WOOLLEY, Mr M V	Wavepower Ltd Carlton House Ringwood Road Woodlands SOUTHAMPTON SO4 2HT

WYPER, Miss Hilary	2 Cameron Cresc CARMUNNOCK
ZAMBONI, Miss F	Energy Studies Unit University of Strathclyde 100 Montrose St GLASGOW

FURTHER REGISTRATIONS

PROVEN, Mr. G.	Proven Engineering Products Ltd., Moorfield Industrial Estate, KILMARNOCK
HIMMEL, Mr. M.	Kirchstrasse 12, D-5170 Juelich, Fed. Rep. of Germany, GERMANY
ROXBURGH, Mr. G.	Roxburgh and Partners, 6 Rowan Road, GLASGOW G41

Author – Paper Index

(See participants' list for addresses and institutions)

Subject Index